Yang Lo

Value Distribution Theory

Springer-Verlag Berlin Heidelberg GmbH

Yang Lo
Institute of Mathematics
Academia Sinica
The People's Republic of China

Revised edition of the original Chinese edition published by Science Press Beijing 1982 as the 9th volume in the Series in Pure and Applied Mathematics.

Distribution rights throughout the world, excluding the People's Republic of China, granted to Springer-Verlag Berlin Heidelberg New York London Paris Tokyo Hong Kong Barcelona Budapest

Mathematics Subject Classification (1991): 30D35, 30D30

ISBN 978-3-662-02917-6   ISBN 978-3-662-02915-2 (eBook)
DOI 10.1007/978-3-662-02915-2

© Springer-Verlag Berlin Heidelberg New York 1993
Originally published by Springer-Verlag Berlin Heidelberg New York 1993
Softcover reprint of the hardcover 1st edition 1993

Typesetting: Science Press, Beijing. The People's Republic of China
41/3140-543210 Printed on acid-free paper

Introduction

It is well known that solving certain theoretical or practical problems often depends on exploring the behavior of the roots of an equation such as

$$f(z) = a, \tag{1}$$

where $f(z)$ is an entire or meromorphic function and a is a complex value. It is especially important to investigate the number $n(r, f = a)$ of the roots of (1) and their distribution in a disk $|z| \le r$, each root being counted with its multiplicity. It was the research on such topics that raised the curtain on the theory of value distribution of entire or meromorphic functions.

In the last century, the famous mathematician E. Picard obtained the pathbreaking result: Any non-constant entire function $f(z)$ must take every finite complex value infinitely many times, with at most one exception. Later, E. Borel, by introducing the concept of the order of an entire function, gave the above result a more precise formulation as follows. An entire function $f(z)$ of order $\lambda(0 < \lambda < \infty)$ satisfies

$$\varlimsup_{r \to \infty} \frac{\log n(r, f = a)}{\log r} = \lambda$$

for every finite complex value a, with at most one exception. This result, generally known as the Picard-Borel theorem, lay the foundation for the theory of value distribution and since then has been the source of many research papers on this subject.

At the beginning of this century, P. Montel introduced the concept of normality for a family of holomorphic or meromorphic functions in a region, which in some sense corresponds to the concept of compactness for this family, and established the Montel Criterion. This theorem connects the normality of a family with the value assignment of every function of this family: Let $\mathcal{F}$ be a family of holomorphic functions defined in a region D, and a, b two distinct finite complex values. If none of the functions in $\mathcal{F}$ takes either a or b, then the family $\mathcal{F}$ is normal in D. Later, by applying the Montel Criterion to an arbitrary transcendental entire function $f(z)$, G. Julia proved that there exists a ray $J : \arg z = \theta_0$ such that in any angular domain $|\arg z - \theta_0| < \varepsilon,\ \ \varepsilon > 0$, the function $f(z)$ takes every finite complex value, except for at most one value. Such a ray is known as a Julia direction of $f(z)$, and it symbolizes the beginning of the research on the theory of angular distribution.

It was R. Nevanlinna who made the decisive contribution to the development of the theory of value distribution. Before him, the principal object and tool of the theory were the class of entire functions and the maximum modulus, respectively. It was Nevanlinna who elevated the theory of meromorphic functions to a new level by introducing the characteristic function $T(r, f)$ for a meromorphic function $f(z)$ in a domain $|z| < R(R \leq \infty, 0 < r < R)$, as an efficient tool. His theory can be summed up as follows.

Let $f(z)$ be a meromorphic function in the finite plane and a a complex value. Then the equality

$$N(r, f = a) + m(r, f = a) + O(1) = T(r, f) \qquad (2)$$

holds as $r \to \infty$, where

$$N(r, f = a) = \int_0^r \frac{n(t, f = a)}{t} dt$$

and

$$m(r, f = a) = \frac{1}{2\pi} \int_0^{2\pi} \log^+ \frac{1}{|f(re^{i\theta}) - a|} d\theta.$$

Moreover, if $a_j (j = 1, \ldots, q)$ are q distinct finite complex numbers, then the inequality

$$(q - 2)T(r, f) < \sum_{j=1}^{q} N(r, f = a_j) - N_1(r) + S(r, f) \qquad (3)$$

holds, where $N_1(r)$ is a non-negative term given essentially by the multiple roots of the equation $f(z) = a$ and $S(r, f)$ is the error term, which increases much more slowly in general than other terms. It is easy to see that the Picard-Borel Theorem is then a direct consequence of Nevanlinna theory. Furthermore, define the deficiency of a with respect to f to be

$$\delta(a, f) = \varliminf_{r \to \infty} \frac{m(r, f = a)}{T(r, f)} = 1 - \varlimsup_{r \to \infty} \frac{N(r, f = a)}{T(r, f)}.$$

Then the set $\{a | \delta(a, f) > 0\}$ of the deficient values of a transcendental meromorphic function $f(z)$ in the finite plane is at most countable and

$$\sum \delta(a, f) \leq 2, \tag{4}$$

where the sum is taken over all the deficient values.

In this volume we intend to present a systematic and comprehensive survey of the theory of value distribution of meromorphic functions, including both the classical results as well as more recent achievements, especially the developments in the past two decades.

Chapters 1 and 2 afford a brief introduction to Nevanlinna theory and the normal family, respectively, though the last section of Chapter 1 is devoted to a recent important theorem due to Osgood and Steinmetz, which improves the deficiency relation (4) as

$$\sum \delta(a(z), f) \leq 2,$$

where the sum is taken over all meromorphic functions $a(z)$ with $T(r, a(z)) = o(T(r, f))$ as $r \to \infty$.

In Chapter 3 we prove that every meromorphic function $f(z)$ of order $\lambda (0 < \lambda < \infty)$ in the finite plane must have a Borel direction $\arg z = \theta_0$ such that for any positive number ε and any complex number a we have

$$\varlimsup_{r \to \infty} \frac{\log n(r, \theta_0, \varepsilon, f = a)}{\log r} = \lambda,$$

except for at most two values of a, where the symbol $n(r, \theta_0, \varepsilon, f = a)$ denotes the number of zeros of $f(z) - a$ in the region $(|z| \leq r) \cap (|\arg z - \theta_0| \leq \varepsilon)$, each zero being counted with its multiplicity.

In Chapter 4 we discuss the value distribution of a meromorphic function together with its derivatives. In this connection Hayman obtained an important inequality, where the characteristic function $T(r, f)$ can be bounded by just two counting functions $N(r, f = 0)$ and $N(r, f^{(k)} = 1)$ with $k \geq 1$. Corresponding to Hayman's inequality, we prove an alternative theorem: If $f(z)$ is meromorphic in $|z| < 1$ with $f(z) \neq 0$ and

$f^{(k)}(z) \neq 1$ for a positive integer k, then either $|f(z)| < 1$ or $|f(z)| > C_k > 0$ uniformly in $|z| < 1/32$, where C_k does not depend on f. Then Gu Yong-xing's criterion for normality follows immediately. Moreover, the author obtained the following general result: Let $\mathcal{F}$ be a family of meromorphic functions in a region D and k a positive integer. If for each function $f(z)$ in $\mathcal{F}$, neither $f(z)$ nor $f^{(k)}(z)$ has a fixed point in D, then $\mathcal{F}$ is normal. Finally, we shall also discuss a precise estimate of the total deficiency of $f^{(k)}(z)$ due to the author.

In Chapter 5, we first establish the following theorem on the distribution of Borel directions: If λ is a positive number and E is a non-empty closed set of real numbers (mod 2π), then there exists a meromorphic function $f(z)$ of order λ such that all of its Borel directions constitute exactly the set $\{\arg z = \theta | \theta \in E\}$. Next, we give a simple proof to an important result of H. Milloux which asserts that for an entire function $f(z)$ of order $\lambda(0 < \lambda < \infty)$, every Borel direction of order λ of $f'(z)$ is also a Borel direction of order λ of $f(z)$ itself. Then it is natural to ask if the inverse of Milloux's theorem is true or not. In this connection, Zhang Qing-de and the author proved that if $f(z)$ is a meromorphic function of order $\lambda(0 < \lambda < \infty)$ in the finite plane, then there exists a direction $\arg z = \theta_0 (0 \leq \theta_0 < 2\pi)$ such that for any positive number ε, any positive integer k and any two finite complex numbers a and b with $b \neq 0$, we have

$$\varlimsup_{r \to \infty} \frac{\log\{n(r, \theta_0, \varepsilon, f = a) + n(r, \theta_0, \varepsilon, f^{(k)} = b)\}}{\log r} = \lambda.$$

Consequently, if the meromorphic function $f(z)$ has a finite exceptional value, then all its Borel directions will be conserved under the derivation.

Chapter 6 deals with the relationship between the number of deficient values and the number of Borel directions, due to Zhang Guang-hou and the author. Given any meromorphic function of a finite positive order in the finite plane, the total number of its deficient values does not exceed the total number of its Borel directions. Moreover, if the function is entire, then the total number of its finite deficient values does not exceed half the total number of its Borel directions.

In Chapter 7 we discuss Baernstein's function T^* and the spread relation. As applications we present Fuchs' theorem, the ellipse theorem and a solution of the deficiency problem in the case of lower order less than one.

Most results in Chapters 4, 5, 6 and 7 were obtained in the past two decades and have not, as yet, been included in any book. On the other hand, Chapters 1, 2 and 3 provide the requisite knowledge of Nevanlinna theory, normal family and Borel direction for the reading of the successive chapters. Therefore, this volume is self-contained and should be useful to both researchers and graduate students.

The contents of this volume are the result of a series of lectures by the author at the Graduate School of Chinese Academy of Sciences several years ago, and most recently at the University of Notre Dame. The author wishes to thank Professors D. Drasin, G.Frank, W.H.J. Fuchs, W.K. Hayman, W. Stoll, A. Weitsman, H. Wu and C.C Yang for discussions and encouragement. Special thanks go to Professor Jiang Jia-he and Madam K. Weltin who read the manuscripts most carefully and provided many valuable comments. Finally, the author acknowledges the partial support provided by the National Science Foundation of China.

Lo Yang

Contents

Chapter 1

Essentials of Nevanlinna Theory

In 1925, R. Nevanlinna[1] established two fundamental theorems; in one stroke he initiated the modern research on the theory of value distribution, and laid down the foundation for its development ever since. Therefore, the first chapter will be devoted to a brief introduction to Nevanlinna theory[1], and the last section of the chapter, as an illustration of the development, will discuss an important theorem by Osgood[1] and Steinmetz[1].

1.1 The Poisson-Jensen Formula

1.1.1 The Poisson–Jensen formula.   In Nevanlinna theory, the following Poisson-Jensen formula plays a very important role.

Theorem 1.1   *Suppose $f(\zeta)$ is meromorphic in $|\zeta| \leq R$ ($0 < R < \infty$) and that a_μ ($\mu = 1, 2, \cdots, M$) and b_ν ($\nu = 1, 2, \cdots, N$) are the zeros and poles of $f(\zeta)$ in $|\zeta| < R$, respectively. If $z = re^{i\theta}$ is a point in $|\zeta| < R$, distinct from a_μ and b_ν, then*

$$\log\left|f(z)\right| = \frac{1}{2\pi} \int_0^{2\pi} \log\left|f(Re^{i\varphi})\right| \frac{R^2 - r^2}{R^2 - 2Rr\cos(\theta - \varphi) + r^2} d\varphi$$

$$+ \sum_{\mu=1}^{M} \log\left|\frac{R(z - a_\mu)}{R^2 - \bar{a}_\mu z}\right| - \sum_{\nu=1}^{N} \log\left|\frac{R(z - b_\nu)}{R^2 - \bar{b}_\nu z}\right|.$$

(1.1.1)

---

1) cf. Nevanlinna [1, 2, 3], Hayman [2], Tsuji [1], Goldberg & Ostrowski [1] and Ozawa [1].

Proof. First of all, we consider the special case where $f(\zeta)$ has neither zero nor pole on $|\zeta| \le R$. Then $\log f(\zeta)$ is regular on $|\zeta| \le R$. Thus

$$\log f(0) = \frac{1}{2\pi i} \int_{|\zeta|=R} \log f(\zeta) \frac{d\zeta}{\zeta} = \frac{1}{2\pi} \int_0^{2\pi} \log f(Re^{i\varphi}) d\varphi. \qquad (1.1.2)$$

Taking the real part of both sides in (1.1.2), we prove (1.1.1) in the case of $z = 0$. In the general case, set $w = \dfrac{R(\zeta - z)}{R^2 - \bar{z}\zeta}$, which maps $|\zeta| \le R$ onto $|w| \le 1$ and $\zeta = z$ to $w = 0$. Its inverse is $\zeta = R(Rw+z)/(R+\bar{z}w)$. Letting $F(w) = f(R(RW + z)/(R + \bar{z}w))$, we see that $F(w)$ is meromorphic on $|w| \le 1$ and has neither zero nor pole there. According to (1.1.2), we have

$$\log F(0) = \frac{1}{2\pi i} \int_{|w|=1} \log F(w) \frac{dw}{w}.$$

Since

$$\frac{dw}{w} = d(\log w) = \frac{d\zeta}{\zeta - z} + \frac{\bar{z}d\zeta}{R^2 - \bar{z}\zeta} = \frac{(R^2 - |z|^2)d\zeta}{(R^2 - \bar{z}\zeta)(\zeta - z)},$$

we deduce

$$\log f(z) = \frac{1}{2\pi i} \int_{|\zeta|=R} \log f(\zeta) \frac{R^2 - |z|^2}{(R^2 - \bar{z}\zeta)(\zeta - z)} d\zeta. \qquad (1.1.3)$$

Therefore, on $|\zeta| = R$, we have $\zeta = Re^{i\varphi}$, $d\zeta = iRe^{i\varphi} d\varphi$, and

$$(R^2 - \bar{z}\zeta)(\zeta - z) = (R^2 - Rre^{i(\varphi-\theta)})(Re^{i\varphi} - re^{i\theta})$$

$$= Re^{i\varphi}\left\{R^2 - 2Rr\cos(\varphi - \theta) + r^2\right\}.$$

Substituting these quantities into (1.1.3) and taking the real parts, we obtain

$$\log |f(z)| = \frac{1}{2\pi} \int_0^{2\pi} \log |f(Re^{i\varphi})| \frac{R^2 - r^2}{R^2 - 2Rr\cos(\varphi - \theta) + r^2} d\varphi. \qquad (1.1.4)$$

Next, when $f(\zeta)$ has a finite number of zeros and poles on $|\zeta| = R$ and none in $|\zeta| < R$, we consider the domain D_ε consisting of all points of $|\zeta| \le R$ of distance greater than ε from every pole and zero of $f(\zeta)$. For small ε, the boundary of D_ε is denoted by Γ_ε. The function

$$\frac{\log f(\zeta)(R^2 - |z|^2)}{(R^2 - \bar{z}\zeta)(\zeta - z)}$$

is holomorphic in D_ε, except for the point $\zeta = z$, which is a pole with residue $\log f(z)$. Thus

$$\log f(z) = \frac{1}{2\pi i} \int_{\Gamma_\varepsilon} \log f(\zeta) \frac{R^2 - |z|^2}{(R^2 - \bar{z}\zeta)(\zeta - z)} d\zeta.$$

Since the length of every small arc is less than $2\pi\varepsilon$ and the integrand on it is $O(\log 1/\varepsilon)$, the corresponding integral tends to zero with ε. Thus we also obtain (1.1.4) in this case.

Finally, when $f(\zeta)$ has zeros and poles in $|\zeta| < R$, denoted by $a_\mu (\mu = 1, 2, \cdots, M)$ and $b_\nu (\nu = 1, 2, \cdots, N)$, respectively, the function

$$g(\zeta) = f(\zeta) \frac{\displaystyle\prod_{\nu=1}^{N} \frac{R(\zeta - b_\nu)}{R^2 - \bar{b}_\nu \zeta}}{\displaystyle\prod_{\mu=1}^{M} \frac{R(\zeta - a_\mu)}{R^2 - \bar{a}_\mu \zeta}}$$

is meromorphic on $|\zeta| \leq R$ and has neither zero nor pole in $|\zeta| < R$, and therefore, according to the above result, we have

$$\log |g(z)| = \frac{1}{2\pi} \int_0^{2\pi} \log |g(Re^{i\varphi})| \frac{R^2 - r^2}{R^2 - 2Rr\cos(\theta - \varphi) + r^2} d\varphi.$$

Since

$$|g(Re^{i\varphi})| = |f(Re^{i\varphi})|, \quad 0 < \varphi < 2\pi,$$

(1.1.1) can be derived immediately. $\square$

1.1.2 Corollaries

Corollary 1.   *Under the assumptions of Theorem 1.1, if $f(\zeta)$ has neither zeros nor poles on $|\zeta| \leq R$, then for every point z with $|z| = r < R$, we have*

$$\log |f(z)| = \frac{1}{2\pi} \int_0^{2\pi} \log |f(Re^{i\varphi})| \frac{R^2 - r^2}{R^2 - 2Rr\cos(\theta - \varphi) + r^2} d\varphi. \quad (1.1.5)$$

This is Poisson's formula.

Corollary 2.   *Under the assumptions of Theorem 1.1, if $f(0) \neq 0$, ∞, then we have Jensen's formula as follows*

$$\log|f(0)| = \frac{1}{2\pi} \int_0^{2\pi} \log|f(Re^{i\varphi})|d\varphi - \sum_{\mu=1}^M \log\frac{R}{|a_\mu|} + \sum_{\nu=1}^N \log\frac{R}{|b_\nu|}. \quad (1.1.6)$$

When $f(0) = 0$, denote by $n(0, f = 0)$ its multiplicity; when $f(0) = \infty$, denote by $n(0, f = \infty)$ the corresponding multiplicity. (If $f(0) \neq 0$, then $n(0, f = 0) = 0$. Similarly, if $f(0) \neq \infty$, then $n(0, f = \infty) = 0$.) Setting $\tau = n(0, f = 0) - n(0, f = \infty)$, we have the following expansion in a neighborhood of the origin:

$$f(\zeta) = c_\tau \zeta^\tau + \cdots, \quad c_\tau \neq 0.$$

Write

$$g(\zeta) = \begin{cases} f(\zeta)\left(\dfrac{R}{\zeta}\right)^\tau, & \zeta \neq 0; \\ c_\tau R^\tau, & \zeta = 0. \end{cases}$$

Clearly $g(\zeta)$ is meromorphic on $|\zeta| \leq R$ and $g(0) \neq 0, \infty$. Applying Jensen's formula (1.1.6) and noting $|g(Re^{i\varphi})| = |f(Re^{i\varphi})|$, we obtain

$$\log|c_\tau| + \tau\log R$$

$$= \frac{1}{2\pi} \int_0^{2\pi} \log|f(Re^{i\varphi})|d\varphi - \sum_{0<|a_\mu|<R} \log\frac{R}{|a_\mu|} + \sum_{0<|b_\nu|<R} \log\frac{R}{|b_\nu|}.$$

Thus

$$\log|c_\tau| + n(0, f = 0)\log R = \frac{1}{2\pi} \int_0^{2\pi} \log|f(Re^{i\varphi})|d\varphi - \sum_{0<|a_\mu|<R} \log\frac{R}{|a_\mu|}$$

$$+ \sum_{0<|b_\nu|<R} \log\frac{R}{|b_\nu|} + n(0, f = \infty)\log R.$$

$$(1.1.7)$$

This is the general form of Jensen's formula.

1.2 Characteristic Functions and the First Fundamental Theorem

1.2.1 Characteristic function

Definition 1.1 *For $x \geq 0$, we define*

$$\log^+ x = \max(\log x, 0) = \begin{cases} \log x, & x \geq 1; \\ 0, & 0 \leq x < 1. \end{cases} \tag{1.2.1}$$

It is easy to see

$$\log x = \log^+ x - \log^+ \frac{1}{x}$$

holds for any positive number x. If $f(z)$ is meromorphic in $|z| < R$, and $0 < r < R$, then

$$\frac{1}{2\pi} \int_0^{2\pi} \log |f(re^{i\varphi})| d\varphi = \frac{1}{2\pi} \int_0^{2\pi} \log^+ |f(re^{i\varphi})| d\varphi$$

$$- \frac{1}{2\pi} \int_0^{2\pi} \log^+ \frac{1}{|f(re^{i\varphi})|} d\varphi.$$

On the other hand, denote by $n(r, f)$ the number of poles of $f(z)$ on $|z| \leq r$, each pole being counted with its proper multiplicity. Denote by $n(0, f)$ the multiplicity of the pole of $f(z)$ at the origin. (If $f(0) \neq \infty$, then $n(0, f) = 0$.) The poles of $f(z)$ in $0 < |z| \leq r$ are denoted by $b_1, b_2, \cdots, b_N$, each pole occuring with its multiplicity. Thus

$$\sum_{\nu=1}^N \log \frac{r}{|b_\nu|} = \int_0^r \log \frac{r}{t} d(n(t, f) - n(0, f)).$$

By integration by parts, we obtain

$$\sum_{\nu=1}^N \log \frac{r}{b_\nu} = \int_0^r \frac{n(t, f) - n(0, f)}{t} dt.$$

Similarly, the notation $n(r, 1/f)$ denotes the number of zeros of $f(z)$ on $|z| \leq r$, each zero being counted with its proper multiplicity, and $n(0, 1/f)$ the number of times $f(z)$ vanishes at the origin. Denote by $a_1, a_2, \cdots, a_M$

the zeros of $f(z)$ in $0 < |z| \leq r$, each zero occuring with its multiplicity.
Thus

$$\sum_{\mu=1}^{M} \log \frac{r}{|a_\mu|} = \int_0^r \frac{n(t, 1/f) - n(0, 1/f)}{t} dt.$$

Therefore Jensen's formula can be written as

$$\log |c_r| + \frac{1}{2\pi} \int_0^{2\pi} \log^+ \frac{1}{|f(re^{i\varphi})|} d\varphi$$

$$+ \int_0^r \frac{n(t, 1/f) - n(0, 1/f)}{t} dt + n\left(0, \frac{1}{f}\right) \log r$$

$$= \frac{1}{2\pi} \int_0^{2\pi} \log^+ |f(re^{i\varphi})| d\varphi + \int_0^r \frac{n(t, f) - n(0, f)}{t} dt + n(0, f) \log r.$$

Definition 1.2

$$\left. \begin{array}{l} m(r, f) = \dfrac{1}{2\pi} \displaystyle\int_0^{2\pi} \log^+ |f(re^{i\varphi})| d\varphi, \\[4mm] m\left(r, \dfrac{1}{f-a}\right) = \dfrac{1}{2\pi} \displaystyle\int_0^{2\pi} \log^+ \dfrac{1}{|f(re^{i\varphi}) - a|} d\varphi, \quad a \neq \infty. \end{array} \right\} \tag{1.2.2}$$

$m(r, f)$, also expressed as $m(r, f = \infty)$ or $m(r, \infty)$, means the average of the positive logarithm of $f(z)$ on $|z| = r$. $m(r, 1/(f - a))$ is also denoted by $m(r, f = a)$ or $m(r, a)$.

Definition 1.3

$$\left. \begin{array}{l} N(r, f) = \displaystyle\int_0^r \dfrac{n(t, f) - n(0, f)}{t} dt + n(0, f) \log r, \\[4mm] N\left(r, \dfrac{1}{f-a}\right) = \displaystyle\int_0^r \dfrac{n(t, 1/(f-a)) - n(0, 1/(f-a))}{t} dt \\[4mm] \qquad\qquad + n\left(0, 1/(f-a)\right) \log r, \quad a \neq \infty. \end{array} \right\} \tag{1.2.3}$$

$n(t, 1/(f - a))$, sometimes expressed as $n(t, f = a)$ or $n(t, a)$, denotes the number of zeros of $f(z) - a$ on $|z| \leq t$, each zero being counted with its proper multiplicity. $n(0, 1/(f - a))$, or $n(0, f = a)$ or $n(0, a)$, is the multiplicity of the zero of $f(z) - a$ at the origin. $N(r, f)$, sometimes expressed as $N(r, f = \infty)$ or $N(r, \infty)$, is called the counting function of poles of $f(z)$. $N(r, 1/(f - a))$ is sometimes denoted by $N(r, f = a)$ or $N(r, a)$.

Definition 1.4

$$T(r, f) = m(r, f) + N(r, f).\qquad(1.2.4)$$

$T(r, f)$ is usually called the characteristic function of $f(z)$. Obviously $T(r, f)$ is non-negative.

Thus Jensen's formula can be written as

$$\log |c_\tau| + T\left(r, \frac{1}{f}\right) = T(r, f).\qquad(1.2.5)$$

(1.2.5) is also known as the Jensen-Nevanlinna formula.

1.2.2 Characteristic functions of products and sums; Examples.
In order to give upper bounds of the characteristic functions of the products and sums of a finite number of meromorphic functions, we note first the following properties of the positive logarithm.

If a_ν $(\nu = 1, 2, \cdots, p)$ are p finite complex numbers, then we have

$$\log^+ \left| \prod_{\nu=1}^p a_\nu \right| \leq \log^+ \left\{ \prod_{\nu=1}^p \max\left(1, |a_\nu|\right) \right\} = \log \left\{ \prod_{\nu=1}^p \max(1, |a_\nu|) \right\}$$

$$= \sum_{\nu=1}^p \log \left\{ \max(1, |a_\nu|) \right\} = \sum_{\nu=1}^p \log^+ |a_\nu|,$$

and

$$\log^+ \left| \sum_{\nu=1}^p a_\nu \right| \leq \log^+ \left\{ \sum_{\nu=1}^p |a_\nu| \right\} \leq \log^+ \left\{ p \left(\max_{1 \leq \nu \leq p} |a_\nu| \right) \right\}$$

$$\leq \sum_{\nu=1}^p \log^+ |a_\nu| + \log p.$$

Thus, if $f_\nu(z)$ $(\nu = 1, 2, \cdots, p)$ are meromorphic in $|z| < R$, then we have

$$m\left(r, \prod_{\nu=1}^p f_\nu\right) \leq \sum_{\nu=1}^p m(r, f_\nu),$$

$$m\left(r, \sum_{\nu=1}^p f_\nu\right) \leq \sum_{\nu=1}^p m(r, f_\nu) + \log p,$$

whenever $0 < r < R$. On the other hand, it is clear that

$$n\left(r, \prod_{\nu=1}^{p} f_{\nu}\right) \leq \sum_{\nu=1}^{p} n(r, f_{\nu}),$$

$$n\left(r, \sum_{\nu=1}^{p} f_{\nu}\right) \leq \sum_{\nu=1}^{p} n(r, f_{\nu}).$$

Therefore we obtain

$$T\left(r, \prod_{\nu=1}^{p} f_{\nu}\right) \leq \sum_{\nu=1}^{p} T(r, f_{\nu}), \tag{1.2.6}$$

$$T\left(r, \sum_{\nu=1}^{p} f_{\nu}\right) \leq \sum_{\nu=1}^{p} T(r, f_{\nu}) + \log p. \tag{1.2.7}$$

Examples.

1) $f(z) \equiv C$.

In this case, $m(r, f) = \log^+ |C|$ and $N(r, f) = 0$. Thus $T(r, f) = \log^+ |C|$.

2) $f(z) = \dfrac{a_p z^p + a_{p-1} z^{p-1} + \cdots + a_0}{b_q z^q + b_{q-1} z^{q-1} + \cdots + b_0}, \quad a_p, b_q \neq 0.$

When r is sufficiently large,

$$m(r, f) = \begin{cases} (p - q) \log r + O(1), & p > q, \\ O(1), & p \leq q, \end{cases}$$

$$N(r, f) = q \log r.$$

Hence

$$T(r, f) = \max(p, q) \log r + O(1).$$

3) $f(z) = e^z$.

We have

$$m(r, f) = \frac{1}{2\pi} \int_0^{2\pi} \log^+(e^{r \cos \theta}) d\theta = \frac{1}{2\pi} \int_{-\frac{\pi}{2}}^{\frac{\pi}{2}} r \cos \theta d\theta$$

$$= \frac{r}{\pi}, \qquad N(r, f) \equiv 0.$$

Consequently, $T(r, f) = r/\pi$.

1.2.3 First fundamental theorem; Properties of characteristic functions.   Let us now prove the first fundamental theorem of Nevanlinna.

Theorem 1.2   *Let $f(z)$ be meromorphic in $|z| < R(\leq \infty)$. If a is an arbitrary complex number and $0 < r < R$, then we have*

$$m\left(r, \frac{1}{f-a}\right) + N\left(r, \frac{1}{f-a}\right) = T(r,f) + \log|C_r| + \varepsilon(a,r), \qquad (1.2.8)$$

where C_r is the first non-zero coefficient in the Taylor expansion of $\dfrac{1}{f(z)-a}$ at the origin, and

$$|\varepsilon(a,r)| \leq \log^+|a| + \log 2. \qquad (1.2.9)$$

In fact, applying Jensen's formula (1.2.5) to $f(z) - a$ gives

$$T\left(r, \frac{1}{f-a}\right) = T(r, f-a) + \log|C_r|.$$

Since

$$T(r, f-a) \leq T(r,f) + \log^+|a| + \log 2,$$

and

$$T(r,f) = T(r, f - a + a) \leq T(r, f-a) + \log^+|a| + \log 2,$$

the conclusion of Theorem 1.2 follows immediately.

We now discuss two properties of characteristic functions which will be useful later.

1) Let $f(z)$ be meromorphic in $|z| < R$. If $g(z) = (\alpha f + \beta)/(\gamma f + \delta)$, where $\alpha, \beta, \gamma, \delta$ are constants such that $\alpha\delta - \beta\gamma \neq 0$, then

$$T(r,g) = T(r,f) + O(1), \quad 0 < r < R. \qquad (1.2.10)$$

Since $f(z)$ can be expressed as

$$f(z) = \frac{-\delta g + \beta}{\gamma g - \alpha},$$

we need only to prove

$$T(r,g) \leq T(r,f) + O(1), \quad 0 < r < R. \qquad (1.2.11)$$

As a matter of fact,

$$T(r,g) = T\left(r, \frac{\alpha}{\gamma} + \frac{\beta - (\alpha\delta)/\gamma)}{\gamma f + \delta}\right)$$

$$\leq \log^+\left|\frac{\alpha}{\gamma}\right| + \log^+\left|\beta - \frac{\alpha\delta}{\gamma}\right| + T\left(r, \frac{1}{\gamma f + \delta}\right) + \log 2. \tag{1.2.12}$$

Clearly

$$T\left(r, \frac{1}{\gamma f + \delta}\right) = T(r, \gamma f + \delta) + O(1)$$

$$\leq T(r, f) + \log^+|\gamma| + \log^+|\delta| + \log 2 + O(1). \tag{1.2.13}$$

Substituting (1.2.13) into (1.2.12), we obtain (1.2.11).

2) If $f(z)$ is holomorphic on $|z| \leq R$ and $M(r, f) = \max_{|z| \leq r} |f(z)|$, then we have

$$T(r, f) \leq \log^+ M(r, f) \leq \frac{R+r}{R-r} T(R, f), \tag{1.2.14}$$

whenever $0 < r < R$.

Since $f(z)$ is holomorphic, the first inequality of (1.2.14) is obvious from the definition of $m(r, f)$.

If $M(r, f) \leq 1$, the second inequality of (1.2.14) holds automatically. If $M(r, f) > 1$, we can assume $|f(z_0)| = M(r, f)$, where $z_0 = re^{i\varphi}$. Applying the Poisson-Jensen formula and noting $|(R(z - a_\mu))/(R^2 - \bar{a}_\mu z)| < 1$, we have

$$\log^+ M(r, f) = \log|f(z_0)|$$

$$\leq \frac{1}{2\pi} \int_0^{2\pi} \log|f(Re^{i\varphi})| \frac{R^2 - r^2}{R^2 - 2Rr\cos(\theta - \varphi) + r^2} d\varphi$$

$$\leq \frac{R+r}{R-r} \cdot \frac{1}{2\pi} \int_0^{2\pi} \log^+\left|f(Re^{i\varphi})\right| d\varphi$$

$$= \frac{R+r}{R-r} T(R, f).$$

1.2.4 Cartan's identity; Order.   In order to introduce an identity of H. Cartan which will be used for discussing further properties of characteristic functions, the following lemma is needed.

Lemma 1.1   *If a is an arbitrary finite complex number, then we have*

$$\frac{1}{2\pi} \int_0^{2\pi} \log |a - e^{i\theta}| d\theta = \log^+ |a|. \tag{1.2.15}$$

We may assume that $a \neq 0$, since (1.2.15) holds obviously for $a = 0$. When $|a| > 1$, the function $a - z$ has neither zero nor pole in $|z| < 1$, so that

$$\log |a| = \frac{1}{2\pi} \int_0^{2\pi} \log |a - e^{i\theta}| d\theta.$$

When $|a| \leq 1$, $a - z$ has one zero, namely a and no pole in $|z| < 1$. Thus

$$\log |a| = \frac{1}{2\pi} \int_0^{2\pi} \log |a - e^{i\theta}| d\theta - \log \frac{1}{|a|}.$$

Therefore, (1.2.15) holds in each case.

Theorem 1.3   *If f(z) is meromorphic in $|z| < R$, then we have*

$$T(r, f) = \frac{1}{2\pi} \int_0^{2\pi} N(r, f = e^{i\theta}) d\theta + \log^+ |f(0)|, \tag{1.2.16}$$

whenever $0 < r < R$.

(1.2.16) is usually referred to as Cartan's identity.

Proof.   Applying Jensen's formula to $f(z) - e^{i\theta}$ yields

$$\log |f(0) - e^{i\theta}| = \frac{1}{2\pi} \int_0^{2\pi} \log |f(re^{i\varphi}) - e^{i\varphi}| d\varphi + N(r, f) - N(r, f = e^{i\theta}).$$

Integrating both sides with respect to θ and changing the order of integration in the resulting double integral on the right–hands side, we obtain

$$\frac{1}{2\pi} \int_0^{2\pi} \log |f(0) - e^{i\theta}| d\theta = \frac{1}{2\pi} \int_0^{2\pi} \left\{ \frac{1}{2\pi} \int_0^{2\pi} \log \left| f(re^{i\varphi}) - e^{i\theta} \right| d\theta \right\} d\varphi$$

$$+ N(r, f) - \frac{1}{2\pi} \int_0^{2\pi} N(r, f = e^{i\theta}) d\theta.$$

From Lemma 1.1, the left-hand side of the above equality equals $\log^+ |f(0)|$ and the first term of the right-hand side is

$$\frac{1}{2\pi} \int_0^{2\pi} \log^+ |f(re^{i\varphi})| d\varphi, \quad \text{i.e. } m(r, f).$$

This gives (1.2.16).   □

Corollary 1. $T(r, f)$ *is a non-decreasing function of* r.

Corollary 2. $T(r, f)$ *is a convex function of* $\log r$.

In fact,

$$\frac{dT(r, f)}{d \log r} = \frac{1}{2\pi} \int_0^{2\pi} n(t, f = e^{i\theta}) d\theta.$$

Since $n(r, f = e^{i\theta})$ is a non-decreasing function of r for every $\theta (0 \leq \theta < 2\pi)$, $T(r, f)$ is a convex function of $\log r$.

Next we introduce a definition of order.

Definition 1.5   *Let* $S(r)$ *be a real function in* (r_0, ∞), *where* $r_0 \geq 0$. *If* $S(r)$ *is non-negative and non-decreasing in this interval, then its order* λ *and lower order* μ *are defined respectively by*

$$\lambda = \varlimsup_{r \to \infty} \frac{\log^+ S(r)}{\log r}, \quad \mu = \varliminf_{r \to \infty} \frac{\log^+ S(r)}{\log r}.$$

It is obvious that $0 \leq \mu \leq \lambda \leq \infty$.

If $f(z)$ is meromorphic in the finite plane[1], then $T(r, f)$ is non-negative and non-decreasing in $(0, \infty)$, and therefore by the Definition 1.5, the order and the lower order of $T(r, f)$ are defined.

Definition 1.6   *Suppose* $f(z)$ *is meromorphic in the finite plane. The order* λ *and the lower order* μ *of* $f(z)$ *are defined as the order and the lower order of* $T(r, f)$, *respectively* :

$$\lambda = \varlimsup_{r \to \infty} \frac{\log^+ T(r, f)}{\log r}, \quad \mu = \varliminf_{r \to \infty} \frac{\log^+ T(r, f)}{\log r}.$$

From Theorem 1.2, we can see that the order of $N(r, 1/(f - a))$ does not exceed the order of $f(z)$. Since

$$n\left(r, \frac{1}{f - a}\right) \leq \frac{1}{\log 2} \int_r^{2r} \frac{n\left(t, \frac{1}{f - a}\right)}{t} dt \leq \frac{1}{\log 2} N\left(2r, \frac{1}{f - a}\right),$$

we have the following conclusion.

---

1) This means $f(z)$ is meromorphic in $|z| < \infty$.

If $f(z)$ is meromorphic in the finite plane, then the order of $n(r, 1/(f - a))$ does not exceed the order of $f(z)$ for every complex number a.

A very nontrivial fact is that the order of $n(r, 1/(f-a))$ equals exactly the order of $f(z)$, except for at most two values of the complex number a. This is a consequence of Nevanlinna's second fundamental theorem which will be introduced in the next section.

1.3 The Second Fundamental Theorem

1.3.1   A simple form of the second fundamental theorem.   Before proving the second fundamental theorem of Nevanlinna, we note first a simple fact.

Lemma 1.2   If $f_1(z)$ and $f_2(z)$ are meromorphic in $|z| < R$ $(R \leq \infty)$, then we have

$$N(r, f_1 f_2) - N\left(r, \frac{1}{f_1 f_2}\right) = N(r, f_1) + N(r, f_2) - N\left(r, \frac{1}{f_1}\right) - N\left(r, \frac{1}{f_2}\right),$$
$$(1.3.1)$$

whenever $0 < r < R$.

Proof.   We may assume without loss of generality that $f_1(0)$ and $f_2(0)$ are finite and non-zero, otherwise it would be sufficient to replace $f_1(0)$ or $f_2(0)$ by the first non-zero coefficient of the corresponding Taylor expansions at the origin. Then we have

$$N(r, f_1 f_2) - N\left(r, \frac{1}{f_1 f_2}\right)$$

$$= \frac{1}{2\pi} \int_0^{2\pi} \log \frac{1}{|f_1(re^{i\varphi}) f_2(re^{i\varphi})|} d\varphi + \log |f_1(0) f_2(0)|$$

$$= \left\{ \frac{1}{2\pi} \int_0^{2\pi} \log \frac{1}{|f_1(re^{i\varphi})|} d\varphi + \log |f_1(0)| \right\}$$

$$+ \left\{ \frac{1}{2\pi} \int_0^{2\pi} \log \frac{1}{|f_2(re^{i\varphi})|} d\varphi + \log |f_2(0)| \right\}$$

$$= N(r, f_1) - N\left(r, \frac{1}{f_1}\right) + N(r, f_2) - N\left(r, \frac{1}{f_2}\right). \quad \square$$

Now let us prove a simple form of the second fundamental theorem of

Nevanlinna, involving three counting functions. This is the original result of R. Nevanlinna [1] in 1925.

Theorem 1.4   Let $f(z)$ be meromorphic in $|z| < R(\leq \infty)$. If $f(0) \neq 0, 1, \infty$ and $f'(0) \neq 0$, then we have for $r \in (0, R)$

$$T(r, f) \leq N(r, f) + N\left(r, \frac{1}{f}\right) + N\left(r, \frac{1}{f-1}\right) - N_1(r) + S(r, f), \quad (1.3.2)$$

where

$$N_1(r) = \left(2N(r, f) - N(r, f')\right) + N\left(r, \frac{1}{f'}\right) \qquad (1.3.3)$$

and

$$S(r, f) = m\left(r, \frac{f'}{f}\right) + m\left(r, \frac{f'}{f-1}\right) + \log\left|\frac{f(0)(f(0) - 1)}{f'(0)}\right| + \log 2. \quad (1.3.4)$$

Proof.   The identity

$$\frac{1}{f} \equiv 1 - \frac{f'}{f} \cdot \frac{f-1}{f'}$$

gives

$$m\left(r, \frac{1}{f}\right) \leq m\left(r, \frac{f'}{f}\right) + m\left(r, \frac{f-1}{f'}\right) + \log 2. \qquad (1.3.5)$$

We apply Jensen's formula to the term $m(r, 1/f))$ and obtain

$$m\left(r, \frac{1}{f}\right) = T(r, f) - N\left(r, \frac{1}{f}\right) + \log\frac{1}{|f(0)|}. \qquad (1.3.6)$$

Similarly we have

$$m\left(r, \frac{f-1}{f'}\right) = m\left(r, \frac{f'}{f-1}\right) + \log\left|\frac{f(0) - 1}{f'(0)}\right|$$

$$+ \left\{N\left(r, \frac{f'}{f-1}\right) - N\left(r, \frac{f-1}{f'}\right)\right\}.$$

By Lemma 1.2,

$$m\left(r, \frac{f-1}{f'}\right) = m\left(r, \frac{f'}{f-1}\right) + \log\left|\frac{f(0) - 1}{f'(0)}\right|$$

$$+ \left\{N(r, f') + N\left(r, \frac{1}{f-1}\right) - N(r, f) - N\left(r, \frac{1}{f'}\right)\right\}.$$
$$(1.3.7)$$

Substituting (1.3.6) and (1.3.7) into (1.3.5), we obtain (1.3.2), where $N_1(r)$ and $S(r,f)$ are given by (1.3.3) and (1.3.4), respectively. $\square$

Remark. In Theorem 1.4, the condition that $f(0) \neq 0, 1, \infty$ and $f'(0) \neq 0$ is not an essential restriction and was used merely to justify the application of Jensen's formula. When it is not satisfied, we need only to make a suitable change in the term $\log|(f(0)(f(0)-1))/(f'(0))|$ of $S(r,f)$.

1.3.2 A general form of the second fundamental theorem.

In the following, we shall derive a general form of Nevanlinna's second fundamental theorem, as obtained by Collingwood [1] and Littlewood.

Theorem 1.5 *Suppose $f(z)$ is meromorphic and nonconstant in $|z| < R$. Suppose furthermore that $a_\nu (\nu = 1, 2, \cdots, q)$ are $q(\geq 2)$ finite complex numbers such that $\min_{1 \leq \nu_1 \leq \nu_2 \leq q} |a_{\nu_1} - a_{\nu_2}| \geq \delta > 0$. If $f(0) \neq 0$, ∞ and $f'(0) \neq 0$, then we have for every $r \in (0, R)$*

$$m(r, f) + \sum_{\nu=1}^{q} m(r, a_\nu) \leq 2T(r, f) - N_1(r) + S(r, f), \qquad (1.3.8)$$

where $N_1(r)$ is defined by (1.3.3) and

$$S(r, f) = m\left(r, \frac{f'}{f}\right) + m\left(r, \sum_{\nu=1}^{q} \frac{f'}{f - a_\nu}\right) + q \log^+ \frac{2q}{\delta} + \log 2 + \log \frac{1}{|f'(0)|}.$$
$$(1.3.9)$$

Proof. Let

$$F(z) = \sum_{\nu=1}^{q} \frac{1}{f(z) - a_\nu}.$$

For every fixed value of r, set

$$E_j = \left\{ \theta : 0 \leq \theta < 2\pi \text{ and } |f(re^{i\theta}) - a_j| < \frac{\delta}{2q} \right\}, \quad j = 1, 2, \cdots, q.$$

When $\nu \neq j$ and $\theta \in E_j$, it is clear that

$$|f(re^{i\theta}) - a_\nu| \geq |a_\nu - a_j| - |f(re^{i\theta}) - a_j| \geq \delta\left(1 - \frac{1}{2q}\right).$$

Since

$$F(re^{i\theta}) = \frac{1}{f(re^{i\theta}) - a_j}\left\{1 + \sum_{\nu \neq j} \frac{f(re^{i\theta}) - a_j}{f(re^{i\theta}) - a_\nu}\right\},$$

we have

$$|F(re^{i\theta})| > \left|\frac{1}{f(re^{i\theta}) - a_j}\right| \left\{1 - (q-1)\frac{\delta/2q}{\delta(1 - (1/2q))}\right\} > \frac{1}{2|f(re^{i\theta}) - a_j|}.$$

When $\theta \in E_j$, we have $\theta \bar{\in} E_\nu (\nu \neq j)$, hence

$$\log^+ |F(re^{i\theta})| > \log^+ \frac{1}{|f(re^{i\theta}) - a_j|} - \log 2$$

$$\geq \sum_{\nu=1}^q \log^+ \frac{1}{|f(re^{i\theta}) - a_\nu|} - q \log^+ \frac{2q}{\delta} - \log 2. \tag{1.3.10}$$

The inequality (1.3.10) is obvious for $\theta \bar{\in} \cup_{j=1}^q E_j$. Thus (1.3.10) holds for $0 \leq \theta < 2\pi$ and

$$m(r, F) \geq \sum_{\nu=1}^q m\left(r, \frac{1}{f - a_\nu}\right) - q \log^+ \frac{2q}{\delta} - \log 2. \tag{1.3.11}$$

Moreover, we have an appropriate upper bound of $m(r, F)$:

$$m(r, F) \leq m(r, f'F) + m\left(r, \frac{1}{f'}\right)$$

$$\leq m(r, f'F) + T(r, f') - N\left(r, \frac{1}{f'}\right) + \log \frac{1}{|f'(0)|}. \tag{1.3.12}$$

Clearly then

$$T(r, f') = m(r, f') + N(r, f') \leq m(r, f) + m\left(r, \frac{f'}{f}\right) + N(r, f')$$

$$= T(r, f) + m\left(r, \frac{f'}{f}\right) + \left\{N(r, f') - N(r, f)\right\}. \tag{1.3.13}$$

Substituting (1.3.13) into (1.3.12) and comparing the resulting inequality with (1.3.11), we obtain

$$m(r, f) + \sum_{\nu=1}^q m\left(r, \frac{1}{f - a_\nu}\right) \leq 2T(r, f)$$

$$-\left\{2N(r, f) - N(r, f') + N\left(r, \frac{1}{f'}\right)\right\} + m\left(r, \sum_{\nu=1}^q \frac{f'}{f - a_\nu}\right) \tag{1.3.14}$$

$$+m\left(r, \frac{f'}{f}\right) + q \log^+ \frac{2q}{\delta} + \log 2 + \log \frac{1}{|f'(0)|},$$

which is the desired result. $\square$

1.3.3 The fundamental lemma on the logarithmic derivative.

For further applications, we should prove that $S(r, f)$ in Theorems 1.4 and 1.5 grows more slowly than $T(r, f)$ when r tends to R. For a given function $f(z)$, the last two terms of $S(r, f)$ are constants and the first two terms of $S(r, f)$ have the same form $m(r, f'/f)$ and $m(r, (f-1)'/(f-1))$, and therefore we should study the growth of $m(r, f'/f)$.

As a matter of fact, for some special functions, $m(r, f'/f)$ grows more slowly than $T(r, f)$. For instance, if $f(z)$ is a polynomial $P(z)$ of degree k, then $T(r, p) = k \log r + O(1)$ and $m(r, p'/p) = 0(r > r_0)$. If $f(z) = e^z$, then $T(r, e^z) = r/\pi$ and $m(r, (e^z)'/e^z) = 0$. Nevanlinna's fundamental lemma, a very deep result, states that the above assertion holds for an arbitrary meromorphic function. This lemma is also called the lemma on the logarithmic derivative and plays a key role in applications of the second fundamental theorem. The following form of the lemma is an improved form given by G. Valiron [3].

Lemma 1.3   *Let $f(z)$ be meromorphic in $|z| < R(\leq \infty)$. If $f(0) \neq 0, \infty$, then we have*

$$m\left(r, \frac{f'}{f}\right) < 10 + 4\log^+ \log^+ \frac{1}{|f(0)|} + 2\log^+ \frac{1}{r}$$

$$+3\log^+ \frac{1}{\rho - r} + 4\log^+ \rho + 4\log^+ T(\rho, f),$$

(1.3.15)

whenever $0 < r < \rho < R$.

Proof.   Since $f'/f = (\log f)'$, we can start from the Poisson-Jensen formula for $\log f$ to get an estimate of $m(r, f'/f)$ by taking the derivative of every term and estimating its modulus.

Setting $z = re^{i\theta}$, the Poisson-Jensen formula has the form

$$\log|f(z)| = \frac{1}{2\pi} \int_0^{2\pi} \log|f(\rho e^{i\varphi})| \frac{\rho^2 - r^2}{\rho^2 - 2\rho r \cos(\varphi - \theta) + r^2} d\varphi$$

$$- \sum_{|a_\mu| \leq \rho} \log\left|\frac{\rho^2 - \bar{a}_\mu z}{\rho(z - a_\mu)}\right| + \sum_{|b_\nu| \leq \rho} \log\left|\frac{\rho^2 - \bar{b}_\nu z}{\rho(z - b_\nu)}\right|.$$

Because

$$\frac{\rho^2 - r^2}{\rho^2 - 2\rho r \cos(\varphi - \theta) + r^2} = Re\left(\frac{\rho e^{i\varphi} + z}{\rho e^{i\varphi} - z}\right),$$

we have

$$\log f(z) = \frac{1}{2\pi} \int_0^{2\pi} \log |f(\rho e^{i\varphi})| \frac{\rho e^{i\varphi} + z}{\rho e^{i\varphi} - z} d\varphi$$

$$- \sum \log \frac{\rho^2 - \bar{a}_\nu z}{\rho(z - a_\mu)} + \sum \log \frac{\rho^2 - \bar{b}_\nu z}{\rho(z - b_\nu)} + iC, \tag{1.3.16}$$

hence

$$\frac{f'(z)}{f(z)} = \frac{1}{2\pi} \int_0^{2\pi} \log |f(\rho e^{i\varphi})| \frac{2\rho e^{i\varphi}}{(\rho e^{i\varphi} - z)^2} d\varphi$$

$$- \sum \frac{|a_\mu|^2 - \rho^2}{(z - a_\mu)(\rho^2 - \bar{a}_\mu z)} + \sum \frac{|b_\nu|^2 - \rho^2}{(z - b_\nu)(\rho^2 - \bar{b}_\nu z)}.$$

When $|z| = r$, it is clear that

$$\left| \frac{|a_\mu|^2 - \rho^2}{(z - a_\mu)(\rho^2 - \bar{a}_\mu z)} \right| = \frac{\rho(\rho^2 - |a_\mu|^2)}{|\rho^2 - \bar{a}_\mu z|^2} \left| \frac{\rho^2 - \bar{a}_\mu z}{\rho(z - a_\mu)} \right|$$

$$\leq \frac{\rho^3}{(\rho^2 - \rho r)^2} \left| \frac{\rho^2 - \bar{a}_\mu z}{\rho(z - a_\mu)} \right|$$

$$= \frac{\rho}{(\rho - r)^2} \left| \frac{\rho^2 - \bar{a}_\mu z}{\rho(z - a_\mu)} \right|.$$

Similarly we have

$$\left| \frac{|b_\nu|^2 - \rho^2}{(z - b_\nu)(\rho^2 - \bar{b}_\nu z)} \right| \leq \frac{\rho}{(\rho - r)^2} \left| \frac{\rho^2 - \bar{b}_\nu z}{\rho(z - b_\nu)} \right|.$$

Thus

$$\left| \frac{f'(z)}{f(z)} \right| \leq \frac{2\rho}{(\rho - r)^2} \left\{ \frac{1}{2\pi} \int_0^{2\pi} |\log |f(\rho e^{i\varphi})|| d\varphi \right.$$

$$\left. + \sum \left| \frac{\rho^2 - \bar{a}_\mu z}{\rho(z - a_\mu)} \right| + \sum \left| \frac{\rho^2 - \bar{b}_\nu z}{\rho(z - b_\nu)} \right| \right\}.$$

Since

$$\frac{1}{2\pi} \int_0^{2\pi} \left| \log |f(\rho e^{i\varphi})| \right| d\varphi$$

$$= m(\rho, f) + m\left(\rho, \frac{1}{f}\right) \leq 2T(\rho, f) + \log \frac{1}{|f(0)|},$$

we obtain

$$\log^+\left|\frac{f'(z)}{f(z)}\right| \le \log^+\frac{2\rho}{(\rho-r)^2} + \log^+ 2T(\rho, f)$$

$$+ \log^+\log^+\frac{1}{|f(0)|} + \sum\log^+\left|\frac{\rho^2-\bar{a}_\mu z}{\rho(z-a_\mu)}\right|$$

$$+ \sum\log^+\left|\frac{\rho^2-\bar{b}_\nu z}{\rho(z-b_\nu)}\right| + \log\left\{n(\rho, f) + n\left(\rho, \frac{1}{f}\right) + 2\right\}.$$

For the function $(\rho^2 - \bar{a}_\mu z)/(\rho(z-a_\mu))$, Jensen's formula gives

$$\log\frac{\rho}{|a_\mu|} = m\left(r, \frac{\rho^2-\bar{a}_\mu z}{\rho(z-a_\mu)}\right) + \log^+\frac{r}{|a_\mu|},$$

so

$$\sum m\left(r, \frac{\rho^2-\bar{a}_\mu z}{\rho(z-a_\mu)}\right) = \sum\log\frac{\rho}{|a_\mu|} - \sum\log^+\frac{r}{|a_\mu|}$$

$$= N\left(\rho, \frac{1}{f}\right) - N\left(r, \frac{1}{f}\right).$$

Similarly we have

$$\sum m\left(r, \frac{\rho^2-\bar{b}_\nu z}{\rho(z-b_\nu)}\right) = N(\rho, f) - N(r, f).$$

Therefore

$$m\left(r, \frac{f'}{f}\right) < 2\log 2 + \log^+\rho + 2\log^+\frac{1}{\rho-r}$$

$$+ \log^+ T(\rho, f) + \log^+\log^+\frac{1}{|f(0)|} + N(\rho, f) \qquad (1.3.17)$$

$$- N(r, f) + N\left(\rho, \frac{1}{f}\right) - N\left(r, \frac{1}{f}\right)$$

$$+ \log\left\{n(\rho, f) + n\left(\rho, \frac{1}{f}\right) + 2\right\}.$$

In order to estimate $n(\rho) = n(\rho, f) + n(\rho, 1/f)$, we denote its corresponding counting function by $N(\rho)$. Choosing ρ' such that $\rho < \rho' < R$, we have

$$N(\rho') \ge \int_\rho^{\rho'}\frac{n(t)dt}{t} \ge n(\rho)\frac{\rho'-\rho}{\rho'},$$

so

$$n(\rho) \le \frac{\rho'}{\rho'-\rho}N(\rho') \le \frac{\rho'}{\rho'-\rho}\left\{2T(\rho', f) + \log^+\frac{1}{|f(0)|}\right\}.$$

Thus

$$\log^+\{n(\rho)+2\} \leq \log^+ \rho' + \log^+ \frac{1}{\rho'-\rho} + \log^+ \log^+ \frac{1}{|f(0)|}$$

$$+ \log^+ T(\rho',f) + 4\log 2. \tag{1.3.18}$$

Next, for an estimate of $N(\rho) - N(r)$. Since $N(t)$ is a convex function of $\log t$,

$$\frac{N(\rho) - N(r)}{\log \rho - \log r} \leq \frac{N(\rho') - N(r)}{\log \rho' - \log r},$$

i.e.

$$N(\rho) - N(r) \leq \frac{\log \frac{\rho}{r}}{\log \frac{\rho'}{r}} N(\rho'),$$

whenever $0 < r < \rho < \rho' < R$.

By

$$\log \frac{\rho}{r} = \int_r^\rho \frac{dt}{t} \leq \frac{\rho-r}{r}$$

and

$$\log \frac{\rho'}{r} = \int_r^{\rho'} \frac{dt}{t} \geq \frac{\rho'-r}{\rho'},$$

we obtain

$$N(\rho) - N(r) \leq \frac{\rho'}{r} \cdot \frac{\rho-r}{\rho'-r}\left\{2T(\rho',f) + \log^+ \frac{1}{|f(0)|}\right\}.$$

Choosing ρ such that

$$\rho = r + \frac{r(\rho'-r)}{2\rho'\left\{T(\rho',f) + \log^+ \frac{1}{|f(0)|} + 1\right\}}, \tag{1.3.19}$$

we have $0 < r < \rho < \rho' < R$ and

$$N(\rho) - N(r) < 1. \tag{1.3.20}$$

Furthermore, (1.3.19) gives

$$\log^+ \frac{1}{\rho-r} \leq \log^+ \frac{1}{r} + \log^+ \frac{1}{\rho'-r} + \log^+ \rho'$$

$$+ \log^+ T(\rho',f) + \log^+ \log^+ \frac{1}{|f(0)|} + \log 6. \tag{1.3.21}$$

Finally,

$$\rho - r < \frac{\rho' - r}{2}$$

implies

$$\rho' - \rho = (\rho' - r) - (\rho - r) > \frac{\rho' - r}{2},$$

hence

$$\log^+ \frac{1}{\rho' - \rho} < \log 2 + \log^+ \frac{1}{\rho' - r}. \tag{1.3.22}$$

Then, substituting (1.3.18), (1.3.20) and (1.3.21) into (1.3.17) and noting (1.3.22), we obtain

$$m\left(r, \frac{f'}{f}\right) < 9\log 2 + 2\log 3 + 1 + 4\log^+ \log^+ \frac{1}{|f(0)|} + 2\log^+ \frac{1}{r}$$

$$+ 3\log^+ \frac{1}{\rho' - r} + 4\log^+ \rho' + 4\log^+ T(\rho', f).$$

This proves Lemma 1.3. $\square$

1.3.4 The Borel Lemma.

In order to deal with the term $\log^+ T(\rho, f)$ which appears in the estimate of $m(r, f'/f)$ in Lemma 1.3, we need a lemma on monotone functions. This lemma is essentially due to Borel [1].

Lemma 1.4 (1) *If $T(r)$ is a continuous, non-decreasing function in $[r_0, \infty)$ and $T(r_0) \geq 1$, then we have*

$$T\left(r + \frac{1}{T(r)}\right) < 2T(r), \tag{1.3.23}$$

except on a set E_0 of r with linear measure less than or equal to 2.

(2) *If $T(r)$ is a continuous, non-decreasing function in $[r_0, R)$ and $T(r_0) \geq 1$, then we have*

$$T\left(r + \frac{R - r}{eT(r)}\right) < 2T(r), \tag{1.3.24}$$

except on a set E_0 of r with $\int_{E_0} (dr)/(R - r) \leq 2$. In particular, if $r_0 < r_1 < r_2 < R$ and $R - r_2 < (R - r_1)/(e^2)$, then there exist some values of r in (r_1, r_2) satisfying (1.3.24).

Proof.   (1) Denote by E_0 the set of $r(\geq r_0)$ such that (1.3.23) is not satisfied. Since $T(r)$ is continuous, E_0 is a closed set. Let

$$r_0' = r_0,$$

$$r_1 = \min\left\{E_0 \cap [r_0', \infty)\right\}, \qquad r_1' = r_1 + \frac{1}{T(r_1)},$$

$$\cdots\cdots$$

$$r_\nu = \min\left\{E_0 \cap [r_{\nu-1}', \infty)\right\}, \qquad r_\nu' = r_\nu + \frac{1}{T(r_\nu)},$$

$$\cdots\cdots$$

If $E_0 \cap [r_\nu', \infty) \neq \emptyset$ for every $\nu = 0, 1, 2, \cdots$, then we have an infinite sequence $\{r_\nu\}$ with $r_\nu < r_\nu' \leq r_{\nu+1}(\nu = 1, 2, \cdots)$. It is easy to see that $\lim_{\nu\to\infty} r_\nu = \tau < \infty$ implies $\lim_{\nu\to\infty} r_\nu' = \tau$, hence $0 = \lim_{\nu\to\infty}(r_\nu' - r_\nu) = \lim_{\nu\to\infty} 1/(T(r_\nu)) \geq 1/(T(\tau)) > 0$, a contradiction. Therefore we have $\lim_{\nu\to\infty} r_\nu = \infty$, implying $E_0 \subset \cup_{\nu=1}^\infty [r_\nu, r_\nu']$, hence mes $E_0 \leq \sum_{\nu=1}^\infty (r_\nu' - r_\nu) = \sum_{\nu=1}^\infty 1/(T(r_\nu))$. Since $r_\nu \in E_0(\nu = 1, 2, \cdots)$, we have

$$T(r_\nu) \geq T(r_{\nu-1}') \geq 2T(r_{\nu-1}) \geq \cdots \geq 2^{\nu-1}T(r_1) \geq 2^{\nu-1}.$$

So we obtain finally

$$\text{mes } E_0 \leq \sum_{\nu=1}^\infty \frac{1}{2^{\nu-1}} = 2.$$

If $E_0 \cap [r_{\tilde\nu}', \infty) = \emptyset$ for some $\tilde\nu = 0, 1, 2, \cdots$, then $E_0 \subset \cup_{\nu=1}^{\tilde\nu}[r_\nu, r_\nu']$, and therefore mes $E_0 \leq 2$ in view of the above argument.

(2) We shall change the variable and reduce this case to (1). In fact, by setting

$$\rho = \log\frac{1}{R-r}, \qquad r = R - e^{-\rho}, \qquad \rho_0 = \log\frac{1}{R-r_0},$$

then $T_1(\rho) = T(R - e^{-\rho})$ is defined in $\rho_0 \leq \rho < \infty$. According to the case (1), the linear measure of the set E of the values $\rho(\geq \rho_0)$ such that the inequality

$$T_1\left(\rho + \frac{1}{T_1(\rho)}\right) \geq 2T_1(\rho)$$

holds does not exceed 2. If the set E_0 of r corresponds to E, then

$$\int_{E_0} \frac{dr}{R-r} = \int_E d\rho \leq 2.$$

When $r \bar{\in} E_0$, we have $T(r') < 2T(r)$, where

$$\log \frac{1}{R - r'} = \log \frac{1}{R - r} + \frac{1}{T(r)}.$$

This gives

$$r' = R - (R - r)e^{-\frac{1}{T(r)}} = r + (R - r)\left\{1 - e^{-\frac{1}{T(r)}}\right\}.$$

Noting that

$$1 - e^{-\frac{1}{T(r)}} = \int_0^{\frac{1}{T(r)}} \frac{dx}{e^x} \geq \frac{\frac{1}{T(r)}}{e^{\frac{1}{T(r)}}} \geq \frac{1}{eT(r)},$$

we obtain

$$r' \geq r + \frac{R - r}{eT(r)}.$$

When $R - r_2 < (R - r_1)/e^2$, it is clear that

$$\int_{r_1}^{r_2} \frac{dt}{R - t} = \log \frac{R - r_1}{R - r_2} > 2.$$

Thus some r can be found in (r_1, r_2) such that $r \bar{\in} E_0$, hence

$$T\left(r + \frac{R - r}{eT(r)}\right) < 2T(r). \quad \square$$

1.3.5 Error terms in the second fundamental theorem. We now discuss the error terms in the second fundamental theorem by using Lemma 1.3 and Lemma 1.4.

Theorem 1.6 *Suppose $f(z)$ is meromorphic in the finite plane and non-degenerate into a constant, and $S(r, f)$ is expressed by (1.3.4) of Theorem 1.4 or (1.3.9) of Theorem 1.5. If the order of $f(z)$ is finite, we have*

$$S(r, f) = O(\log r), \quad r \to \infty. \tag{1.3.25}$$

If the order of $f(z)$ is infinite, we have

$$S(r, f) = O(\log(rT(r, f))), \quad r \to \infty, \tag{1.3.26}$$

except on a set E with finite linear measure.

Proof.   We consider only the error terms in Theorem 1.5 since those in Theorem 1.4 are even simpler.

If

$$\varphi(z) = \prod_{\nu=1}^{q}(f(z) - a_\nu),$$

then

$$S(r, f) = m\left(r, \frac{f'}{f}\right) + m\left(r, \frac{\varphi'}{\varphi}\right) + O(1).$$

When the order λ of $f(z)$ is finite, we have

$$T(r, f) < r^{\lambda+1}, \quad r > r_0,$$

so

$$T(r, \varphi) \leq \sum_{\nu=1}^{q} T(r, f - a_\nu) \leq qT(r, f) + O(1).$$

Choosing $\rho = 2r$ in Lemma 1.3, we can see that

$$m\left(r, \frac{f'}{f}\right) = O(\log r), \quad m\left(r, \frac{\varphi'}{\varphi}\right) = O(\log r), \quad r \to \infty.$$

Thus

$$S(r, f) = O(\log r), \quad r \to \infty.$$

When the order of $f(z)$ is infinite, there exists r_0 such that $T(r_0, f) \geq 1$ and $T(r_0, \varphi) \geq 1$. Denoting by E_1 and E_2 the exceptional sets illustrated in Lemma 1.4 which correspond to $T(r, f)$ and $T(r, \varphi)$, respectively, we have mes $E_j \leq 2(j = 1, 2)$. When $r\bar{\in}(E_1 \cup E_2)$, choosing $\rho = r + 1/(T(r, f))$, we have

$$m\left(r, \frac{f'}{f}\right) = O\{\log(rT(r, f))\},$$

$$m\left(r, \frac{\varphi'}{\varphi}\right) = O\{\log(rT(r, f))\}, \quad r \to \infty.$$

Hence

$$S(r, f) = O\{\log(rT(r, f))\}, \quad r \to \infty,$$

except on a set with linear measure less than or equal to 4.   □

If a_ν is finite and $f(0) \neq a_\nu$, then

$$m\left(r, \frac{1}{f - a_\nu}\right) = T(r, f - a_\nu) - N\left(r, \frac{1}{f - a_\nu}\right) + \log \frac{1}{|f(0) - a_\nu|}$$

$$\leq T(r, f) + \log^+ |a_\nu| + \log 2$$

$$-N\left(r, \frac{1}{f - a_\nu}\right) + \log \frac{1}{|f(0) - a_\nu|}.$$

Therefore Theorem 1.5 can be expressed as follows.

Theorem 1.5′   *Suppose $f(z)$ is meromorphic in the finite plane and non-degenerate into a constant. If $a_\nu (\nu = 1, 2, \cdots, q)$ are $q(\geq 3)$ distinct complex numbers (one of them may be infinity), then*

$$(q - 2)T(r, f) < \sum_{\nu=1}^{q} N\left(r, \frac{1}{f - a_\nu}\right) - N_1(r) + S(r, f),$$

where $N_1(r)$ is expressed by (1.3.3) and $S(r, f)$ has the properties formulated in Theorem 1.6.

1.4 Applications of the Second Fundamental Theorem

1.4.1 The Picard-Borel theorem.   Before introducing the Picard theorem, we note the following lemma.

Lemma 1.5   *If $f(z)$ is a transcendental meromorphic function in the finite plane[1], then we have*

$$\lim_{r \to \infty} \frac{T(r, f)}{\log r} = \infty. \tag{1.4.1}$$

Proof.   Suppose the conclusion is not true.   Then there exist a large positive integer M and a sequence r_ν tending to infinity such that $\lim_{\nu \to \infty} (T(r_\nu, f))/(\log r_\nu) < M$.   By the first fundamental theorem of Nevanlinna, we have

$$\lim_{\nu \to \infty} \frac{N(r_\nu, f)}{\log r_\nu} < M, \quad \lim_{\nu \to \infty} \frac{N\left(r_\nu, \frac{1}{f}\right)}{\log r_\nu} < M.$$

---

[1] A meromorphic function in the finite plane, non-degenerate into a rational function, is said to be a transcendental meromorphic function.

Thus both the number of zeros and the number of poles of $f(z)$ in the finite plane are less than M. Constructing a rational function $R(z)$ which has the same zeros and poles with the same multiplicities as $f(z)$, we have

$$\frac{f(z)}{R(z)} = e^{g(z)},$$

where $g(z)$ is entire. Since $R(z)$ is rational, we have

$$T(r, R) = O(\log r).$$

Thus

$$T\left(r_\nu, \frac{f}{R}\right) \leq T(r_\nu, f) + T(r_\nu, R) + O(1) = O(\log r_\nu),$$

so

$$\log M\left(r_\nu, \frac{f}{R}\right) \leq \frac{2r_\nu + r_\nu}{2r_\nu - r_\nu} T\left(2r_\nu, \frac{f}{R}\right) = O(\log r_\nu).$$

Therefore we have

$$\operatorname{Reg}(z) = \log\left|\frac{f(z)}{R(z)}\right| = O(\log r_\nu)$$

for every point z on $|z| = r_\nu$. By the generalized Liouville theorem (See Titchmarsh [1, 86–87]), $g(z)$ must be a constant. Consequently, $f(z)$ is rational, which contradicts the hypothesis of Lemma 1.5.

Now let us prove the following Picard theorem.

Theorem 1.7   *If $f(z)$ is transcendental meromorphic in the finite plane, then $f(z)$ takes every complex number infinitely many times, except for at most two values.*

Proof.   Suppose that the conclusion of Theorem 1.7 is not true. Then there exist three distinct complex numbers $a_\nu (\nu = 1, 2, 3)$ such that $f(z)$ takes every a_ν only finitely many times. Thus

$$N(r, a_\nu) = O(\log r), \quad (\nu = 1, 2, 3).$$

By defining

$$g(z) = \frac{f(z) - a_1}{f(z) - a_3} \cdot \frac{a_2 - a_3}{a_2 - a_1}, \tag{1.4.2}$$

it is clear that

$$N\left(r, \frac{1}{g}\right) + N\left(r, \frac{1}{g-1}\right) + N(r, g) = O(\log r).$$

Then, applying Theorem 1.4 and Theorem 1.6 to $g(z)$, we have

$$T(r, g) = O(\log(rT(r, g))),$$

except on a set with linear measure less than or equal to 4. Thus

$$\lim_{r \to \infty} \frac{T(r, g)}{\log r} < \infty.$$

By Lemma 1.5, $g(z)$ is rational, and therefore so is $f(z)$. This contradicts the hypothesis of Theorem 1.7. $\square$

Definition 1.7   Let $f(z)$ be meromorphic in the finite plane and let a be a complex number. a is called an exceptional value of $f(z)$ in the sense of Picard, if $f(z) - a$ has no zeros.

By Theorem 1.7, the number of exceptional values of a transcendental meromorphic function in the sense of Picard is at most equal to 2. Clearly, the upper bound 2 is precise. For instance, e^z has two Picard exceptional values, 0 and ∞. Moreover, $e^z - a$ has infinitely many zeros for any finite non-zero complex number a.

Definition 1.8   Let $f(z)$ be a meromorphic function of order $\lambda(0 < \lambda < \infty)$ in the finite plane. A complex number a is called an exceptional value of $f(z)$ in the sense of Borel, if

$$\varlimsup_{r \to \infty} \frac{\log n(r, f = a)}{\log r} < \lambda. \tag{1.4.3}$$

Lemma 1.6   If $f(z)$ is meromorphic and of order $\lambda(0 < \lambda < \infty)$ in the finite plane, then a necessary and sufficient condition that $f(z)$ have an exceptional value a in the sense of Borel is

$$\varlimsup_{r \to \infty} \frac{\log N(r, f = a)}{\log r} < \lambda.$$

Proof.   From

$$n(r, f = a) \leq \frac{n(r, f = a)}{\log 2} \int_r^{2r} \frac{dt}{t} \leq \frac{1}{\log 2} N(2r, f = a), \quad r \geq 1,$$

and

$$N(r, f = a) - N(r_0, f = a)$$

$$= \int_{r_0}^{r} \frac{n(t, f = a)dt}{t} \leq n(r, f = a) \log \frac{r}{r_0},$$

we can see that

$$\varlimsup_{r \to \infty} \frac{\log N(r, f = a)}{\log r} = \varlimsup_{r \to \infty} \frac{\log n(r, f = a)}{\log r}.$$

Thus the conclusion of Lemma 1.6 follows from Definition 1.8.   □

Now let us prove the following Borel theorem by use of the second fundamental theorem of Nevanlinna.

Theorem 1.8   *If $f(z)$ is meromorphic and of order $\lambda(0 < \lambda < \infty)$ in the finite plane, then $f(z)$ has at most two exceptional values in the sense of Borel.*

Proof.   Suppose the conclusion is not true. Then there are three distinct exceptional values $a_\nu(\nu = 1, 2, 3)$ in the sense of Borel. By Lemma 1.6, we have

$$\varlimsup_{r \to \infty} \frac{\log N(r, f = a_\nu)}{\log r} < \lambda, \quad \nu = 1, 2, 3.$$

We can assume that the $a_\nu(\nu = 1, 2, 3)$ are 0, 1, ∞, for otherwise, it would be sufficient to make the transformation (1.4.2). Applying Theorems 1.4 and 1.6 to $f(z)$, we have

$$T(r, f) < \sum_{\nu=1}^{3} N(r, f = a_\nu) + O(\log r).$$

Thus

$$\varlimsup_{r \to \infty} \frac{\log T(r, f)}{\log r} < \lambda,$$

which contradicts the supposition that the order of $f(z)$ equals λ.   □

Clearly, an exceptional value in the sense of Picard is also one in the sense of Borel. Thus the Borel theorem generalizes the Picard theorem for functions of finite and nonzero order.

1.4.2 Deficiency relation.   The second fundamental theorem of Nevan-linna, in addition to implying the Picard-Borel theorem, also leads to some new concepts such as deficient value, deficiency and deficiency relation.

Definition 1.9   Let $f(z)$ be transcendental meromorphic in the finite plane. The deficiency of a complex number a with respect to $f(z)$ is defined by

$$\delta(a, f) = \varliminf_{r \to \infty} \frac{m\left(r, \dfrac{1}{f - a}\right)}{T(r, f)} = 1 - \varlimsup_{r \to \infty} \frac{N\left(r, \dfrac{1}{f - a}\right)}{T(r, f)}. \tag{1.4.4}$$

It is easy to see $0 \le \delta(a, f) \le 1$.

Definition 1.10   A complex number a is called a deficient value of $f(z)$, if its deficiency is positive. A deficient value is also called an exceptional value in the sense of Nevanlinna.

Theorem 1.9   If $f(z)$ is transcendental meromorphic in the finite plane, then the set of deficient values of $f(z)$ is at most countable, and

$$\sum \delta(a, f) \le 2. \tag{1.4.5}$$

Proof.   Suppose $a_\nu (\nu = 1, 2, \cdots, q)$ are distinct complex numbers. Theorem 1.5 yields

$$\varliminf_{r \to \infty} \left\{ \sum_{\nu=1}^{q} \frac{m\left(r, \dfrac{1}{f - a_\nu}\right)}{T(r, f)} \right\} \le \varlimsup_{r \to \infty} \frac{2T(r, f)}{T(r, f)} + \varliminf_{r \to \infty} \frac{S(r, f)}{T(r, f)}.$$

Thus

$$\sum_{\nu=1}^{q} \delta(a_\nu, f) \le \varliminf_{r \to \infty} \left\{ \frac{\displaystyle\sum_{\nu=1}^{q} m\left(r, \dfrac{1}{f - a_\nu}\right)}{T(r, f)} \right\} \le 2.$$

Consequently, for each positive integer j, the number of deficient values with deficiencies larger than $1/j$ is less than $2j$. Since

$$E\{a : \delta(a, f) > 0\} = \bigcup_{j=1}^{\infty} E\left\{a : \delta(a, f) > \frac{1}{j}\right\},$$

the set of deficient values is at most countable. Moreover, for any finite q, we have $\sum_{\nu=1}^{q} \delta(a_\nu, f) \leq 2$, so that the sum of all the deficiencies does not exceed 2.   □

The upper bound 2 may be attained. For instance, the exponential function e^z satisfies

$$\delta(0, e^z) + \delta(\infty, e^z) = 2.$$

When $f(z)$ is a transcendental entire function, it is clear that $\delta(\infty, f) = 1$. Thus

$$\sum_{a \neq \infty} \delta(a, f) \leq 1. \tag{1.4.6}$$

Obviously, Theorem 1.9 implies Theorem 1.7.

Remark. The deficient value, though actually introduced by Collingwood, is usually called the exceptional value in the sense of Nevanlinna, because it was introduced on the basis of Theorem 1.5, a generalization of the second fundamental theorem of Nevanlinna (Theorem 1.4).

In Theorems 1.7 and 1.9, we assume that $f(z)$ is transcendental. When $f(z)$ is rational, the corresponding conclusions[1]   can be obtained by a similar method. As a matter of fact, if $f(z) = P(z)/Q(z)$, then

$$m\left(r, \frac{f'}{f}\right) \leq m\left(r, \frac{P'}{P}\right) + m\left(r, \frac{Q'}{Q}\right) = 0, \quad r \to \infty,$$

so $S(r, f) = O(1)$. The rest of the proof is the same.

Furthermore the following conclusion can be obtained by a direct computation.

If $f(z)$ is rational, then $f(z)$ takes any complex number, except for at most one value . Moreover, $f(z)$ has a unique deficient value.

1.4.3 Multiple value.   Now we derive another form of the second fundamental theorem of Nevanlinna.

---

1) The conclusion of Theorem 1.7 should be revised by saying that $f(z)$ can take every complex value, except for at most two values.

Let $f(z)$ be meromorphic in $|z| < R(\leq \infty)$ and a be a finite complex number. For $0 < r < R$, denote by $\bar{n}(r, f = a)$, sometimes by $\bar{n}(r, 1/(f - a))$ or $\bar{n}(r, a)$, the number of zeros of $f(z) - a$ in $|z| \leq r$, each zero being counted only once. Moreover, let

$$\bar{n}(0, f = a) = \begin{cases} 0, & \text{if } f(0) \neq a, \\ 1, & \text{if } f(0) = a. \end{cases}$$

Introduce the notation

$$\overline{N}(r, f = a) = \int_0^r \frac{\bar{n}(t, f = a) - \bar{n}(0, f = a)}{t} dt + \bar{n}(0, f = a) \log r,$$

which is sometimes denoted by $\overline{N}(r, 1/(f - a))$ or $\overline{N}(r, a)$, and called the reduced counting function of $f(z) - a$. Similarly we can define $\bar{n}(r, f = \infty)$ (sometimes denoted by $\bar{n}(r, f)$ or $\bar{n}(r, \infty)$) and $\overline{N}(r, f = \infty)$ (sometimes denoted by $\overline{N}(r, f)$ or $\overline{N}(r, \infty)$).

If $a_\nu (\nu = 1, 2, \cdots, q)$ are distinct complex numbers, then we have from Theorem 1.5'

$$(q - 2)T(r, f) < \sum_{\nu=1}^{q} N\left(r, \frac{1}{f - a_\nu}\right) - N_1(r) + S_1(r, f),$$

where $S_1(r, f)$ has the properties formulated in Theorem 1.6.

Now let us discuss the properties of

$$N_1(r) = 2N(r, f) - N(r, f') + N\left(r, \frac{1}{f'}\right).$$

Since

$$2N(r, f) - N(r, f') = \int_0^r \frac{n_1(t, f = \infty) - n_1(0, f = \infty)}{t} dt$$
$$+ n_1(0, f = \infty) \log r,$$

where $n_1(t, f = \infty) = 2n(t, f) - n(t, f')$, it is a counting function of multiple poles. If $f(z)$ has a pole of order k at z_0, then $n_1(t, f = \infty)$ is counted $k - 1$ times there. When a is a finite complex number and $f(z) - a$ has a zero of order k at z_0', then $n(t, 1/f')$ is also counted $k - 1$ times at z_0'. Thus

$$N_1(r) = \int_0^r \frac{n_1(t) - n_1(0)}{t} dt + n_1(0) \log r$$

is a counting function of multiple values (finite or infinite), where

$$n_1(t) = n_1(t, f = \infty) + n\left(t, \frac{1}{f'}\right),$$

so

$$\sum_{\nu=1}^{q} N\left(r, \frac{1}{f - a_\nu}\right) - N_1(r) \leq \sum_{\nu=1}^{q} \overline{N}\left(r, \frac{1}{f - a_\nu}\right).$$

Therefore the second fundamental theorem of Nevanlinna can be written as

$$(q - 2)T(r, f) < \sum_{\nu=1}^{q} \overline{N}\left(r, \frac{1}{f - a_\nu}\right) + S_1(r, f). \qquad (1.4.7)$$

Definition 1.11   *Suppose $f(z)$ is a transcendental meromorphic function in the plane and a is a complex number. We define*

$$\Theta(a, f) = 1 - \overline{\lim_{r \to \infty}} \frac{\overline{N}(r, a)}{T(r, f)},$$

$$\theta(a, f) = \lim_{r \to \infty} \frac{N(r, a) - \overline{N}(r, a)}{T(r, f)}.$$

From (1.4.7) and

$$\delta(a, f) + \theta(a, f) = \lim_{r \to \infty} \frac{m(r, a)}{T(r, f)} + \lim_{r \to \infty} \frac{N(r, a) - \overline{N}(r, a)}{T(r, f)}$$

$$\leq \lim_{r \to \infty} \frac{m(r, a) + N(r, a) - \overline{N}(r, a)}{T(r, f)} = \Theta(a, f),$$

we have Theorem 1.10.

Theorem 1.10   *If $f(z)$ is transcendental meromorphic in the finite plane, then the set of values a for which $\Theta(a, f) > 0$ is at most countable and*

$$\sum_{a} \left\{\delta(a, f) + \theta(a, f)\right\} \leq \sum_{a} \Theta(a, f) \leq 2. \qquad (1.4.8)$$

Definition 1.12   *a is called a complete multiple value of $f(z)$, if all the zeros of $f(z) - a$ (or all the poles of $f(z)$, when $a = \infty$) are multiple.*

Corollary 1.   *Under the hypothesis of Theorem 1.10, the number of complete multiple values of $f(z)$ does not exceed 4.*

In fact, if a is a complete multiple value, then

$$\Theta(a, f) = 1 - \varlimsup_{r\to\infty} \frac{\overline{N}(r, a)}{T(r, f)} \geq 1 - \varlimsup_{r\to\infty} \frac{\overline{N}(r, a)}{N(r, a)} \geq \frac{1}{2}. \tag{1.4.9}$$

Thus (1.4.8) yields the required conclusion.

Corollary 2.   *If $f(z)$ is a transcendental entire function, then $f(z)$ has at most two complete multiple values.*

In fact, we have

$$\sum_{a\neq\infty} \Theta(a, f) \leq 1. \tag{1.4.10}$$

Comparing (1.4.10) with (1.4.9), the conclusion of Corollary 2 is immediate.

The upper bounds of Corollaries 1 and 2 are precise. For instance, the Weierstrass elliptic function $\mathcal{B}(z)$ is meromorphic in the finite plane and satisfies

$$\mathcal{B}'(z)^2 = \big\{\mathcal{B}(z) - a_1\big\}\big\{\mathcal{B}(z) - a_2\big\}\big\{\mathcal{B}(z) - a_3\big\}, \tag{1.4.11}$$

where a_1, a_2, and a_3 are distinct finite complex numbers. Obviously $\mathcal{B}'(z) = 0$ when $\mathcal{B}(z) = a_\nu$ $(\nu = 1, 2, 3)$. Thus the a_ν $(\nu = 1, 2, 3)$ are complete multiple values of $\mathcal{B}(z)$. Moreover, ∞ is another complete multiple value of $\mathcal{B}(z)$, since $\mathcal{B}(z)$ has only poles of order 2 by (1.4.11).

It is easy to see that $\sin z$ is an entire function with two complete multiple values 1 and -1.

Corollary 3.   *Let $f(z)$ be transcendental meromorphic in the finite plane. Suppose a_ν $(\nu = 1, 2, \cdots, p)$ are distinct complex numbers and l_ν $(\nu = 1, 2, \cdots, p)$ are integers greater than 1. If all the zeros of $f(z) - a_\nu$ $(\nu = 1, 2, \cdots, p)$ have multiplicities $\geq l_\nu$, then we have*

$$\sum_{\nu=1}^{p} \left(1 - \frac{1}{l_\nu}\right) \leq 2. \tag{1.4.12}$$

1.5 Generalizations of the Second Fundamental Theorem

In this section, we discuss some generalizations of the second fundamental theorem by replacing complex numbers with "small" meromorphic functions.

1.5.1 A special result.   R. Nevanlinna himself obtained a generalization involving three small functions.

Theorem 1.11   *If $f(z)$ and $\varphi_\nu(z)$ ($\nu = 1, 2, 3$) are meromorphic functions in the finite plane such that*

$$T(r, \varphi_\nu) = o\{T(r, f)\}, \quad \nu = 1, 2, 3,$$

then we have

$$\{1 - o(1)\}T(r, f) < \sum_{\nu=1}^{3} N\left(r, \frac{1}{f - \varphi_\nu}\right) + S(r, f), \qquad (1.5.1)$$

where $S(r, f)$ has the properties formulated in Theorem 1.6.

Define:
$$g(z) = \frac{f(z) - \varphi_1(z)}{f(z) - \varphi_3(z)} \cdot \frac{\varphi_2(z) - \varphi_3(z)}{\varphi_2(z) - \varphi_1(z)}. \qquad (1.5.2)$$

Then from Theorem 1.4 we have

$$T(r, g) < N(r, g) + N\left(r, \frac{1}{g}\right) + N\left(r, \frac{1}{g - 1}\right) + S(r, g). \qquad (1.5.3)$$

Next, since (1.5.2) can be written as

$$\frac{1}{f - \varphi_3} = \frac{1}{\varphi_3 - \varphi_1}\left(\frac{\varphi_2 - \varphi_1}{\varphi_2 - \varphi_3}g - 1\right),$$

we have
$$T(r, f) \leq T(r, f - \varphi_3) + T(r, \varphi_3) + \log 2$$
$$\leq T\left(r, \frac{1}{f - \varphi_3}\right) + o\big(T(r, f)\big)$$
$$\leq T(r, g) + o\big(T(r, f)\big).$$

Thus

$$\bigl(1 - o(1)\bigr)T(r, f) < T(r, g).$$

From (1.5.2), it is easy to see that

$$N(r, g) + N\Bigl(r, \frac{1}{g}\Bigr) + N\Bigl(r, \frac{1}{g-1}\Bigr)$$

$$\leq \sum_{\nu=1}^{3} N\Bigl(r, \frac{1}{f - \varphi_\nu}\Bigr) + N\Bigl(r, \frac{1}{\varphi_2 - \varphi_1}\Bigr)$$

$$+ N\Bigl(r, \frac{1}{\varphi_2 - \varphi_3}\Bigr) + N\Bigl(r, \frac{1}{\varphi_1 - \varphi_3}\Bigr)$$

$$\leq \sum_{\nu=1}^{3} N\Bigl(r, \frac{1}{f - \varphi_\nu}\Bigr) + o\{T(r, f)\}.$$

(1.5.1) is then derived from (1.5.3) by the above inequalities.
Furthermore, since

$$g(z) = \frac{\varphi_2 - \varphi_3}{\varphi_2 - \varphi_1}\Bigl\{\frac{\varphi_3 - \varphi_1}{f - \varphi_3} + 1\Bigr\},$$

we have

$$T(r, g) \leq \bigl\{1 + o(1)\bigr\}T(r, f).$$

When the order of $f(z)$ is finite, so is the order of $g(z)$. Hence $S(r, g) = O(\log r)$ in (1.5.3). The case of infinite order is similar.

Furthermore, R. Nevanlinna [2] asked if complex numbers can be replaced by small functions in the general form of the second fundamental theorem of Nevanlinna (Theorem 1.5). Chuang Chi-tai [3] introduced the Wronskian and obtained the first nontrivial result as follows.

Theorem 1.12   Let $f(z)$ and $a_j(z)(j = 1, 2, \cdots, q)$ be mero-morphic in the finite plane. If the $a_j(z)(j = 1, 2, \cdots, q)$ are distinct and satisfy

$$T(r, a_j) = o(T(r, f)), \quad j = 1, 2, \cdots, q,$$

then we have

$$\bigl\{q - 1 - o(1)\bigr\}T(r, f) < \sum_{\nu=1}^{q} N\Bigl(r, \frac{1}{f - a_j}\Bigr) + q\overline{N}(r, f) + S(r, f), \quad (1.5.4)$$

where $S(r, f) = O(\log(rT(r, f)))$, except on possibly a sequence of exceptional intervals with finite total length, when the order of $f(z)$ is infinite.

Following to the definition of deficiency (Definition 1.9), it is now natural to define

$$\delta(a(z), f) = \varliminf_{r \to \infty} \frac{m\left(r, \dfrac{1}{f - a(z)}\right)}{T(r, f)} = 1 - \varlimsup_{r \to \infty} \frac{N\left(r, \dfrac{1}{f - a(z)}\right)}{T(r, f)},$$

$$\tag{1.5.5}$$

where $a(z)$ is a meromorphic function with the condition

$$T(r, a(z)) = o(T(r, f)). \tag{1.5.6}$$

As a consequence of Theorem 1.12, the set of deficient functions of every transcendental entire function is countable and the total sum of the corresponding deficiencies does not exceed 1.

Next, Yang Lo [4] proved a result in the general case of meromorphic functions.

Theorem 1.12′   *If $f(z)$ is meromorphic and of finite lower order μ, then the set of its deficient functions is countable and the total sum of the corresponding deficiencies does not exceed*

$$\min\left\{[2\mu] + 1, \max\left(1, \frac{\sqrt{2}}{2}\mu\pi\right)\right\}.$$

Subsequently, Frank and Weissenborn [1] obtained the following result by using an elegant method.

Theorem 1.12″   *Let $f(z)$ be a transcendental meromorphic function in the finite plane and $a_j(z)(j = 1, 2, \cdots, q)$ be q rational functions. If ε is a positive number, then we have*

$$m(r, f) + \sum_{j=1}^{q} m\left(r, \frac{1}{f - a_j(z)}\right) < (2 + \varepsilon)T(r, f),$$

except on a set E_ε of r with finite linear measure.

As a corollary of Theorem 1.12", we have

$$\delta(\infty, f) + \sum \delta(a(z), f) \leq 2,$$

where the summation is taken over all the rational functions.

Recently, Osgood [1] and Steinmetz [1] independently settled Nevanlinna's problem completely.

Theorem 1.13   *Let $f(z)$ be a transcendental meromorphic function in the finite plane and $a_j(z)$ be q meromorphic functions such that*

$$T(r, a_j(z)) = o(T(r, f)).$$

If ε is a positive number, then we have

$$m(r, f) + \sum_{j=1}^{q} m\left(r, \frac{1}{f - a_j(z)}\right) < (2 + \varepsilon)T(r, f), \qquad (1.5.7)$$

except on a set E_ε of r with finite linear measure.

1.5.2 Proof of Theorem 1.13.   In order to introduce Steinmetz's elegant proof of Theorem 1.13, we need some preparation.

Denote by $L(s)$ the linear space which is generated by the functions of the form

$$a_1^{n_1} \cdots a_q^{n_q}; \quad n_1, \cdots, n_q \geq 0, \quad n_1 + \cdots + n_q = s.$$

For a fixed positive integer s, let $\beta_1, \cdots, \beta_n$ be a basis of $L(s)$ and $b_1, \cdots, b_k$ be a basis of $L(s + 1)$.

We recall that the Wronskian of functions $g_1, g_2, \cdots, g_m$ is defined by

$$W(g_1, g_2, \cdots, g_m) \equiv \begin{vmatrix} g_1 & g_2 & \cdots g_m \\ g_1' & g_2' & \cdots g_m' \\ g_1'' & g_2'' & \cdots g_m'' \\ & \cdots & \\ g_1^{(m-1)} & g_2^{(m-1)} & \cdots g_m^{(m-1)} \end{vmatrix}.$$

Now we consider the properties of the Wronskian

$$P(f) = W(b_1, \cdots, b_k, \beta_1 f, \cdots, \beta_n f). \qquad (1.5.8)$$

It is clear that $P(f)$ can not be identically zero. For every $a_j (j = 1, 2, \cdots, q)$, since $\beta_1 a_j, \cdots, \beta_n a_j$ belong to $L(s+1)$ and $\{b_1, \cdots, b_k\}$ is a basis of $L(s+1)$, we have

$$P(f - a_j) = P(f).$$

Lemma 1.7   *If $f(z)$ is transcendental meromorphic in the finite plane, then*

$$m\left(r, \frac{f^{(k)}}{f}\right) = S(r, f) \qquad (1.5.9)$$

for every positive integer k, where $S(r, f)$ has the properties formulated in Theorem 1.6.

When $k = 1$, Lemma 1.7 is true by Lemma 1.3 and Theorem 1.6. The general conclusion can be easily proved by induction.

Lemma 1.8 *Notation as above, we have*

$$T(r, P(f)) \leq n \cdot m(r, f) + (n + k)N(r, f) + S(r, f). \qquad (1.5.10)$$

Proof.   Since

$$P(f) = f^n Q\left(\frac{f'}{f}\right),$$

where Q is a differential polynomial with coefficients in small meromorphic functions satisfying (1.5.6), we have

$$m(r, P(f)) = nm(r, f) + S(r, f) \qquad (1.5.11)$$

from Lemma 1.7. By direct computation, it is easy to see that

$$W(gg_1, gg_2, \cdots, gg_m) = g^m W(g_1, g_2, \cdots, g_m). \qquad (1.5.12)$$

Thus

$$P(f) = f^{k+n} W\left(\frac{b_1}{f}, \cdots, \frac{b_k}{f}, \beta_1, \cdots, \beta_n\right),$$

whence

$$N(r, P(f)) \leq (k+n)N(r, f) + S(r, f). \qquad (1.5.13)$$

Lemma 1.8 follows from the sum of (1.5.11) and (1.5.13). □

Now we prove Theorem 1.13. Let

$$d(z) = \frac{1}{2} \min_{0 \leq i \neq j \leq q} |a_i(z) - a_j(z)|.$$

Denote by $E_j(j = 1, 2, \cdots, q)$ the set of points z for which the inequality

$$|f(z) - a_j(z)| < d(z)$$

holds.

From

$$P(f) \equiv P(f - a_j) = (f - a_j)^n Q\left(\frac{(f - a_j)'}{f - a_j}\right),$$

we have

$$\log^+ \frac{1}{|f(z) - a_j(z)|} \leq \frac{1}{n} \log^+ \frac{1}{|P(f(z))|} + \frac{1}{n} \log^+ \left|Q\left(\frac{f' - a_j'}{f - a_j}\right)\right|.$$

For $z \in E_j$ and $k \neq j$, it is clear that

$$|f(z) - a_k(z)| \geq |a_k(z) - a_j(z)| - |a_j(z) - f(z)|$$

$$\geq 2d(z) - d(z) = d(z),$$

so that

$$\log^+ \frac{1}{|f(z) - a_k(z)|} \leq \log^+ \frac{1}{d(z)}.$$

Thus

$$\sum_{j=1}^{q} \log^+ \frac{1}{|f(z) - a_j(z)|}$$

$$\leq \frac{1}{n} \log^+ \frac{1}{|P(f(z))|} + n \log^+ \frac{1}{d(z)} + \frac{1}{n} \sum_{j=1}^{q} \log^+ \left|Q\left(\frac{f' - a_j'}{f - a_j}\right)\right|$$

for $z \in E_j (j = 1, 2, \cdots, q)$. The inequality is obvious, when z does not belong to $\cup_{j=1}^{q} E_j$.

Therefore,

$$\sum_{j=1}^{q} m\left(r, \frac{1}{f - a_j}\right) \leq \frac{1}{n} m\left(r, \frac{1}{P(f)}\right) + nm\left(r, \frac{1}{d}\right) + S(r, f). \quad (1.5.14)$$

Using Lemma 1.8,

$$m\left(r, \frac{1}{P(f)}\right) \leq T(r, P(f)) + O(1)$$
$$\leq n \cdot m(r, f) + (n + k)N(r, f) + S(r, f). \quad (1.5.15)$$

It follows from the definition of $d(z)$ that

$$\frac{1}{d(z)} \leq 2 \sum_{1 \leq i < j \leq q} \frac{1}{|a_i(z) - a_j(z)|},$$

so

$$m\left(r, \frac{1}{d(z)}\right) \leq 2 \sum_{1 \leq i < j \leq q} m\left(r, \frac{1}{a_i - a_j}\right) = S(r, f). \quad (1.5.16)$$

Substituting (1.5.15) and (1.5.16) into (1.5.14), we obtain

$$m(r, f) + \sum_{j=1}^{q} m\left(r, \frac{1}{f - a_j}\right)$$
$$\leq 2m(r, f) + \left(1 + \frac{k}{n}\right)N(r, f) + S(r, f).$$

Now we claim that there is at least one positive integer s such that $\dim L(s + 1) < (1 + \varepsilon) \dim L(s)$. In fact, if this assertion is not true, then we have

$$\dim L(s) \geq (1 + \varepsilon) \dim L(s - 1) \geq (1 + \varepsilon)^2 \dim L(s - 2)$$
$$\geq \cdots \geq (1 + \varepsilon)^{s-1} \dim L(1) \geq (1 + \varepsilon)^{s-1}.$$

On the other hand, there are at most $(s + 1)^q$ different elements of $a_1^{n_1} \cdots a_q^{n_q}, 0 \leq n_1, n_2, \cdots, n_q \leq s$. Thus

$$(s + 1)^q \geq (1 + \varepsilon)^{s-1},$$

and this is impossible for sufficiently large s. The proof of Theorem 1.13 is complete. □

Corollary.
$$\delta(\infty, f) + \sum \delta(a(z), f) \leq 2,$$

where the summation is taken over all the functions $a(z)$ satisfying condition (1.5.6).

Chapter 2

Normal Families

In this chapter, we shall see that the study of the value assignment of a function in a neighborhood of an isolated essential singularity can be reduced to that of the properties of a family of functions in a fixed region. For this purpose, an introduction to the concept of normal family and some criteria for normality would be appropriate. Nevanlinna theory plays an important role in this discussion.

2.1 Normal Families of Holomorphic Functions

2.1.1 Definition and fundamental properties

Definition 2.1 *A family $\mathcal{F}$ of holomorphic functions in a region D is said to be normal if for each sequence (f_μ) of $\mathcal{F}$, we can extract a subsequence (f_{μ_ν}) converging uniformly in every bounded closed subregion of D to a limit function which may be identical to infinity.*

Let P be an arbitrary point of D. $\mathcal{F}$ is normal at P if there exists a neighborhood $U(P)$ of P such that $\mathcal{F}$ is normal in $U(P)$.

The concept of normal family was introduced by P. Montel [1] at the beginning of this century. He wrote a series of papers on this topic.

Lemma 2.1 *A family $\mathcal{F}$ is normal in a region D if and only if $\mathcal{F}$ is normal at every point of D.*

Proof. The necessity of the condition is obvious.

For the proof of sufficiency, let

$$E_j = \left\{ z \in D : |z| \leq j \text{ and } \min_{\zeta \in Bd(D)} |z - \zeta| \geq \frac{1}{j} \right\}, \quad j = 1, 2, \cdots,$$

where $Bd(D)$ denotes the boundary of D. It is clear that every E_j is a bounded closed set in D and each bounded closed set in D is contained in some E_j.

According to the hypothesis, every point $p \in D$ has a neighborhood $U(P)$ such that $\mathcal{F}$ is normal in it. Let $V(p)$ be a smaller neighborhood of p with $\overline{V(p)} \subset U(p)$. By the Heine-Borel theorem, from the open covering $\{V(p)|p \in E_j\}$ of E_j for every fixed $j = 1, 2, \cdots$, we can extract a finite subcovering $\{V(p_i)\ i = 1, 2, \cdots, L\}$. By the normality of $\mathcal{F}$ in every $U(p)$, for an arbitrary sequence $\{f_\mu\}$ in $\mathcal{F}$, there is a subsequence $\{f_{\mu_1}^1\}$ of $\{f_\mu\}$, which converges in $\bar{V}(p_1)$. Thus there is a subsequence $\{f_{\mu_2}^2\}$ of $\{f_{\mu_1}^1\}$, converging uniformly in $\bar{V}(p_1) \cup \bar{V}(p_2)$. By this procedure, we find finally a subsequence of $\{f_\mu\}$, converging uniformly in $\cup_{i=1}^l \bar{V}_i$, hence in E_j.

So, according to the above result, for an arbitrary sequence $\{f_\nu\}$ in $\mathcal{F}$, there is a subsequence $\{f_{\nu_1}^1|\nu_1 = 1, 2, \cdots\}$ converging uniformly in E_1, and a subsequence $\{f_{\nu_2}^2 : \nu_2 = 1, 2, \cdots\}$ of $\{f_{\nu_1}^1\}$, converging uniformly in E_2; this procedure can then be continued infinitely. Therefore, the diagonal subsequence $\{f_i^i|i = 1, 2, \cdots\}$ of $\{f_\nu\}$ converges uniformly in every E_j, and hence in every bounded closed set in D. This proves the sufficiency of Lemma 2.1. $\square$

2.1.2 Criteria for normality. Let $\mathcal{F}$ be a family of holomorphic functions in a region D. Under what conditions, can $\mathcal{F}$ become a normal family? Such conditions are often referred to as the criteria for normality. Naturally, it is an important problem to find these criteria.

In this section, we discuss a simple and fundamental criterion due to Montel.

Definition 2.2 *A family $\mathcal{F}$ of functions is uniformly bounded on a set E, if there exists a positive number M such that $|f(z)| \leq M$ holds for every function $f(z)$ of $\mathcal{F}$ and every point of E.*

Definition 2.3 *A family $\mathcal{F}$ of functions is equicontinuous on a set E, if, for any positive number ε, there exists a corresponding number $\delta = \delta(\varepsilon)$ such that $|f(z_1) - f(z_2)| < \varepsilon$ for every function $f(z)$ of $\mathcal{F}$ and every pair of points z_1 and z_2 in E, provided that $|z_1 - z_2| < \delta$.*

Theorem 2.1   *Let $\mathcal{F}$ be a family of holomorphic functions in a region D. If $\mathcal{F}$ is uniformly bounded in D, then $\mathcal{F}$ is normal there.*

Proof. By Lemma 2.1, we need only prove that $\mathcal{F}$ is normal at every point z_0 of D. We choose a small positive number d such that $|z - z_0| \le 2d$ is contained in D. If z_1, z_2 is a pair of points in $U(z_0) : |z - z_0| \le d$, then, for every function f of $\mathcal{F}$, the Cauchy formula gives us:

$$f(z_1) - f(z_2) = \frac{1}{2\pi i} \int_{|z-z_0|=2d} \frac{f(\zeta)d\zeta}{\zeta - z_1} - \frac{1}{2\pi i} \int_{|z-z_0|=2d} \frac{f(\zeta)d\zeta}{\zeta - z_2}$$

$$= \left(\frac{1}{2\pi i} \int_{|z-z_0|=2d} \frac{f(\zeta)d\zeta}{(\zeta - z_1)(\zeta - z_2)} \right)(z_1 - z_2).$$

Since $\mathcal{F}$ is uniformly bounded in D, there is a positive number M such that $|f(z)| < M (f \in \mathcal{F}, z \in D)$. Thus

$$|f(z_1) - f(z_2)| \le \frac{1}{2\pi} \cdot \frac{M}{d^2} \cdot 2\pi(2d)|z_1 - z_2| = \frac{2M}{d}|z_1 - z_2|.$$

This means $\mathcal{F}$ is also equicontinuous in $U(z_0)$. We now prove that $\mathcal{F}$ is normal in $U(z_0)$.

We choose a set E which is countable and everywhere dense in $U(z_0)$ (e.g. all the points in $U(z_0)$ which have two rational coordinates). For an arbitrary sequence $(f_\nu(z))$ in $\mathcal{F}$, since $\mathcal{F}$ is uniformly bounded in $U(z_0)$, we can extract by the diagonal method a subsequence $(f_{\nu_i}(z))$ converging at every point of E. From the equicontinuity of $(f_{\nu_i}(z))$, for any positive number ε, there exists a positive number δ such that

$$|f_{\nu_i}(z_1) - f_{\nu_i}(z_2)| < \frac{\varepsilon}{3}, \quad i = 1, 2, \cdots, \tag{2.1.1}$$

for every pair of points z_1, z_2 in $U(z_0)$, provided that $|z_1 - z_2| < \delta$.

Divide the z-plane by a square lattice of length $\delta/2$. In every square intersecting $U(z_0)$, we choose an interior point which also belongs to E. Then finitely many such points $e_j (j = 1, 2, \cdots, J)$ can be found. There exists a positive integer N corresponding to ε such that

$$|f_{\nu_i}(e_j) - f_{\nu_k}(e_j)| < \frac{\varepsilon}{3}, \quad j = 1, 2, \cdots, J, \tag{2.1.2}$$

provided $\nu_i > N$ and $\nu_k > N$.

Let z be an arbitrary point in $U(z_0)$. There is an e_j such that $|z - e_j| < \delta$. Thus, from (2.1.1) and (2.1.2) we have

$$|f_{\nu_i}(z) - f_{\nu_k}(z)| \le |f_{\nu_i}(z) - f_{\nu_i}(e_j)| + |f_{\nu_i}(e_j) - f_{\nu_k}(e_j)|$$

$$+ |f_{\nu_k}(e_j) - f_{\nu_k}(z)| < \varepsilon, \tag{2.1.3}$$

provided $\nu_i > N$ and $\nu_k > N$. Since N does not depend on z, $(f_{\nu_i}(z))$ is uniformly convergent in $U(z_0)$, so that $\mathcal{F}$ is normal in $U(z_0)$. □

Corollary. *If there is a positive number M such that $|f(z)| > M$ holds for every holomorphic function $f(z)$ belonging to a family $\mathcal{F}$ in a region D, then $\mathcal{F}$ is normal there.*

In fact, $\{1/f(z),\ f \in \mathcal{F}\}$ satisfies the condition of Theorem 2.1, so it is normal in D, and therefore so is $\mathcal{F}$.

2.1.3 Marty criterion.   Now we discuss the Marty criterion, a necessary and sufficient condition for normality, which is worthy of note.

Theorem 2.2   *Let $\mathcal{F}$ be a family of holomorphic functions in a region D. For $\mathcal{F}$ to be normal in D, a necessary and sufficient condition is that for every bounded closed subregion D_1 of D there is a positive number M such that*

$$|f'(z)| \le M(1 + |f(z)|^2) \tag{2.1.4}$$

for every $f(z) \in \mathcal{F}$ and every $z \in \overline{D}_1$.

Proof.   First of all, we prove necessity. If it is not true, then there exists a bounded closed subregion D_1 and a sequence $f_\mu(z)(\mu = 1, 2, \cdots)$ from $\mathcal{F}$ such that

$$\lim_{u \to \infty} \left\{ \max_{z \in \overline{D}_1} \frac{|f'_\mu(z)|}{1 + |f_\mu(z)|^2} \right\} = \infty. \tag{2.1.5}$$

Since $\mathcal{F}$ is normal, we can choose a subsequence (f_{μ_ν}) converging uniformly to $\varphi(z)$ on every bounded closed subregion $\overline{D}_1$ of D.
   If $\varphi(z) \not\equiv \infty$, then

$$\frac{|f'_{\mu_\nu}(z)|}{1 + |f_{\mu_\nu}(z)|^2} \longrightarrow \frac{|\varphi'(z)|}{1 + |\varphi(z)|^2}.$$

Hence

$$\max_{z \in \bar{D}_1} \frac{|f'_{\mu_\nu}(z)|}{1 + |f_{\mu_\nu}(z)|^2} \le 1 + \max_{z \in \bar{D}_1} \frac{|\varphi'(z)|}{1 + |\varphi(z)|^2} < \infty,$$

whenever ν is sufficiently large. If $\varphi(z) \equiv \infty$, then $\dfrac{1}{f_{\mu_\nu}(z)}$ converges uniformly to zero on D_1, hence

$$\frac{|f'_{\mu_\nu}(z)|}{1 + |f_{\mu_\nu}(z)|^2} = \frac{\left|\left(\dfrac{1}{f_{\mu_\nu}(z)}\right)'\right|}{1 + \left|\dfrac{1}{f_{\mu_\nu}(z)}\right|^2} \longrightarrow 0.$$

In either case we have a contradiction to (2.1.5), and necessity is proved.

In order to show the sufficiency of the condition, we need only prove that $\mathcal{F}$ is normal at every point of D. Let z_0 be an interior point of D. Then there is a positive number d such that $U(z_0) = \{z : |z - z_0| \le d\}$ is contained in D. By the condition of the theorem, (2.1.4) holds in $U(z_0)$ with a corresponding M. Set $U_1(z_0) = \left\{z : |z - z_0| \le \min\left(d, 1/5M\right)\right\}$.

Let us prove the following two assertions for an arbitrary f in $\mathcal{F}$:

(1) $|f(z_0)| < 1$ implies $|f(z)| < 2$ for every $z \in U_1(z_0)$.

If the assertion is false, then there exists a point $z_1 \in U_1(z_0)$ such that $|f(z_1)| = 2$ and $|f(z)| < 2$ for every $z \in \overline{z_0 z_1}$. Thus we have

$$2 = |f(z_1)| \le |f(z_0)| + \left|\int_{\overline{z_0 z_1}} f'(\zeta)d\zeta\right| < 1 + 5M|z_1 - z_0|$$

from (2.1.4), hence

$$|z_1 - z_0| > \frac{1}{5M},$$

a contradiction.

(2) $|f(z_0)| \ge 1$ implies $|f(z)| > 1/2$ for every $z \in U_1(z_0)$. If the assertion is false, then there exists $z_1 \in U_1(z_0)$ such that $|f(z_1)| = 1/2$ and $|f(z)| > 1/2$ for every $z \in \overline{z_0 z_1}$. Thus for $z \in \overline{z_0 z_1} \subset U_1(z_0)$ we have

$$\frac{\left|\left(\dfrac{1}{f(z)}\right)'\right|}{1 + \left(\dfrac{1}{|f(z)|}\right)^2} = \frac{|f'(z)|}{1 + |f(z)|^2} \le M,$$

hence

$$\left|\left(\frac{1}{f(z)}\right)'\right| < 5M.$$

Therefore

$$2 = \frac{1}{|f(z_1)|} \leq \frac{1}{|f(z_0)|} + \left| \int_{z_0 z_1} \left(\frac{1}{f(z)} \right)' dz \right| < 1 + 5M|z_1 - z_0|,$$

hence

$$|z_1 - z_0| > \frac{1}{5M},$$

a contradiction.

Consequently, for every point z_0 of D, there exists a neighborhood $U_1(z_0)$ such that for every function f of $\mathcal{F}$, either $|f(z)| < 2$ or $1/(|f(z)|) < 2$ uniformly holds in the neighborhood. Clearly $\mathcal{F}$ is normal in D. □

2.2 Montel's Criterion

In this section, we deduce a fundamental criterion for normality due to Montel.

2.2.1 Preliminary Lemmas. The following lemmas will be used frequently later on.

Lemma 2.2 *Given $x \geq 0$ and $A \geq e$, we have*

$$A\log^+ x \leq x + A(\log A - 1). \tag{2.2.1}$$

Proof. When $0 \leq x \leq 1$, the conclusion of Lemma 2.2 is obvious. When $x > 1$, we consider

$$\varphi(x) = x - A\log x$$

which has the minimum $A - A\log A$ at $x = A$. Thus

$$x - A\log x \geq A - A\log A. \quad □$$

Lemma 2.3 *Given $x > 0$ and $A \geq e$, we have*

$$\log x + A\log^+ \log^+ \frac{1}{x} \leq \log^+ x + A(\log A - 1). \tag{2.2.2}$$

Proof.   Lemma 2.2 gives

$$A \log^+ \log^+ \frac{1}{x} \le \log^+ \frac{1}{x} + A(\log A - 1).$$

Since

$$\log x = \log^+ x - \log^+ \frac{1}{x},$$

(2.2.2) can be derived immediately.   □

In many cases, we need the following lemma, due essentially to F. Bureau, though the improved version below is due to Hiong King-lai [2].

Lemma 2.4   *Suppose $T(r)$ is a continuous, non-decreasing and non-negative function, and $a(r)$ is a non-increasing and non-negative function in an interval (r_0, R). Suppose b and c are two constants. If*

$$T(r) < a(r) + b \log^+ \frac{1}{\rho - r} + c \log^+ T(\rho), \tag{2.2.3}$$

whenever $r_0 < r < \rho < R$, then

$$T(r) < 2a(r) + B \log^+ \frac{2}{R - r} + C, \tag{2.2.4}$$

where B and C are constants depending only on b and c.

Proof.   If $T(r) < 1$ in (r_0, R), then the conclusion of Lemma 2.4 is obvious. Otherwise, there exists r_0' such that $r_0 \le r_0' < R$ and $T(r_0') \ge 1$. Then, whenever $r_0' < r < R$, there exists r_1 in $\left(r, (7R + r)/8\right)$ by Lemma 1.4 such that

$$T\left(r_1 + \frac{R - r_1}{eT(r_1)}\right) < 2T(r_1).$$

Since $r_0' < r < r_1 < r_1 + \dfrac{R - r_1}{eT(r_1)} < R$, (2.2.3) yields

$$T(r_1) < a(r_1) + b \log^+ \frac{eT(r_1)}{R - r_1} + c \log^+ T\left(r_1 + \frac{R - r_1}{eT(r_1)}\right)$$

$$< a(r_1) + b \log^+ \frac{e}{R - r_1} + c \log 2 + (b + c) \log^+ T(r_1).$$

Noting that

$$2(b + c) \log^+ T(r_1) \le T(r_1) + 2(b + c)\{\log 2(b + c) - 1\},$$

we deduce that

$$T(r_1) < 2a(r_1) + 2b\log^+ \frac{e}{R - r_1} + 2c\log 2$$

$$+2(b + c)\{\log 2(b + c) - 1\},$$

so

$$T(r) < 2a(r) + 2b\log^+ \frac{8e}{R - r} + 2c\log 2$$

$$+2(b + c)\{\log 2(b + c) - 1\}.$$

The lemma is proved.   □

2.2.2 Theorems of bounded type. In this section, we establish two theorems of bounded type, from which the Montel criterion for normality is easily derived.

Theorem 2.3   *Let $f(z)$ be holomorphic in $|z| < R$. If $f(z) \neq 0, 1$ there and $f'(0) \neq 0$, then for every $r \in (0, R)$ we have*

$$T(r, f) < C\left\{ \log^+ |f(0)| + \log^+ \frac{1}{R|f'(0)|} + \log \frac{2R}{R - r} \right\}, \qquad (2.2.5)$$

where C is a positive constant.

Proof.   Applying the second fundamental theorem of Nevanlinna (Theorem 1.4) to $f(z)$, we have

$$T(r, f) < m\left(r, \frac{f'}{f}\right) + m\left(r, \frac{f'}{f - 1}\right) + \log\left| \frac{f(0)(f(0) - 1)}{f'(0)} \right| + \log 2.$$

By Nevanlinna's Lemma (Lemma 1.3), the inequality

$$m\left(r, \frac{f'}{f}\right) < 10 + 4\log^+ \log^+ \frac{1}{|f(0)|} + 2\log^+ \frac{1}{r}$$

$$+3\log^+ \frac{1}{\rho - r} + 4\log^+ \rho + 4\log^+ T(\rho, f)$$

holds whenever $0 < r < \rho < R$. Similarly,

$$m\left(r, \frac{f'}{f - 1}\right) < 10 + 4\log^+ \log^+ \frac{1}{|f(0) - 1|} + 2\log^+ \frac{1}{r}$$

$$+3\log^+ \frac{1}{\rho - r} + 4\log^+ \rho + 4\log^+ T(\rho, f) + 4\log 2.$$

Thus

$$T(r,f) < 20 + 5\log 2 + \left(\log|f(0)| + 4\log^+\log^+\frac{1}{|f(0)|}\right)$$

$$+\left(\log|f(0)-1| + 4\log\log\frac{1}{|f(0)-1|}\right) + \log\frac{1}{|f'(0)|}$$

$$+4\log^+\frac{1}{r} + 6\log^+\frac{1}{\rho-r} + 8\log^+\rho + 8\log^+ T(\rho,f).$$

From Lemma 2.3, we have the following estimates:

$$\log|f(0)| + 4\log^+\log^+\frac{1}{|f(0)|} \leq \log^+|f(0)| + 4(\log 4 - 1)$$

and

$$\log|f(0)-1| + 4\log^+\log^+\frac{1}{|f(0)-1|}$$

$$\leq \log^+|f(0)-1| + 4(\log 4 - 1)$$

$$\leq \log^+|f(0)| + \log 2 + 4(\log 4 - 1).$$

Hence

$$T(r,f) < 12 + 22\log 2 + 2\log^+|f(0)| + \log\frac{1}{|f'(0)|}$$

$$+4\log^+\frac{1}{r} + 6\log^+\frac{1}{\rho-r} + 8\log^+\rho + 8\log^+ T(\rho,f).$$

Choosing r and ρ with $R/2 \leq r < \rho < R$, we obtain

$$T(r,f) < 12 + 26\log 2 + 2\log^+|f(0)| + \log\frac{1}{|f'(0)|}$$

$$+4\log^+\frac{1}{R} + 8\log^+ R + 6\log^+\frac{1}{\rho-r} + 8\log^+ T(\rho,f).$$

Then Lemma 2.4 yields

$$T(r,f) < C\Big\{1 + \log^+|f(0)| + \log^+\frac{1}{|f'(0)|}$$

$$+\log^+\frac{2}{R-r} + \log^+ R + \log^+\frac{1}{R}\Big\}.$$

When $R = 1$ and $1/2 \leq r < 1$, we have the desired inequality:

$$T(r,f) < C\Big\{\log^+|f(0)| + \log^+\frac{1}{|f'(0)|} + \log\frac{2}{1-r}\Big\}.$$

If $0 < r < 1/2$, this inequality is obvious from the fact that

$$T(r, f) \le T(1/2, f).$$

When $R \ne 1$, by setting $f(Rz) = g(z)$, we see that $g(z)$ is holomorphic in $|z| < 1$ and $g(z) \ne 0, 1$ there. Noting $g'(0) = Rf'(0) \ne 0$, we obtain finally

$$T(r, f) = T\left(\frac{r}{R}, g\right) < C\left\{\log^+ |g(0)| + \log^+ \frac{1}{|g'(0)|} + \log \frac{2}{1 - \frac{r}{R}}\right\}$$

$$< C\left\{\log^+ |f(0)| + \log^+ \frac{1}{R|f'(0)|} + \log \frac{2R}{R - r}\right\}. \quad \square$$

Theorem 2.4 *Suppose $f(z)$ is holomorphic in $|z| < R$ and $f(z) \ne 0, 1$ there. If $M(r, f)$ denotes $\max\limits_{|z| \le r} |f(z)|$, then for every $r \in (0, R)$ we have*

$$\log M(r, f) < \frac{CR}{R - r}\left\{\log^+ |f(0)| + \log \frac{2R}{R - r}\right\}, \qquad (2.2.6)$$

where C is a positive constant [1].

Proof. We assume $R = 1$ and distinguish between two cases.

(1) $|f'(0)| \ge 1$.

In this case, Theorem 2.3 gives

$$\log M(r, f) \le \frac{\frac{1+r}{2} + r}{\frac{1+r}{2} - r} T\left(\frac{1+r}{2}, f\right)$$

$$< \frac{4C}{1 - r}\left\{\log^+ |f(0)| + \log \frac{2}{1 - \frac{1+r}{2}}\right\}.$$

(2) $|f'(0)| < 1$.

Choose a point ζ on $|z| = r$ such that $|f(\zeta)| = \max\limits_{|z|=r} |f(z)|$. Two subcases arise.

---

1) Hereinafter we use C to denote a positive constant which need not be the same at every occurance.

(2.1) $|f'(z)| < 1$ holds uniformly on the segment $\overline{O\zeta}$.

In this case, we have

$$|f(\zeta)| \leq |f(0)| + \left| \int_{\overline{O\zeta}} f'(z)dz \right|.$$

Thus

$$\log M(r, f) \leq \log^+ |f(0)| + \log 2.$$

(2.2) There exists a point z_0 on $\overline{O\zeta}$ such that $|f'(z_0)| = 1$ and $|f'(z)| < 1$ in $\overline{Oz_0}$. This gives

$$|f(z_0)| \leq |f(0)| + 1.$$

Let d be the distance between z_0 and ζ. Then $f(z)$ is holomorphic in $U(z_0) = \{z : |z - z_0| < d + 1 - r\}$, $|f'(z_0)| = 1$ and $f(z) \neq 0, 1$ in $U(z_0)$. Theorem 2.3 gives[1]

$$T\left(d + \frac{1-r}{2}, z_0, f\right)$$

$$< C\left\{ \log^+ |f(z_0)| + \log^+ \frac{1}{d+1-r} + \log \frac{2(d+1-r)}{d+1-r-(d+\frac{1-r}{2})} \right\}$$

$$< C\left\{ \log^+ |f(0)| + \log 2 + \log \frac{4}{1-r} \right\}$$

$$< C\left\{ \log^+ |f(0)| + \log \frac{2}{1-r} \right\}.$$

Thus

$$\log M(r, f) = \log M(d, z_0, f)$$

$$\leq \frac{d + \frac{1-r}{2} + d}{d + \frac{1-r}{2} - d} C\left\{ \log^+ |f(0)| + \log \frac{2}{1-r} \right\}$$

$$< \frac{C}{1-r}\left\{ \log^+ |f(0)| + \log \frac{2}{1-r} \right\}.$$

When $R = 1$, we consider $g(z) = f(Rz)$. We see that $g(z)$ is holomor-

---

1) $T(\tau.z_0, f)$ denotes the characteristic function of $f(z)$ with respect to $\{z : |z - z_0| \leq \tau\}$. $M(\tau, z_0, f)$ is the maximum modulus of $f(z)$ on this disk.

phic in $|z| < 1$ and $g(z) \neq 0, 1$ there. Therefore

$$\log M(r, f) = \log M\left(\frac{r}{R}, g\right) < \frac{C}{1 - \frac{r}{R}}\left\{\log^+ |g(0)| + \log \frac{2}{1 - \frac{r}{R}}\right\}$$

$$= \frac{CR}{R - r}\left\{\log^+ |f(0)| + \log \frac{2R}{R - r}\right\}. \quad \square$$

Theorem 2.4 is the well-known Schottky theorem. It was obtained by Schottky with a different method and in a different formulation.

2.2.3 Montel's criterion

Theorem 2.5 *A family $\mathcal{F}$ of holomorphic functions in a region D is normal if $f(z) \neq 0$ or 1 in D for every $f(z)$ in $\mathcal{F}$.*

Proof. By Lemma 2.1, we need only prove the normality of $\mathcal{F}$ at every point of D. Given $z_0 \in D$, there exists a positive number d such that $U(z_0) = \{z : |z - z_0| < d\} \subset D$.

Let (f_μ) be an arbitrary sequence in $\mathcal{F}$. If there is a subsequence (f_{μ_ν}) such that $(|f_{\mu_\nu}(z_0)|)$ is bounded, then (f_{μ_ν}) is uniformly bounded on $U_1(z_0) = \{z : |z - z_0| \leq d/2\}$ by Theorem 2.4. Thus $\mathcal{F}$ is normal at z_0. If $\lim\limits_{\mu \to \infty} |f_\mu(z_0)| = \infty$, then we can consider the sequence $\left\{1/f_\mu\right\}$. Since every $1/f_\mu$ is holomorphic in $U(z_0)$ and does not take 0 or 1 there, Theorem 2.4 asserts that $1/f_\mu$ $(\mu = 1, 2, \cdots)$ is uniformly bounded on $U_1(z_0)$. Consequently there is a subsequence $1/f_{\mu_\nu}$ $(\nu = 1, 2, \cdots)$ which converges uniformly in the interior of $U_1(z_0)$ to a limit function $F(z)$. When $F(z) \not\equiv 0$, since $1/f_{\mu_\nu}$ has no zero on $U_1(z_0)$, neither does $F(z)$. Thus $f_{\mu_\nu}(z)$ converges uniformly in the interior of $U_1(z_0)$. When $F(z) \equiv 0$, then $f_{\mu_\nu}(z)$ converges uniformly to infinity. In either case, $\mathcal{F}$ is normal at z_0. $\quad \square$

Theorem 2.5′ *If every $f(z)$ in $\mathcal{F}$ is holomorphic and takes neither of the two finite distinct complex values a and b in a domain D, then $\mathcal{F}$ is normal there.*

In fact, we consider the family $\mathcal{F}_1 : \left\{(f(z) - a)/(b - a)\right\}$, where no function $(f(z) - a)/(b - a)$ takes 0 or 1 in D. Thus $\mathcal{F}_1$ is normal, and hence so is $\mathcal{F}$.

We can see that Montel's criterion (Theorem 2.5) is an important development of his original result (Theorem 2.1). Moreover this criterion is closely connected with the value assignment problem of a function.

2.3 Montel Cycle, Normal Families of Meromorphic Functions

2.3.1 Montel cycle.   A series of important theorems surround the Montel criterion for normality (Theorem 2.5), thereby constituting the so-called Montel cycle. The famous Schottky theorem (Theorem 2.4) is one of them.

The following Landau theorem also belongs to the Montel cycle.

Theorem 2.6   Let $f(z)$ be holomorphic in $|z| < R$ and $f(z) \neq 0$ or 1 there. If $f'(0) \neq 0$, then we have

$$R \leq \frac{C\max(1, |f(0)|^C)}{|f'(0)|}. \tag{2.3.1}$$

Proof.   The Cauchy inequality gives

$$|f'(0)| \leq \frac{M(R/2, f)}{R/2},$$

and Theorem 2.4 gives

$$\log M\left(\frac{R}{2}, f\right) < C\left\{\log^+ |f(0)| + 1\right\}.$$

Thus we have $R \leq \dfrac{2e^{C\{\log^+ |f(0)|+1\}}}{|f'(0)|}$. Hence

$$R \leq \frac{C}{|f'(0)|}$$

when $|f(0)| \leq 1$; and

$$R \leq \frac{C|f(0)|^C}{|f'(0)|}$$

when $|f(0)| > 1$. □

From the Landau theorem, it is easy to obtain the Picard theorem again.

Theorem 2.7   *A nonconstant entire function $f(z)$ must assume every finite complex number, with at most one exception.*

Proof.   Assuming the contrary, then $f(z)$ takes neither of the two distinct finite complex numbers a and b. Since $f(z)$ is nonconstant, there is a point z_0 such that $f'(z_0) \neq 0$. Given a positive number R, we see that $g(z) = (f(z) - a)/(b - a)$ is holomorphic in $|z - z_0| < R$, $g(z) \neq 0, 1$ and $g'(z_0) \neq 0$. By Theorem 2.6, R must have an upper bound. This contradicts the fact that $g(z)$ is entire.   □

Furthermore, we can obtain the so-called Picard big theorem from the Montel criterion for normality.

Theorem 2.8   *Let $f(z)$ be holomorphic in $D = \{z : 0 < |z - z_0| < R\}$. If z_0 is an isolated essential singularity, then $f(z)$ must take every finite complex number in D, with at most one exception.*

Proof.   Without loss of generality, we may assume $z_0 = 0$. Let us prove that the family

$$f_\mu(z) = f\left(\frac{z}{2^\mu}\right), \quad (\mu = 1, 2, \cdots) \tag{2.3.2}$$

is not normal in D.

If that is not true, then the following two cases arise.

(1) There exists a subsequence (f_{μ_ν}) converging uniformly to a holomorphic function in every closed subregion of D. Thus

$$|f_{\mu_\nu}(z)| < M$$

on $|z| = R/2$, i.e.,

$$|f(z)| < M \tag{2.3.3}$$

whenever $|z| = R/2^{\mu_\nu + 1}$ ($\nu = 1, 2, \cdots$). Consequently (2.3.3) holds whenever $0 < |z| < R/(2^{\mu_1 + 1})$ and z_0 cannot be an isolated essential singularity.

(2) (f_μ) has no subsequence converging uniformly to a holomorphic function, but it has a subsequence converging uniformly to infinity in every closed subregion of D.

In this case, there exists a positive number ε such that $f(z)$ has no zero in $0 < |z| < \varepsilon$. In fact, if there is a sequence $z_\nu \to 0$ with $f(z_\nu) = 0$, then we can choose positive integers (μ_ν) such that $z_\nu = 2^{-\mu_\nu} z'_\nu$ with $R/4 \le |z'_\nu| \le R/2$. Thus (f_{μ_ν}) contains a subsequence tending uniformly to a holomorphic function in every closed subregion of D. This contradicts the hypothesis of case (2).

Therefore, when μ is sufficiently large, $\varphi_\mu = 1/f_\mu$ is holomorphic in $0 < |z| < R$, and (φ_μ) contains a subsequence converging uniformly to 0 in every closed subregion of D. Thus $1/f \to 0$ when $z \to 0$. This is impossible since z_0 is an isolated essential singularity. Therefore (f_μ) is not normal in D.

If $f(z)$ takes neither of the two distinct finite complex numbers a and b, then (f_μ) is normal in D. This contradicts the above assertion. $\square$

Theorem 2.8 can be strengthened to the following Julia theorem.

Theorem 2.9   Let $f(z)$ be holomorphic in $D = \{z : 0 < |z-z_0| < R\}$ and z_0 its isolated essential singularity. Then there exists a ray $J = \{z : \arg(z - z_0) = \theta_0\}$ such that for every $\varepsilon > 0$, $f(z)$ takes every finite complex number in the region $\{z : |\arg(z - z_0) - \theta_0| < \varepsilon\} \cap D$, with at most one exception.

Such a ray J is called a Julia direction of $f(z)$. Its introduction by Julia in 1919, symbolized the beginning of the research on angular distribution theory.

Proof.   Set $z_0 = 0$. By the proof of Theorem 2.8, the family

$$f_\mu(z) = f\left(\frac{z}{2^\mu}\right), \quad \mu = 1, 2, \cdots,$$

is not normal in D and thus there exists a point z_1 in D such that (f_μ) is not normal at z_1. We now show that the ray $\overline{Oz_1}$ can be taken to be the J asserted in the theorem. In fact, given $\varepsilon > 0$, the intersection Ω of $\{z : |\arg z - \theta_0| < \varepsilon\}$ and $\{z : 0 < |z| < \varepsilon\}$ contains $U(z_1) = \{z : |z - z_1| < d\}$. Because (f_μ) is not normal in $U(z_1)$, there exists a subsequence (f_{μ_ν})

taking every finite complex value a in $U(z_1)$, with the possible exception of one value. Thus $f(z)$ takes every finite complex value a in

$$\left| z - \frac{z_1}{2^{\mu_\nu}} \right| < \frac{d}{2^{\mu_\nu}}, \quad \nu = 1, 2, 3, \cdots,$$

with the possible exception of one value. Since these small disks are contained in Ω, Theorem 2.9 follows immediately.   □

2.3.2 Normal families of meromorphic functions.   Let (f_μ) be a sequence of meromorphic functions in a region D and $f(z)$ their limit function which may be constant, including infinity. (f_μ) is said to converge uniformly to $f(z)$ at a point z_0 of D, if there exists a neighborhood $U(z_0)$ such that: then $\{f_\mu\}$ are holomorphic and converge uniformly to f in $U(z_0)$, if $f(z_0) \neq \infty$, and if $f(z_0) = \infty$ then $\{1/f_\mu\}$ are holomorphic for sufficiently large μ and converge uniformly to $1/f$ in $U(z_0)$.

A family $\mathcal{F}$ of meromorphic functions in D is said to be normal, if for each sequence (f_μ) of $\mathcal{F}$, we can extract a subsequence (f_{μ_ν}) converging uniformly at every point of D.

Theorem 2.10   *Let $\mathcal{F}$ be a family of meromorphic functions in D and a, b and c three distinct complex values. If for every function $f(z)$ in $\mathcal{F}$, $f(z)$ does not take a, b or c in D, then $\mathcal{F}$ is normal there.*

Proof.   When one of a, b and c is equal to ∞, Theorem 2.10 is reduced to Theorem 2.5'.

When a, b and c are finite, we consider

$$g(z) = \frac{1}{f(z) - a}$$

for every $f(z)$ in $\mathcal{F}$. Obviously $g(z)$ is holomorphic and does not take either of the two complex values $1/(b - a)$ and $1/(c - a)$ in D. Thus (g) is normal in D and hence so is $\mathcal{F}$.   □

Chapter 3

Borel Directions

The present chapter is another application of the fundamental theorems of Nevanlinna. After G. Julia had proved the existence of a Julia direction using the theory of normal family, investigations were made into the behavior of an entire or meromorphic function near a ray by H. Milloux "filling disks" [1], A. Ostrowski [1], and others. This fieled of study is refered to as angular distribution theory. In 1928, starting from Nevanlinna theory, G. Valiron [2] proved the existence of a Borel direction of a meromorphic function, thus promoting the development of angular distribution theory.

3.1 Preliminaries

3.1.1 Boutroux-Cartan theorem.   The following theorem, first investigated by P. Boutroux and then completed by H. Cartan, will be frequently used later.

Theorem 3.1 *If $a_\mu(\mu = 1, 2, \cdots, n)$ are n points in the finite plane and h is a positive number, then the set of points z satisfying the inequality*

$$\prod_{\mu=1}^{n} |z - a_\mu| \leq \left(\frac{h}{e}\right)^n \tag{3.1.1}$$

can be covered by a collection of disks[1] whose number does not exceed n and the sum of whose radii does not exceed $2h$.

---

1) These disks are called the Boutroux-Cartan exceptional disks about $a_\mu(\mu = 1, 2, \cdots, n)$ and h.

Proof. Let λ_1 be the largest integer such that there exists a disk C_1 having radius $\lambda_1 h/n$ and containing exactly λ_1 points of $\{a_\mu\}$. It is obvious that $\lambda_1 \leq n$. If there is a disk having radius h and containing all the points a_μ, then $\lambda_1 = n$; otherwise, we shall have $\lambda_1 \leq n-1$. In this case, if there is a disk having the radius $(n-1)h/n$ and containing exactly $n-1$ points of $\{a_\mu\}$, then $\lambda_1 = n-1$; otherwise we shall have $\lambda_1 \leq n-2$. Continuing this procedure, we shall finally have $\lambda_1 \geq 1$. If this assertion is false, i.e. no integer of $1, 2, \cdots, n$ is the required λ_1, then the disk $|z - a_1| \leq h/n$ must contain k_1 $(1 < k_1 < n)$ points of $\{a_\mu\}$, the disk $|z - a_1| < k_1 h/n$ must contain k_2 $(k_1 < k_2 < n)$ points of $\{a_\mu\}$, the disk $|z - a_1| < k_2 h/n$ must contain $k_3 (k_2 < k_3 < n)$ points of $\{a_\mu\}$, and finally, the disk $|z - a_1| < k_{n-1} h/n$ must contain $k_n (k_{n-1} < k_n \leq n)$ points of $\{a_\mu\}$. Then

$$n \geq k_n > k_{n-1} > k_{n-2} > \cdots > k_2 > k_1 > 1.$$

Since the $k_j (j = 1, 2, \cdots, n)$ are positive integers, we have clearly $k_n \geq n+1$. This contradiction proves $1 \leq \lambda_1 \leq n$.

Now let us call the λ_1 points of $\{a_\mu\}$ in C_1 "points in the λ_1-range". When $\lambda_1 < n$, we consider the other $n - \lambda_1$ points, and let λ_2 be the largest positive integer with the property that there is a disk C_2 having radius $\lambda_2 h/n$ and containing exactly λ_2 points of the other $n - \lambda_1$ points. Let us call the λ_2 points in C_2 "points in the λ_2-range". It is clear that $\lambda_2 \leq \min\{\lambda_1, n-\lambda_1\}$. When $\lambda_1 + \lambda_2 < n$, we consider the other $n - \lambda_1 - \lambda_2$ points, and define correspondingly the points in the λ_3-range in the same way. Continue this procedure until the equality $\sum_{\nu=1}^p \lambda_\nu = n$ holds, so that $\lambda_p \leq \cdots \leq \lambda_2 \leq \lambda_1 \leq n$ and the disk C_ν contains the λ_ν points of $\{a_\mu\}$ in the λ_ν-range $(\nu = 1, 2, \cdots, p)$.

Let $\lambda \leq n$ be a positive integer. We claim that if a disk S has radius $\lambda h/n$ and contains at least λ points of $\{a_\mu\}$, then one of them must be in a range $\geq \lambda$. In fact, we assume that S contains $\lambda' (\geq \lambda)$ points of $\{a_\mu\}$, each being in a range $< \lambda$. Since each of the λ' points is in a disk C_ν with $\lambda_\nu < \lambda$, the concentric disk S' with radius $\lambda' h/n$ must contain $\lambda'' (> \lambda')$ points of $\{a_\mu\}$, each being in a range $< \lambda$, and then the concentric disk S'' with radius $\lambda'' h/n$ must also contain $\lambda''' (> \lambda'')$ points of $\{a_\mu\}$, each being in a range $< \lambda$. Continuing this procedure, we finally obtain a concentric disk S^* having radius $\lambda^* h/n$ and containing $\lambda^* (> \lambda)$ points of $\{a_\mu\}$, each being in a range $< \lambda$. This contradicts the above definition.

For every C_ν $(\nu = 1, 2, \cdots, p)$, let Γ_μ denote the concentric disk with radius $2\lambda_\nu h/n$. Clearly, the number of Γ_ν's $(\nu = 1, 2, \cdots, p)$ does not exceed n and the total sum of the radii is equal to $2h$. Then, nothing remains but to prove that (3.1.1) is false outside $\cup_{\nu=1}^p \Gamma_\nu$. In fact, if $z_0 \in \cup_{\nu=1}^p \Gamma_\nu$ and $1 \leq \lambda_0 \leq n$, then the disk $S = \{z : |z - z_0| < \lambda_0 h/n\}$ does not meet C_ν, whenever $\lambda_\nu \geq \lambda_0$. Thus each of the points of $\{a_\nu\}$ in S is in a range $< \lambda_0$, so their number does not exceed $\lambda_0 - 1$. Let us rearrange a_μ $(\mu = 1, 2, \cdots, n)$ according to their distance from z_0, i.e. $|z_0 - a_1| \leq |z_0 - a_2| \leq \cdots$. Then the distance between z_0 and a_μ must be $\geq \mu h/n$ $(\mu = 1, 2, \cdots, n)$. Consequently we have

$$\prod_{\mu=1}^n |z_0 - a_\mu| \geq \frac{h}{n} \cdot \frac{2h}{n} \cdots \frac{nh}{n} = \left(\frac{h}{n}\right)^n \cdot n! > \left(\frac{h}{e}\right)^n. \quad \square$$

3.1.2 Spherical distance.

In order to introduce the Borel direction of a meromorphic function, we need to use the concept of spherical distance and its properties.

We take the real and imaginary axes of the complex plane as two axes of a rectangular coordinate system in space. The third axis is perpendicular to the complex plane. The Riemann Sphere with center at $(0, 0, 1/2)$ and radius $1/2$ is tangent to the complex plane. Denote by N the north pole $(0, 0, 1)$. Given a point z in the complex plane, the point of intersection w of the segment zN and S is called the spherical image of z. In particular, the spherical image of infinity is N.

Definition 3.1 *Given two complex numbers z_1 and z_2, the distance between their spherical images w_1 and w_2 is called the spherical distance of z_1 and z_2, denoted by $|z_1, z_2|$.*

By definition, we have $0 \leq |z_1, z_2| \leq 1$.

By Definition 3.1 and elementary computations it is easy to verify that the spherical image w of $z = x + iy$ is

$$\left(\frac{x}{1+|z|^2}, \frac{y}{1+|z|^2}, \frac{|z|^2}{1+|z|^2}\right),$$

and hence

$$|z, \infty| = \frac{1}{(1+|z|^2)^{\frac{1}{2}}}; \tag{3.1.2}$$

the spherical distance of two finite complex numbers z_1 and z_2 is

$$|z_1, z_2| = \frac{|z_1 - z_2|}{(1 + |z_1|^2)^{\frac{1}{2}}(1 + |z_2|^2)^{\frac{1}{2}}}. \tag{3.1.3}$$

The following properties will be used later on:

(1) If z is a complex number satisfying $|z, \infty| > d > 0$, then $|z| < 1/d$.

(2) If z_1 and z_2 are two finite complex numbers, then

$$\log^+ |z_1| + \log^+ |z_2| + \log \frac{1}{|z_1 - z_2|} \le \log \frac{1}{|z_1, z_2|}. \tag{3.1.4}$$

In fact, we have from (3.1.3)

$$\log \frac{1}{|z_1, z_2|} = \log \frac{(1 + |z_1|^2)^{1/2}(1 + |z_2|^2)^{1/2}}{|z_1 - z_2|}$$

$$= \log(1 + |z_1|^2)^{1/2} + \log(1 + |z_2|^2)^{1/2} + \log \frac{1}{|z_1 - z_2|}$$

$$\ge \log^+ |z_1| + \log^+ |z_2| + \log \frac{1}{|z_1 - z_2|}.$$

Definition 3.2 *Let a_0 be a complex number and $r(\le 1)$ a positive number. $\{z : |z, a_0| < r\}$ is called a spherical disk with center a_0 and radius r.*

Using a proof similar to that of Theorem 3.1, we can prove Theorem 3.1′.

Theorem 3.1′ *Let a_μ ($\mu = 1, 2, \cdots, n$) be n complex numbers and h a positive number. Then the points which satisfy the inequality*

$$\prod_{\mu=1}^{n} |z, a_\mu| \le \left(\frac{h}{e}\right)^n$$

can be covered by a collection of spherical disks whose number does not exceed n and the sum of whose radii does not exceed $2h$.

3.1.3 Theorem of bounded type. In order to derive the existence of a Borel direction, we shall use the second fundamental theorem of Nevanlinna with an appropriate estimate of the error terms.

Theorem 3.2   *Let $f(z)$ be meromorphic in $|z| \leq R$. If $f(0) \neq 0, 1, \infty$ and $f'(0) \neq 0$, then we have*

$$T(r, f) < 2\{N(R, 0) + N(R, 1) + N(R, \infty)\} + 191 + 4\log^+ |f(0)|$$
$$+ 2\log \frac{1}{R|f'(0)|} + 12\log \frac{R}{R-r},$$

$$(3.1.5)$$

whenever $0 < r < R$.

Proof.   Starting from the Nevanlinna second fundamental theorem (Theorem 1.4) we have

$$T(r, f) < N(r, 0) + N(r, 1) + N(r, \infty) + m\left(r, \frac{f'}{f}\right) + m\left(r, \frac{f'}{f-1}\right)$$
$$+ \log \left| \frac{f(0)(f(0) - 1)}{f'(0)} \right| + \log 2.$$

In a way similar to the proof of Theorem 2.3, we deduce by using Lemma 1.3

$$T(r, f) < N(r, 0) + N(r, 1) + N(r, \infty) + 12 + 22\log 2$$
$$+ 4\log^+ \frac{1}{r} + 6\log^+ \frac{1}{\rho - r} + 8\log^+ \rho$$
$$+ 8\log^+ T(\rho, f) + 2\log^+ |f(0)| + \log^+ \frac{1}{|f'(0)|}. \qquad (3.1.6)$$

If $T(r, f) < 1$ holds uniformly in $(0, R)$, then (3.1.5) is already proved. Otherwise, there exists r_0, $0 \leq r_0 < R$ with $T(r_0, f) = 1$. Using Lemma 1.4 to $r_0 < r < R$, there exists τ in $\left[r, \dfrac{7R + r}{8}\right]$ such that

$$T\left(\tau + \frac{R - \tau}{eT(\tau, f)}, f\right) < 2T(\tau, f).$$

Choose

$$\tau' = \tau + \frac{R - \tau}{eT(\tau, f)}.$$

Clearly $0 < r \le \tau < \tau' < R$. The inequality (3.1.6) yields

$$T(\tau, f) < N(R, 0) + N(R, 1) + N(R, \infty)$$

$$+ 28 + 4\log^+ \frac{1}{\tau} + 6\log^+ \frac{1}{\tau' - \tau} + 8\log^+ \tau'$$

$$+ 8\log^+ T(\tau', f) + 2\log^+ |f(0)| + \log^+ \frac{1}{|f'(0)|}.$$

Since

$$\log^+ T(\tau', f) \le \log T(\tau, f) + \log 2$$

and

$$\log^+ \frac{1}{\tau' - \tau} \le \log^+ \frac{1}{R - \tau} + 1 + \log^+ T(\tau, f)$$

$$\le \log^+ \frac{8}{R - r} + 1 + \log^+ T(\tau, f),$$

we obtain

$$T(\tau, f) < N(R, 0) + N(R, 1) + N(R, \infty) + 28 + 4\log^+ \frac{1}{r}$$

$$+ 6\log^+ \frac{1}{R - r} + 6\log 8 + 6 + 6\log^+ T(\tau, f)$$

$$+ 8\log^+ R + 8\log^+ T(\tau, f) + 8\log 2 + 2\log^+ |f(0)|$$

$$+ \log^+ \frac{1}{|f'(0)|}$$

$$< N(R, 0) + N(R, 1) + N(R, \infty) + 53$$

$$+ 2\log^+ |f(0)| + \log^+ \frac{1}{|f'(0)|} + 4\log^+ \frac{1}{r}$$

$$8\log^+ R + 6\log^+ \frac{1}{R - r} + 14\log^+ T(\tau, f).$$

Noting that

$$14\log^+ T(\tau, f) \le \frac{1}{2}T(\tau, f) + 14(\log 28 - 1),$$

we have

$$T(r, f) \le T(\tau, f)$$

$$< 2\{N(R, 0) + N(R, 1) + N(R, \infty)\}$$

$$+ 176 + 4\log^+ |f(0)| + 2\log^+ \frac{1}{|f'(0)|}$$

$$+ 8\log^+ \frac{1}{r} + 16\log^+ R + 12\log^+ \frac{1}{R - r}.$$

When $r \geq R/2$, we have

$$\log^+ \frac{1}{r} \leq \log^+ \frac{1}{R} + \log 2$$

and therefore by

$$\log^+ \frac{1}{R-r} \leq \log \frac{R}{R-r} + \log^+ \frac{1}{R},$$

we deduce that

$$T(r,f) < 2\{N(R,0) + N(R,1) + N(R,\infty)\} + 182 + 4\log^+ |f(0)|$$

$$+2\log^+ \frac{1}{|f'(0)|} + 20\log^+ \frac{1}{R} + 16\log^+ R + 12\log \frac{R}{R-r}.$$

Moreover, when $r < R/2$, we obtain by the above result

$$T(r,f) \leq T\left(\frac{R}{2}, f\right)$$

$$< 2\{N(R,0) + N(R,1) + N(R,\infty)\}$$

$$+182 + 4\log^+ |f(0)| + 2\log^+ \frac{1}{|f'(0)|}$$

$$+20\log^+ \frac{1}{R} + 16\log^+ R + 12\log 2.$$

Thus

$$T(r,f) < 2\{N(R,0) + N(R,1) + N(R,\infty)\}$$

$$+191 + 4\log^+ |f(0)| + 2\log^+ \frac{1}{|f'(0)|}$$

$$+20\log^+ \frac{1}{R} + 16\log^+ R + 12\log \frac{R}{R-r}$$

holds for all r in $(0, R)$.

Taking $R = 1$, we get

$$T(r,f) < 2\{N(1,0) + N(1,1) + N(1,\infty)\} + 191$$

$$+4\log^+ |f(0)| + 2\log^+ \frac{1}{|f'(0)|} + 12\log \frac{1}{1-r}. \qquad (3.1.7)$$

When $R \neq 1$, set $z = R\zeta$ and $f(z) = f(R\zeta) = g(\zeta)$. Clearly, $g(\zeta)$ is meromorphic in $|\zeta| \leq 1$, $g(0) = f(0)$ and $g'(0) = R\ f'(0)$. Thus

$$T(r,f) = T\left(\frac{r}{R}, g\right) < 2\{N(R,0) + N(R,1) + N(R,\infty)\} + 191$$

$$+4\log^+ |f(0)| + 2\log^+ \frac{1}{R|f'(0)|} + 12\log \frac{R}{R-r}. \qquad \square$$

3.2 Fundamental Theorems

3.2.1 Valiron's fundamental theorem.

From Theorems 3.2 and 3.1, we shall prove the following Valiron fundamental theorem which plays a key role in the derivation of the existence of a Borel direction.

Theorem 3.3   *Let $f(z)$ be meromorphic in $|z| < 1$, and*

$$N = n(1, f = 0) + n(1, f = 1) + n(1, f = \infty).$$

Then, for every $r \in (0,1)$ and every complex number a,

$$n(r, f = a) < \frac{C}{(1-r)^2}\Big\{(N+1)\log\frac{2}{1-r} + \log\frac{1}{|f(z_0), a|}\Big\}, \qquad (3.2.1)$$

where C is a constant and z_0 is a point in $|z| < 1$.

Proof.   Let a_ν $(\nu = 1, 2, \cdots, N)$ be all the zeros, poles and 1-points of $f(z)$ in $|z| < 1$, and let $h = \dfrac{1-r}{40e}$. The Boutroux-Cartan theorem states that

$$\prod_{\nu=1}^{N} |z - a_\nu| > h^N,$$

except for the points in the union Γ of N exceptional disks where the total sum of their radii is $2eh = \dfrac{1-r}{20}$.

In $|z| < \dfrac{1-r}{5}$, we choose a point $z_0 \bar\in \Gamma$. Without loss of generality we can assume that $|f(z_0)| \leq 1$. Otherwise we can consider $\dfrac{1}{f(z)}$ which also satisfies the hypothesis.

Clearly there is a circle $|z - z_0| = \rho$ in

$$r + \frac{2}{5}(1 - r) \leq |z - z_0| \leq r + \frac{3}{5}(1 - r)$$

such that $(|z - z_0| = \rho) \cap \Gamma = \emptyset$. Set

$$|f(\zeta)| = \max_{|z-z_0|=\rho} |f(z)|,$$

and we join z_0 and ζ by a segment. Then the intersection of this segment and Γ is replaced by the corresponding arcs, and thus we obtain a curve

L whose length does not exceed

$$r + \frac{3}{5}(1 - r) + \pi\frac{1 - r}{20} < 1.$$

Now let us consider two mutually exclusive cases.
(1) $|f'(z)| < 1$ holds uniformly on L.
In this case, we have

$$|f(\varsigma)| \leq |f(z_0)| + \left|\int_L f'(z)dz\right| < 2.$$

Thus[1)]

$$m(\rho, z_0, f) < \log 2.$$

Let a_ν ($\nu = 1, 2, \cdots, N_1$) be the poles of $f(z)$ on $|z - z_0| \leq \rho$. Since $z_0 \bar{\in}(\gamma)$, we have

$$\left(\prod_{\nu=1}^{N_1} |z_0 - a_\nu|\right)\left(\prod_{\nu=N_1+1}^{N} |z_0 - a_\nu|\right) > h^N,$$

so

$$\prod_{\nu=1}^{N_1} |z_0 - a_\nu| > \left(\frac{h}{2}\right)^N.$$

Hence

$$N(\rho, z_0, f) < N\log\frac{2}{h} = N\log\frac{80e}{1 - r},$$

and

$$T(\rho, z_0, f) < N\log\frac{80e}{1 - r} + \log 2. \tag{3.2.2}$$

For any complex number a, we have

$$n\left(r + \frac{1 - r}{5}; z_0, f = a\right) = \frac{n\left(r + \frac{1 - r}{5}, z_0, f = a\right)}{\log\dfrac{\rho}{r + \dfrac{1 - r}{5}}}\int_{r + \frac{1-r}{5}}^{\rho}\frac{dt}{t}$$

$$\leq \frac{1}{\log\dfrac{\rho}{r + \dfrac{1 - r}{5}}}N\left(\rho, z_0, \frac{1}{f - a}\right).$$

---

1) Hereinafter we denote by $m(\rho, z_0, f)$, $n(\rho, z_0, f)$, $N(\rho, z_0, f)$ and $T(\rho, z_0, f)$ the corresponding m, n, N and T with respect to the disk $|z - z_0| < \rho$.

From

$$\log \frac{\rho}{r+\frac{1-r}{5}} \geq \log \frac{r+\frac{2(1-r)}{5}}{r+\frac{1-r}{5}} = \int_{r+\frac{1-r}{5}}^{r+\frac{2(1-r)}{5}} \frac{dt}{t} > \frac{1-r}{5}$$

and

$$N(\rho, z_0, \frac{1}{f-a}) \leq T(\rho, z_0, \frac{1}{f-a})$$

$$= T(\rho, z_0, f-a) + \log \frac{1}{|f(z_0)-a|}$$

$$\leq T(\rho, z_0, f) + \log^+ |a| + \log 2 + \log \frac{1}{|f(z_0)-a|},$$

we deduce that

$$n\left(r+\frac{1-r}{5}, z_0, f=a\right)$$

$$< \frac{5}{1-r}\left\{T(\rho, z_0, f) + \log^+ |a| + \log \frac{1}{|f(z_0)-a|} + \log 2\right\}.$$

Substituting (3.2.2) into this inequality and noting (3.1.4), we have

$$n\left(r+\frac{1-r}{5}, z_0, f=a\right) < \frac{5}{1-r}\left\{N\log \frac{80e}{1-r} + \log \frac{1}{|f(z_0), a|} + 2\log 2\right\}.$$

Consequently,

$$n(r, f=a) \leq n\left(r+\frac{1-r}{5}, z_0, f=a\right)$$

$$< \frac{C}{1-r}\left\{(N+1)\log \frac{2}{1-r} + \log \frac{1}{|f(z_0), a|}\right\}.$$

(2) There exists a point z_1 on L with

$$|f'(z_1)| \geq 1, \tag{3.2.3}$$

such that $|f'(z)| < 1$ holds uniformly in L from z_0 to z_1. When $|f'(z_0)| \geq 1$, the point z_0 can be taken as z_1.

It is easy to see that

$$|f(z_1)| \leq 2 \tag{3.2.4}$$

and $|\zeta| < r + 4(1-r)/5$. Thus the function $f(z)$ is meromorphic in

$$|z-z_1| \leq |\zeta-z_1| + \frac{1-r}{10},$$

with $f(z_1) \neq 0, 1$ and $f'(z_1) \neq 0$. Theorem 3.2 gives

$$T\left(|\zeta - z_1| + \frac{1-r}{20}, z_1, f\right) < 2\left\{N(|\zeta - z_1| + \frac{1-r}{10}, z_1, f = 0)\right.$$

$$+ N\left(|\zeta - z_1| + \frac{1-r}{10}, z_1, f = 1\right)$$

$$\left. + N(|\zeta - z_1| + \frac{1-r}{10}, z_1, f = \infty)\right\}$$

$$+ 191 + 4\log^+ |f(z_1)| + 2\log^+ \frac{1}{(|\zeta - z_1| + \frac{1-r}{10})|f'(z_1)|}$$

$$+ 12\log^+ \frac{|\zeta - z_1| + \frac{1-r}{10}}{\left(|\zeta - z_1| + \frac{1-r}{10}\right) - \left(|\zeta - z_1| + \frac{1-r}{20}\right)}.$$

In a way similar to case (1), we have

$$N\left(|\zeta - z_1| + \frac{1-r}{10}, z_1, f = 0\right) + N\left(|\zeta - z_1| + \frac{1-r}{10}, z_1, f = 1\right)$$

$$+ N\left(|\zeta - z_1| + \frac{1-r}{10}, z_1, f = \infty\right) < N\log\frac{2}{h}.$$

From (3.2.3), (3.2.4) and the fact that

$$\frac{1-r}{10} < |\zeta - z_1| + \frac{1-r}{10} < 1,$$

we obtain

$$T\left(|\zeta - z_1| + \frac{1-r}{20}, z_1, f\right)$$

$$< 2N\log\frac{2}{h} + 191 + 4\log 2 + 2\log\frac{10}{1-r} + 12\log\frac{20}{1-r}$$

$$< 2N\log\frac{2}{h} + 238 + 14\log\frac{1}{1-r}. \tag{3.2.5}$$

Applying the Poisson-Jensen formula (Theorem 1.1) to $f(z)$ and the point ζ on the disk $|z - z_1| \leq |\zeta - z_1| + \frac{1-r}{20}$, we have

$$\log|f(\zeta)| \leq \frac{1}{2\pi}\int_0^{2\pi} \log|f(z_1 + R_2 e^{i\varphi})|\frac{R_2^2 - R_1^2}{R_2^2 - 2R_2 R_1\cos(\theta - \varphi) + R_1^2}d\varphi$$

$$+ \sum\log\left|\frac{R_2^2 - \overline{(\beta_\mu - z_1)}(\zeta - z_1)}{R_2(\zeta - \beta_\mu)}\right|,$$

where

$$R_2 = |\zeta - z_1| + \frac{1-r}{20}, \quad R_1 = |\zeta - z_1|, \qquad (3.2.6)$$

and β_μ are the poles of $f(z)$ in $|z - z_1| \le R_2$. Since

$$|R_2^2 - \overline{(\beta_\mu - z_1)}(\zeta - z_1)| \le R_2^2 + R_2^2 < 2R_2,$$

we obtain

$$\log |f(\zeta)| \le \frac{R_2 + R_1}{R_2 - R_1} m(R_2, \; z_1, \; f) + \sum \log \frac{2}{|\zeta - \beta_\mu|}. \qquad (3.2.7)$$

Since $\zeta \bar{\in} (\gamma)$, we have

$$\sum \log \frac{2}{|\zeta - \beta_\mu|} < N \log \frac{4}{h}. \qquad (3.2.8)$$

Substituting (3.2.5), (3.2.6) and (3.2.8) into (3.2.7), we deduce that

$$m(\rho, \; z_0, \; f) \le \log^+ |f(\zeta)| < \frac{82}{1-r} \Big\{ N \log \frac{80e}{1-r} + 119 + 7 \log \frac{1}{1-r} \Big\}.$$

Combining this inequality with $N(\rho, \; z_0, \; f) < N \log \dfrac{80e}{1-r}$, we have

$$T(\rho, \; z_0, \; f) < \frac{83}{1-r} \Big\{ N \log \frac{80e}{1-r} + 119 + 7 \log \frac{1}{1-r} \Big\}. \qquad (3.2.9)$$

Therefore

$$n(r, f = a) \le n\Big(r + \frac{1-r}{5}, \; z_0, \; f = a\Big)$$

$$< \frac{C}{(1-r)^2} \Big\{ (N+1) \log \frac{2}{1-r} + \log \frac{1}{|f(z_0), a|} \Big\}$$

for every complex number a, as shown in case (1). $\square$

3.2.2 Generalizations.   Now let us give two generalizations of Theorem 3.3.

Theorem 3.4   *Let $f(z)$ be meromorphic in $|z| < R$, and let*

$$N = n(R, \; f = \alpha) + n(R, \; f = \beta) + n(R, \; f = \gamma),$$

where α, β and γ are three distinct complex numbers with their spherical distances larger than a positive number d. Then, for every $r \in (0, R)$ and every complex number a,

$$n(r, \ f = a) < \frac{CR^2}{(R-r)^2}\left\{(N+1)\log\frac{2R}{R-r} + \log^+\frac{1}{d} + \log\frac{1}{|f(z_0), a|}\right\},$$

where z_0 is a point in $|z| < R$.

Proof.   Without loss of generality, we can assume that $R = 1$. Otherwise it suffices to consider $F(\zeta) = f(R\zeta) = f(z)$.

We can also assume that $\max(|\alpha|, |\beta|) \leq |\gamma|$. Since the spherical distances among α, β and γ are larger than d, we have

$$|\alpha, \infty| \geq \frac{d}{2}, \quad |\beta, \infty| \geq \frac{d}{2},$$

and hence

$$|\alpha| < \frac{2}{d}, \quad |\beta| < \frac{2}{d}.$$

It is easy to see that

$$g(z) = \frac{f(z) - \alpha}{f(z) - \beta} \cdot \frac{\gamma - \beta}{\gamma - \alpha}$$

is meromorphic in $|z| < 1$ and

$$N = n(1, \ g = 0) + n(1, \ g = 1) + n(1, \ g = \infty).$$

As shown in the proof of Theorem 3.3, we have a union Γ of exceptional disks respect to the zeros, poles and 1-points of $g(z)$ in $|z| < 1$ and a point $z_0 \in \Gamma$ with $|z_0| < (1 - r)/5$. We may assume that $|g(z_0)| \leq 1$, otherwise it suffices to reverse α and β. There exists a circumference $|z - z_0| = \rho$ with $r + 2(1 - r)/5 \leq \rho \leq r + 3(1 - r)/5$, not intersecting Γ. Then by considering two mutually exclusive cases as in the proof of Theorem 3.3, we obtain

$$T(\rho, \ z_0, \ g) < \frac{C(N+1)}{1-r}\log\frac{2}{1-r}.$$

On the other hand,

$$T(\rho, \ z_0, \ f) \leq T(\rho, \ z_0, \ f - \beta) + \log^+ |\beta| + \log 2 = T\left(\rho, \ z_0, \ \frac{1}{f - \beta}\right)$$
$$+ \log |f(z_0) - \beta| + \log^+ |\beta| + \log 2 \leq T\left(\rho, \ z_0, \ \frac{f - \alpha}{f - \beta}\right)$$
$$+ \log^+ \frac{1}{|\beta - \alpha|} + \log 2 + \log |f(z_0) - \beta| + \log^+ |\beta| + \log 2$$
$$\leq T(\rho, \ z_0, \ g) + \log^+ \left|\frac{\gamma - \alpha}{\gamma - \beta}\right| + \log^+ \frac{1}{|\beta - \alpha|}$$
$$+ \log^+ |f(z_0)| + 2 \log^+ |\beta| + 3 \log 2.$$

Since

$$\log^+ \left|\frac{\gamma - \alpha}{\gamma - \beta}\right| = \log^+ \left|1 + \frac{\beta - \alpha}{\gamma - \beta}\right|$$
$$\leq \log^+ |\beta| + \log^+ |\alpha| + 2 \log 2 + \log^+ \frac{1}{|\gamma - \beta|},$$

we have

$$T(\rho, \ z_0, \ f) < T(\rho, \ z_0, \ g) + C\left(\log \frac{1}{d} + 1\right) + \log^+ |f(z_0)|$$
$$< C\left\{\frac{N+1}{1-r} \log \frac{2}{1-r} + \log \frac{1}{d}\right\} + \log^+ |f(z_0)|. \qquad (3.2.10)$$

Thus

$$n(r, \ f = a) \leq n\left(r + \frac{1-r}{5}, \ z_0, \ f = a\right) \leq \frac{5}{1-r} N(\rho, \ z_0, \ f = a)$$
$$< \frac{C}{(1-r)^2}\left\{(N+1) \log \frac{2}{1-r} + \log^+ \frac{1}{d} + \log \frac{1}{|f(z_0), a|}\right\}. \quad \square$$

3.2.3 Rauch Theorem.   Let us give a further generalization of Theorem 3.3, the Rauch theorem, for use later.

Lemma 3.1   *Let $P(z)$ be meromorphic in $|z| < R \ (\leq \infty)$. If*

$$\frac{1}{\pi} \iint_{r_1 < |z| < r_2} \log^+ |P(z)| d\sigma_z < \log \frac{1}{d}, \quad 0 < d < \frac{1}{2},$$

holds for $0 < r_1 < r_2 < R$, then the set of values r in $[r_1, \ r_2]$ satisfying

$$m(r, \ P) \leq \frac{\log 1/d}{r_1(r_2 - r_1)}$$

has a measure not less than $(r_2 - r_1)/2$.

Proof.   If the conclusion is not true, then the set of values r in $[r_1, r_2]$ with

$$m(r, \ P) > \frac{\log \frac{1}{d}}{r_1(r_2 - r_1)}$$

has a measure greater than $(r_2 - r_1)/2$. Thus

$$\int_{r_1}^{r_2} 2m(r, P) r dr > 2 \frac{\log \frac{1}{d}}{r_1(r_2 - r_1)} \cdot r_1 \cdot \frac{r_2 - r_1}{2} = \log \frac{1}{d}.$$

Since the left-hand side of the above inequality equals

$$\frac{1}{\pi} \iint_{r_1 < |z| < r_2} \log^+ |P(z)| d\sigma_2,$$

this contradicts the hypothesis of the lemma.   □

Theorem 3.5   *Let* $f(z)$, $P(z)$, $Q(z)$ *and* $R(z)$ *be meromorphic in* $|z| < 1$, *and let*

$$n(1, \ f = P) + n(1, \ f = Q) + n(1, \ f = R) = N.$$

Suppose the total number of zeros and poles of $P(z)$, $Q(z)$, $R(z)$, $P(z)-Q(z)$, $Q(z)-R(z)$ *and* $R(z)-P(z)$ *in* $|z| < 1$ *does not exceed* p. *If*

$$\frac{1}{\pi} \iint_{|z|<1} \log^+ \left(|P| + |Q| + \frac{1}{|P-Q|} + \frac{1}{|Q-R|} + \frac{1}{|R-P|} \right) d\sigma < \log \frac{1}{d},$$
$$(3.2.11)$$

then, for every $r \in (0, 1)$ *and every complex number* a,

$$n(r, \ f = a) < \frac{C}{(1-r)^2} \left\{ (N+P) \log \frac{2}{1-r} + \log \frac{1}{d} + \log \frac{1}{|f(z_0), a|} \right\},$$
$$(3.2.12)$$

where $|z_0| < 1$.

Proof.   We apply the Boutroux-Cartan theorem to the zeros of $f(z) - P(z)$, $f(z) - Q(z)$ and $f(z) - R(z)$ in $|z| < 1$ with $h = (1 - r)/160e$, and denote by Γ_1 the union of the corresponding exceptional disks, where the sum of their radii does not exceed $2eh = (1 - r)/80$. Then we apply the Boutroux-Cartan theorem to all the zeros and poles of

$P(z)$, $Q(z)$, $R(z)$, $P(z) - Q(z)$, $Q(z) - R(z)$ and $R(z) - P(z)$ in $|z| < 1$ with the same h, and denote by Γ_2 the union of the correspouding exceptional disks, where the sum of their radii does not exceed $2eh = (1-r)/80$. Let Γ be the union of Γ_1 and Γ_2. The total sum of diameters of all the exceptional disks in Γ does not exceed $(1 - r)/20$.

Then

$$g(z) = \frac{f(z) - P(z)}{f(z) - Q(z)} \cdot \frac{R(z) - Q(z)}{R(z) - P(z)}$$

is meromorphic in $|z| < 1$, and

$$n(1,\ g = 0) + n(1,\ g = 1) + n(1,\ g = \infty) \leq N + p.$$

There exists a point $z_0 \bar\in \Gamma$ with $|z_0| < (1 - r)/5$ and $|g(z_0)| \leq 1$. According to the definition of Γ, (3.2.11) and Lemma 3.1, in the angular region

$$r + \frac{2}{5}(1 - r) \leq |z - z_0| \leq r + \frac{3}{5}(1 - r),$$

we can choose a circumference $|z - z_0| = \rho$, not intersecting Γ, such that

$$m(\rho,\ z_0,\ P) + m(\rho,\ z_0,\ Q) + m\left(\rho,\ z_0,\ \frac{1}{P - Q}\right)$$

$$+ m\left(\rho,\ z_0,\ \frac{1}{Q - R}\right) + m\left(\rho,\ z_0,\ \frac{1}{R - P}\right)$$

$$\leq 5\max\left\{m(\rho,\ z_0,\ P),\ m(\rho,\ z_0,\ Q),\ m\left(\rho,\ z_0,\ \frac{1}{P - Q}\right),\right.$$

$$\left. m\left(\rho,\ z_0,\ \frac{1}{Q - R}\right),\ m\left(\rho,\ z_0,\ \frac{1}{R - P}\right)\right\}$$

$$\leq 5m\left(\rho,\ z_0,\ |P| + |Q| + \frac{1}{|P - Q|} + \frac{1}{|Q - R|} + \frac{1}{|R - P|}\right)$$

$$\leq \frac{5 \log \frac{1}{d}}{(r + \frac{2}{5}(1 - r))\frac{1 - r}{5}} \leq \frac{125 \log \frac{1}{d}}{2(1 - r)}.$$

$$(3.2.13)$$

As in the proof of Theorem 3.3, we consider two mutually exclusive cases, to obtain

$$T(\rho,\ z_0,\ g) < \frac{C(N + P + 1)}{1 - r} \log \frac{2}{1 - r}.$$

On the other hand,

$$T(\rho,\ z_0,\ f) \le T(\rho,\ z_0,\ f-Q) + T(\rho,\ z_0,\ Q) + \log 2$$

$$= T\left(\rho,\ z_0,\ \frac{1}{f-Q}\right) + \log|f(z_0) - Q(z_0)| + T(\rho,\ z_0,\ Q) + \log 2$$

$$\le T\left(\rho,\ z_0,\ \frac{f-P}{f-Q}\right) + T\left(\rho,\ z_0,\ \frac{1}{Q-P}\right) + \log 2$$

$$+ \log|f(z_0) - Q(z_0)| + T(\rho,\ z_0,\ Q) + \log 2$$

$$\le T(\rho,\ z_0,\ g) + T\left(\rho,\ z_0,\ \frac{R-P}{R-Q}\right) + T\left(\rho,\ z_0,\ \frac{1}{Q-P}\right)$$

$$+ T(\rho,\ z_0,\ Q) + \log|f(z_0) - Q(z_0)| + 2\log 2.$$

Since $(R-P)/(R-Q) = 1 + (Q-P)/(R-Q)$, implying

$$T\left(\rho,\ z_0,\ \frac{R-P}{R-Q}\right) \le T\left(\rho,\ z_0,\ \frac{1}{R-Q}\right)$$

$$+ T(\rho,\ z_0,\ P) + T(\rho,\ z_0,\ Q) + 2\log 2$$

and

$$\log|f(z_0) - Q(z_0)| \le \log^+|f(z_0)| + \log^+|Q(z_0)| + \log 2$$

$$\le \log^+|f(z_0)| + T(\rho,\ z_0,\ Q) + \log 2,$$

we obtain

$$T(\rho,\ z_0,\ f) \le T(\rho,\ z_0,\ g) + T\left(\rho,\ z_0,\ \frac{1}{R-Q}\right) + T\left(\rho,\ z_0,\ \frac{1}{Q-P}\right)$$

$$+ 3T(\rho,\ z_0,\ Q) + T(\rho,\ z_0, P) + 5\log 2 + \log^+|f(z_0)|.$$

Moreover, the conditions of Theorem 3.5, $z_0 \bar{\in} \Gamma_2$ and (3.2.13), together imply

$$3T(\rho,\ z_0,\ Q) + T(\rho,\ z_0,\ P) + T\left(\rho,\ z_0,\ \frac{1}{R-Q}\right) + T\left(\rho,\ z_0,\ \frac{1}{Q-P}\right)$$

$$\le 3\left\{P\log\frac{2}{h} + \frac{125\log\frac{1}{d}}{2(1-r)}\right\} \le 3\left\{P\log\frac{320e}{1-r} + \frac{125\log\frac{1}{d}}{2(1-r)}\right\}.$$

Thus

$$T(\rho,\ z_0,\ f) < \frac{C}{1-r}\left\{(N+P+1)\log\frac{2}{1-r} + \log\frac{1}{d}\right\} + \log^+|f(z_0)|.$$

Consequently,

$$n(r, \ f = a) \le n\Big(r + \frac{1-r}{5}, \ z_0, \ f = a\Big) \le \frac{5}{1-r}N\Big(\rho, \ z_0, \ \frac{1}{f-a}\Big)$$

$$< \frac{5}{1-r}\Big\{T(\rho, \ z_0, \ f-a) + \log \frac{1}{|f(z_0) - a|}\Big\}$$

$$< \frac{C}{(1-r)^2}\Big\{(N+P+1)\log \frac{2}{1-r} + \log \frac{1}{d} + \log \frac{1}{|f(z_0), a|}\Big\},$$

whatever the $r \in (0,1)$ and the complex number a may be. □

3.3 Filling Disks and Borel Directions

3.3.1 Two lemmas. In order to introduce the concepts of filling disks and Borel direction of a meromorphic function, we need further preparations.

Let $f(z)$ be meromorphic in the finite plane, D a bounded region, $n(D, \ f = a)$ the number of a-points of $f(z)$ in D, and N a positive integer. Suppose that the set E of complex numbers a with $n(D, \ f = a) \ge N$ can not be covered by a collection of spherical disks the total sum of whose radii is equal to $1/2$.

Divide D into p subregions $D_j (j = 1, 2, \cdots, p)$. Let $K_j \ (j = 1, 2, \cdots, p)$ be a disk containing D_j, and K'_j the corresponding concentric disk with radius twice that of K_j. Suppose that for every $j = 1, 2, \cdots, p$,

$$n(K'_j, \ f = \alpha) + n(K'_j, \ f = \beta) + n(K'_j, \ f = \gamma) < n,$$

where α, β and γ are three complex numbers with their spherical distances having a lower bound d and $n > 1$. Then, according to Theorem 3.4, for every complex number a,

$$n(K_j, \ f = a) < C\Big\{n + \log \frac{1}{d} + \log \frac{1}{|a, A_j|}\Big\},$$

where A_j is a constant independent of a. Hence

$$n(D, \ f = a) < C\Big\{np + p\log \frac{1}{d} + \log \frac{1}{\prod_{j=1}^{p} |a, A_j|}\Big\},$$

since $D = \cup_{j=1}^{p} D_j \subset \cup_{j=1}^{p} K_j$.

Moreover, Theorem 3.1' gives us

$$\prod_{j=1}^{p} |a, \ A_j| > h^P,$$

which holds for all a except those in the union Γ of spherical disks the sum of whose radii does not exceed $2eh$. Let us set $h = 1/8e$. Then, according to the above supposition, we have $E \not\subset \Gamma$, i.e. there is a complex number $a_0 \bar{\in} \Gamma$ with $n(D, \ f = a_0) \geq N$. Hence

$$N \leq n(D, \ f = a_0) < C \left\{ np + p \log \frac{1}{d} + p \log 8e \right\}$$

$$< Cnp, \qquad \text{(by choosing } d = e^{-n})^{1)}$$

Therefore an appropriate constant C^* can be found such that $n > C^* N/p$. The above discussion can be summarized as the following lemma.

Lemma 3.2   *Let $f(z)$, D, D_j, K_j and K'_j $(j=1, 2, \cdots, p)$ as defined above. If the set of complex numbers a with $n(D, f=a) \geq N$ cannot be covered by a collection of spherical disks the sum of whose radii is $1/2$ on the Riemann Sphere, then there exists a disk K'_j and a constant C^* such that in K'_j, $f(z)$ takes every complex number at least $C^* N/p$ times, except for those complex numbers contained in the union of two spherical disks each with radius $e^{-C^* N/p}$.*

Let us examine the implications of this lemma.

Let $f(z)$ be a transcendental meromorphic function of finite order in the finite plane. Take the annular region $r < |z| < R$ as the region D in Lemma 3.2. First of all, let us find a lower bound of $n(D, \ f = a)$, i.e. an estimate of $n(R, \ f = a) - n(r, \ f = a)$.

Note that

$$n(R, \ f = a) = \frac{n(R, f = a)}{\log \dfrac{R}{r}} \int_r^R \frac{dt}{t}$$

$$\geq \left[N\left(R, \frac{1}{f-a}\right) - N\left(r, \frac{1}{f-a}\right) \right] \Big/ \log \frac{R}{r},$$

---

1) We recall that C always denotes a constant and need not be the same at every occurrence.

and

$$n(r,\ f = a) = \frac{n(r,\ f = a)}{\log k} \int_r^{kr} \frac{dt}{t} \le \frac{1}{\log k} N(kr,\ f = a)$$

whenever $k > 1$.

For a sufficiently large R, the second fundamental theorem of Nevanlinna (Theorem 1.5') asserts that the set of complex numbers a having the property that

$$N\left(R,\ \frac{1}{f-a}\right) > \frac{T(R,\ f)}{4} \text{ and } |f(0),\ a| > \frac{1}{3} \tag{3.3.1}$$

cannot be covered by a collection of spherical disks the sum of whose radii equals $1/2$. Moreover, by taking R large enough to satisfy

$$T(R,\ f) \ge \max\left\{240,\ \frac{240 \log \frac{R}{r}}{\log k},\ 12T(r,\ f),\ \frac{12T(kr,\ f)}{\log k} \log \frac{R}{r}\right\}, \tag{3.3.2}$$

we have the following estimate

$$n(R,\ f = a) - n(r,\ f = a)$$

$$\ge \frac{N\left(R,\ \frac{1}{f-a}\right)}{\log \frac{R}{r}} - \frac{N\left(r,\ \frac{1}{f-a}\right)}{\log \frac{R}{r}} - \frac{N\left(kr,\ \frac{1}{f-a}\right)}{\log k}$$

$$\ge \frac{\frac{1}{4}T(R,\ f)}{\log \frac{R}{r}} - \frac{T\left(r, \frac{1}{f-a}\right)}{\log \frac{R}{r}} - \frac{T\left(kr,\ \frac{1}{f-a}\right)}{\log k}$$

$$\ge \frac{\frac{1}{4}T(R,\ f)}{\log \frac{R}{r}} - \frac{T(r,f) + \log \frac{1}{|f(0),a|} + \log 2}{\log \frac{R}{r}}$$

$$- \frac{T(kr,\ f) + \log \frac{1}{|f(0),a|} + \log 2}{\log k}$$

$$\geq \frac{T(R, f)}{4\log \dfrac{R}{r}}\Big\{1 - \frac{4T(r, f)}{T(R, f)} - \frac{4\log 6}{T(R, f)}$$

$$-\frac{4T(kr, f)}{T(R, f)}\cdot\frac{\log\dfrac{R}{r}}{\log k} - \frac{4\log 6}{T(R, f)}\cdot\frac{\log\dfrac{R}{r}}{\log k}\Big\}$$

$$\geq \frac{T(R, f)}{15\log \dfrac{R}{r}}.$$

Next, let q be a sufficiently large integer. We divide the annular region $r < |z| < R$ as follows. Choose q rays

$$\arg z = 0, \quad \frac{2\pi}{q}, \quad 2\cdot\frac{2\pi}{q}, \quad \cdots, \quad (q-1)\frac{2\pi}{q}$$

and circles

$$|z| = r, \quad r\Big(1 + \frac{2\pi}{q}\Big), \quad \cdots, \quad r\Big(1 + \frac{2\pi}{q}\Big)^{S},$$

where

$$S = \Big[\frac{\log\dfrac{R}{r}}{\log\Big(1 + \dfrac{2\pi}{q}\Big)}\Big] + 1.$$

Then the region $r < |z| < R$ is divided into p subregions D_j with

$$p \leq q\Big\{\Big[\frac{\log\dfrac{R}{r}}{\log\Big(1 + \dfrac{2\pi}{q}\Big)}\Big] + 1\Big\} < \frac{q^2\log\dfrac{R}{r}}{2}.$$

Let z_j denote the center of D_j. It is easy to see that D_j is contained in $|z - z_j| < 2\pi|z_j|/q$.

Finally, applying Lemma 3.2, we have the following proposition.

Lemma 3.3   *Let $f(z)$ be a transcendental meromorphic function of finite order in the finite plane. Then there exists a point z_j in the region $r < |z| < R$ such that $f(z)$ takes every complex number at least*

$$n = C^*\frac{T(R, f)}{q^2\Big(\log\dfrac{R}{r}\Big)^2} \tag{3.3.3}$$

times in $|z - z_j| < 4\pi|z_j|/q$, except for those complex numbers which can be contained in the union of two spherical disks each with radius e^{-n}, and provided that R and q are sufficiently large to satisfy (3.3.2).

The condition that $f(z)$ is of finite order in Lemma 3.3 is used in (3.3.1). When the order of $f(z)$ is infinite, exceptional intervals in the sense of Theorem 1.6 of Chpater 1 may exist. There is, however, a value R' in $[R, 2R]$, not belonging to the exceptional intervals. Thus the set of complex numbers a such that

$$N(R', \frac{1}{f-a}) > \frac{T(R', f)}{4}, \quad |f(0), a| > \frac{1}{3}$$

cannot be covered by a collection of disks the sum of whose radii is $1/2$. For such complex numbers a, we have

$$n(2R, \ f = a) - n(r, \ f = a)$$

$$\geq \frac{N\left(2R, \ \frac{1}{f-a}\right)}{\log \frac{2R}{r}} - \frac{N\left(r, \ \frac{1}{f-a}\right)}{\log \frac{2R}{r}} - \frac{N\left(kr, \ \frac{1}{f-a}\right)}{\log k}.$$

On the other hand,

$$N\left(2R, \ \frac{1}{f-a}\right) \geq N\left(R', \ \frac{1}{f-a}\right) \geq \frac{1}{4}T(R', \ f) \geq \frac{1}{4}T(R, f).$$

Thus we can obtain the following lemma by a similar deduction shown in the proof of Lemma 3.3.

Lemma 3.4 *If $f(z)$ is transcendental meromorphic in the finite plane, then there exists a point z_j in the region $r < |z| < 2R$ such that $f(z)$ takes every complex number at least*

$$n = C^* \frac{T(R, \ f)}{q^2\left(\log \frac{R}{r}\right)^2}$$

times in $|z - z_j| < 4\pi|z_j|/q$, except for those complex numbers which can be contained in the union of two spherical disks each with radius e^{-n}, and provided that R and q are sufficiently large to satisfy

$$T(R, \ f) \geq \max\left\{240, \ \frac{240\log(2R)}{\log k}, 12T(r, f), \ \frac{12T(kr, \ f)}{\log k}\log \frac{2R}{r}\right\}.$$

$$(3.3.4)$$

3.3.2. Filling disks and Julia directions of a meromorphic function. In Theorem 2.9, we have seen that every transcendental entire function has a Julia direction. For transcendental meromorphic functions, this is however not the case. For instance, A. Ostrowski [1] constructed a transcendental meromorphic function $f(z)$ such that

$$T(r,\ f) = O((\log r)^2), \quad r \to \infty,$$

but this $f(z)$ has no Julia direction.

Nevertheless, we can state the following theorem.

Theorem 3.6 *Let $f(z)$ be meromorphic in the finite plane. If*

$$\varlimsup_{r\to\infty} \frac{T(r,\ f)}{(\log r)^2} = \infty, \tag{3.3.5}$$

then there exists a sequence of disks

$$\Gamma_j : |z - z_j| < \varepsilon_j |z_j|, \ \lim_{j\to\infty} |z_j| = \infty, \ \lim_{j\to\infty} \varepsilon_j = 0, \quad j = 1, 2, \cdots$$

such that for each j, $f(z)$ takes every complex number at least n_j times in Γ_j, except for those complex numbers contained in the union of two spherical disks each with radius e^{-n_j}, where $\lim_{j\to\infty} n_j = \infty$.

The disks with the above property are called filling disks.

Proof.   From (3.3.5), there is a sequence of positive numbers (r_k) tending to infinity and satisfying

$$\lim_{k\to\infty} \frac{T(r_k,\ f)}{(\log r_k)^2} = \infty.$$

Choosing r_{k_1} sufficiently large such that

$$T(r_{k_1},\ f) \geq \max\left\{240, \frac{240 \log(2r_{k_1})}{\log 2}, 12T(1,\ f), \frac{12T(2, f)}{\log 2} \log(2r_{k_1})\right\}$$

and such that

$$q_1 = \frac{T(r_{k_1},\ f)^{\frac{1}{4}}}{(\log r_{k_1})^{\frac{1}{2}}}$$

is also sufficiently large.

Then, by Lemma 3.4, there exists a point z_1 in $1 < |z| < 2r_{k_1}$ such that in $\Gamma_1 : |z - z_1| < 4\pi|z_1|/q_1$, $f(z)$ takes every complex number at least

$$n_1 = C^* \frac{T(r_{k_1}, f)^{\frac{1}{2}}}{\log r_{k_1}}$$

times, except for those complex numbers contained in the union of two spherical disks each with radius e^{-n_1}.

Choose $r_{k_1'} > 3r_{k_1}$ and r_{k_2} sufficiently large such that

$$T(r_{k_2}, f) \geq \max \left\{ 240, \frac{240 \log(2r_{k_2})}{\log 2}, 12T(r_{k_1'}, f), \frac{12T(r_{k_1'}, f)}{\log 2} \log(2r_{k_2}) \right\}$$

and such that

$$q_2 = \frac{T(r_{k_2}, f)^{\frac{1}{4}}}{(\log r_{k_2})^{\frac{1}{2}}}$$

is also sufficiently large.

Then, once more by Lemma 3.4 there exists a point z_2 in $r_{k_1'} < |z| < 2r_{k_2}$ such that, in $\Gamma_2 : |z - z_2| < 4\pi|z_2|/q_2$, $f(z)$ takes every complex number at least

$$n_2 = C^* \frac{T(r_{k_2}, f)^{\frac{1}{2}}}{\log r_{k_2}}$$

times, except for those complex numbers contained in the union of two spherical disks each with radius e^{-n_2}.

Continuing this procedure, we obtain a sequence of disks with the property formulated in Theorem 3.6. □

Corollary. *If $f(z)$ is meromorphic in the finite plane satisfying (3.3.5), then there exists a ray J: arg $z = \theta_0 (0 \leq \theta_0 < 2\pi)$ such that given an arbitrary positive number ε, $f(z)$ takes every complex number infinitely many times in the angular region $|\text{arg } z - \theta_0| < \varepsilon$, with two possible exceptions.*

In fact, according to Theorem 3.6, $f(z)$ has a sequence of filling disks $\Gamma_j : |z - z_j| < \varepsilon_j|z_j|$. The sequence $\{e^{i\arg z_j}, j = 1, 2, \cdots\}$ has an accumulation point $e^{i\theta_0}$ $(0 \leq \theta_0 < 2\pi)$. We claim that the ray J: arg $z = \theta_0$ has the desired property. Otherwise there is an angular region $\Omega : |\text{arg } z - \theta_0| < \varepsilon$ such that $f(z)$ takes none of the three complex numbers a_ν $(\nu = 1, 2, 3)$ in Ω. Note that Ω contains a subsequence of (Γ_j), still to be denoted by (Γ_j) for the brevity. When j is sufficiently large, we have $2e^{-n_j} < \min\limits_{1 \leq \mu \neq \nu \leq 3} \{|a_\mu, a_\nu|\}$. Thus at least one of a_ν $(\nu = 1, 2, 3)$, say a_1, belongs to neither of the two exceptional spherical disks each with radius e^{-n_j}. Therefore $f(z)$ takes a_1 in Γ_j at least n_j times, where $\lim_{j\to\infty} n_j = \infty$. This contradicts the assumption that $f(z)$ does not take a_1 in Ω.

The rays with the property formulated in the corollary are called Julia directions of $f(z)$.

3.3.3 Filling disks and Borel directions of order λ of a meromorphic function.

For meromorphic functions of finite and positive order, we can draw stronger conclusions.

Theorem 3.7 *If $f(z)$ is meromorphic and of order λ $(0 < \lambda < \infty)$ in the finite plane, then there exists a sequence of disks*

$$\Gamma_j : \ |z - z_j| < \varepsilon_j |z_j|, \ \lim_{j\to\infty} \varepsilon_j = 0, \ \lim_{j\to\infty} |z_j| = \infty, \quad j = 1, 2, \cdots$$

such that $f(z)$ takes every complex number at least $|z_j|^{\lambda - \varepsilon_j}$ times in Γ_j, except for those complex numbers contained in the union of two spherical disks each with radius $e^{-|z_j|^{\lambda - \delta_j}}$, where $\lim_{j\to\infty} \delta_j = 0$.

The disks Γ_j are called filling disks of order λ.

Proof. Since the order of $f(z)$ is λ, there is a sequence of positive numbers r_k tending to infinity and satisfying

$$\lim_{k\to\infty} \frac{\log T(r_k, \ f)}{\log r_k} = \lambda.$$

Choose r_{k_1} sufficiently large such that

$$T(r_{k_1}, \ f) \geq \max\left\{240, \ \frac{240 \log r_{k_1}}{\log 2}, 12T(1, f), \ \frac{12T(2, f)}{\log 2} \log r_{k_1}\right\}$$

and such that

$$q_1 = \log r_{k_1}$$

is also sufficiently large. By Lemma 3.3, there exists a point z_1 in $1 < |z| < r_{k_1}$ such that in the disk $\Gamma_1 : \ |z - z_1| < 4\pi|z_1|/\log r_{k_1}$, $f(z)$ takes every complex number at least

$$n_1 = C^* \frac{T(r_{k_1}, f)}{(\log r_{k_1})^4}$$

times, except for those complex numbers contained in the union of two spherical disks each with radius e^{-n_1}.

Take $r_{k'_1} > 2r_{k_1}$, and take r_{k_2} sufficiently large such that

$$T(r_{k_2}, f) \geq \max\left\{240, \frac{240\log r_{k_2}}{\log 2}, 12T(r_{k'_1}, f), \frac{12T(2r_{k'_1}, f)}{\log 2}\log r_{k_2}\right\}$$

and such that $q_2 = \log r_{k_2}$ is also sufficiently large. Then, there is a point z_2 in $r_{k'_1} < |z| < r_{k_2}$ such that in the disk $\Gamma_2 : |z - z_2| < 4\pi|z_2|/\log r_{k_2}$, $f(z)$ takes every complex number at least

$$n_2 = C^* \frac{T(r_{k_2}, f)}{(\log r_{k_2})^4}$$

times, except for those complex numbers contained in the union of two spherical disks each with radius e^{-n_2}.

Continuing this procedure, we obtain a sequence of disks. Since

$$n_j = C^* \frac{T(r_{k_j}, f)}{(\log r_{k_j})^4} > r_{k_j}^{\lambda - \varepsilon'_{k_j}} > |z_{k_j}|^{\lambda - \varepsilon'_{k_j}},$$

this sequence of disks has the property formulated in Theorem 3.7.

Let $f(z)$ be meromorphic in the finite plane, and let $\arg z = \theta_0$ ($0 \leq \theta_0 < 2\pi$) be a ray. For $r > 0$, $\varepsilon > 0$ and an arbitrary complex number a, we denote by $n(r, \theta_0, \varepsilon, f = a)$ or $n(r, \theta_0, \varepsilon, 1/(f - a))$ the number of zeros of $f(z) - a$ in the region $(|z| \leq r) \cap (|\arg z - \theta_0| \leq \varepsilon)$, multiple zeros being counted with their multiplicities. When $a = \infty$, we write $n(r, \theta_0, \varepsilon, f)$.

Theorem 3.8 *If $f(z)$ is meromorphic and of order λ ($0 < \lambda < \infty$) in the finite plane, then there exists a ray B: $\arg z = \theta_0$ ($0 \leq \theta_0 < 2\pi$) such that, given any positive number ε, the equality*

$$\varlimsup_{r\to\infty} \frac{\log n(r, \theta_0, \varepsilon, f = a)}{\log r} = \lambda \tag{3.3.6}$$

holds for every complex number a, with at most two possible exceptions.

Proof. According to Theorem 3.7, $f(z)$ has a sequence of filling disks of order λ, $\Gamma_j : |z - z_j| < \varepsilon_j|z_j|$, such that in every Γ_j, $f(z)$ takes every complex number at least $|z_j|^{\lambda-\varepsilon_j}$ times, except for those complex numbers contained in the union of two spherical disks S_j and S'_j each with radius $e^{-|z_j|^{\lambda-\delta_j}}$, where $\lim_{j\to\infty}\delta_j = 0$.

Let $e^{i\theta_0}$ $(0 \leq \theta_0 < 2\pi)$ be an accumulation point of $\{e^{i\arg z_j}, j = 1,2,\cdots\}$. We claim that $\arg z = \theta_0$ is the desired ray. In fact, if this assertion is not true, there exist an angular region $\Omega_0 : |\arg z - \theta_0| < \varepsilon_0$ and three distinct complex numbers a_ν $(\nu = 1,2,3)$ such that

$$\varlimsup_{r\to\infty} \frac{\log n(r,\ \theta_0,\ \varepsilon_0,\ f = a_\nu)}{\log r} < \lambda, \quad \nu = 1,2,3. \qquad (3.3.7)$$

Ω contains a subsequence of (Γ_j), still to be denoted by (Γ_j). When j is sufficiently large, we have

$$2e^{-|z_j|^{\lambda-\delta_j}} < \max_{1\leq\mu\neq\nu\leq 3}\{|a_\mu,\ a_\nu|\}.$$

Thus at least one of a_ν $(\nu = 1,2,3)$, say a_1, belongs to neither of S_j or S'_j $(j = 1,2,\cdots)$, so that $f(z)$ takes a_1 in Γ_j at least $|z_j|^{\lambda-\delta'_j}$ times. Hence

$$\varlimsup_{r\to\infty} \frac{\log n(r,\ \theta_0,\ \varepsilon,\ f = a_1)}{\log r} \geq \varlimsup_{j\to\infty} \frac{\log n(\Gamma_j,\ f = a_1)}{\log(2|z_j|)}$$

$$\geq \lim_{j\to\infty} \frac{\log(|z_j|^{\lambda-\delta_j})}{\log(2|z_j|)} = \lambda.$$

This inequality contradicts (3.3.7). $\square$

The ray in Theorem 3.8 is usually called a Borel direction of $f(z)$. It is also called a Borel-Valiron direction, since the proof of its existence is due to G. Valiron [2].

Much research was made into Borel directions of entire and meromorphic functions. For instance, M. Biernacki [1] proved the following theorem.

Theorem 3.8'   *If $f(z)$ is meromorphic and of order λ $(0 < \lambda < \infty)$ in the finite plane, then there exists a ray $B: \arg z = \theta_0$ $(0 \leq \theta_0 < 2\pi)$ such that, given any positive number ε, the equality*

$$\varlimsup_{r\to\infty} \frac{\log n(r,\ \theta_0,\ \varepsilon,\ f = \Pi)}{\log r} = \lambda, \qquad (3.3.8)$$

holds for every meromorphic function $\Pi(z)$ of order less than λ (including all the complex numbers), with two possible exceptions at most.

A. Rauch [1] proceeded to generalize the Biernacki theorem. In order to introduce his result, we need the concepts of convergent and divergent classes of meromorphic functions.

Let $f(z)$ be meromorphic and of order λ $(0 < \lambda < \infty)$ in the finite plane. If the integral

$$\int^{\infty} \frac{T(r,f)}{r^{\lambda+1}} dr \tag{3.3.9}$$

is finite, then $f(z)$ belongs to the convergent class of order λ. Otherwise, $f(z)$ belongs to the divergent class of order λ.

Theorem 3.8″   *If $f(z)$ is meromorphic in the finite plane and belongs to the divergent class of order λ, then there exists a ray (B): $\arg z = \theta_0$ $(0 \le \theta_0 < 2\pi)$ such that, given any positive number ε, the equality*

$$\int^{\infty} n(r, \theta_0, \varepsilon, f = \Pi) \frac{dr}{r^{\lambda+1}} = \infty \tag{3.3.10}$$

holds for every meromorphic function $\Pi(z)$ belonging to the convergent class of order λ, with at most two possible exceptional functions.

Theorems 3.8′ and 3.8″ can be proved with the aid of Theorem 3.5. Those readers interested in these results may consult the original papers. We shall introduce some recent results on Borel directions in later chapters.

3.4 Properties of Borel Directions

3.4.1 One lemma. Before introducing the properties of Borel directions, we need to prove a lemma.

Lemma 3.5 *Let $f(z)$ be meromorphic in the finite plane. Suppose r_j $(j = 1, 2, \cdots)$ are the moduli of the poles of $f(z)$ arranged by non-decreasing order and counted with their multiplicities. If σ is a positive number, then the series*

$$\sum r_j^{-\sigma}$$

and the integrals

$$\int^\infty \frac{n(t,f)}{t^{\sigma+1}}dt \quad and \quad \int^\infty \frac{N(t,f)}{t^{\sigma+1}}dt$$

are either simultaneously convergent or simultaneously divergent.

Proof.   Using the Stieltjes integral, it is easy to see that

$$\sum_{j=j_0}^{J} \frac{1}{r_j^\sigma} = \int_{r_0}^R \frac{1}{t^\sigma}dn(t,f) = \frac{n(R,f)}{R^\sigma} - \frac{n(r_0,f)}{r_0^\sigma} + \int_{r_0}^R \frac{\sigma n(t,f)}{t^{\sigma+1}}dt. \quad (3.4.1)$$

When $\Sigma r_j^{-\sigma}$ is convergent, $\int^\infty \frac{n(t,f)}{t^{\sigma+1}}dt$ is also convergent. Conversely, if $\int^\infty \frac{n(t,f)}{t^{\sigma+1}}dt$ converges, then we have

$$1 > \int_R^\infty \frac{n(t,f)}{t^{\sigma+1}}dt \geq n(R,f)\int_R^\infty \frac{dt}{t^{\sigma+1}} = \frac{n(R,f)}{\sigma R^\sigma}$$

for sufficiently large R. Combining this fact and (3.4.1), the series $\Sigma r_j^{-\sigma}$ is also convergent.

Similarly, by using

$$\int_{r_0}^R \frac{n(t,f)}{t^{\sigma+1}}dt = \int_{r_0}^R \frac{1}{t^\sigma}dN(t,f) = \frac{N(R,f)}{R^\sigma} - \frac{N(r_0,f)}{r_0^\sigma} + \int_{r_0}^R \frac{\sigma N(t,f)}{t^{\sigma+1}}dt,$$

$\int^\infty \frac{n(t,f)}{t^{\sigma+1}}dt$ and $\int^\infty \frac{N(t,f)}{t^{\sigma+1}}dt$ are either simultaneously convergent or simultaneously divergent.   □

The following lemma can be proved by the same way.

Lemma 3.5'   *Let $f(z)$ be meromorphic in the finite plane, $0 \leq \theta_0 < 2\pi$, $\eta > 0$, and let a be a complex number. Suppose r_j $(\theta_0, \eta, f = a)$ $(j = 1, 2, \cdots)$ are the moduli of the zeros of $f(z) - a$ in $|\arg z - \theta_0| \leq \eta$, arranged by non-decreasing order and counted with their multiplicities (When $a = \infty$, r_j are the moduli of poles of $f(z)$.), and suppose*

$$N\left(r, \theta_0, \eta, \frac{1}{f-a}\right) = \int_0^r \frac{n\left(t, \theta_0, \eta, \frac{1}{f-a}\right) - n\left(0, \theta_0, \eta, \frac{1}{f-a}\right)}{t}dt$$

$$+ n\left(0, \theta_0, \eta, \frac{1}{f-a}\right)\log r.$$

If σ is positive, then the series

$$\sum \{r_j(\theta_0, \ \eta, \ f = a)\}^{-\sigma}$$

and the integrals

$$\int^\infty \frac{n(t, \theta_0, \eta, \frac{1}{f-a})}{t^{\sigma+1}} dt, \quad \int^\infty \frac{N(t, \theta_0, \eta, \frac{1}{f-a})}{t^{\sigma+1}} dt$$

are either simultaneously convergent or simultaneously divergent.

3.4.2 Properties of meromorphic functions in an angle

Theorem 3.9   Let $f(z)$ be meromorphic in the angle $|\arg z - \theta_0| < \eta$. If there are three distinct complex numbers a_ν $(\nu = 1, 2, 3)$ and a positive number σ such that the series

$$\sum \{r_k \ (\theta_0, \ \eta, \ f = a_\nu)\}^{-\sigma}, \quad \nu = 1, 2, 3 \tag{3.4.2}$$

are convergent, then the series

$$\sum \{r_k(\theta_0, \ \eta - \varepsilon, \ f = a)\}^{-\sigma}$$

converges for any positive number ε and every complex number a, except for a set of a with linear measure zero.

Proof.   Divide the angular domain $|\arg z - \theta_0| < \eta - \varepsilon$ into equal small angles with magnitude not exceeding $\varepsilon/4$. The number of such small angles J satisfies

$$J \leq \left[\frac{2(\eta - \varepsilon)}{\frac{\varepsilon}{4}}\right] + 1.$$

Choosing a sufficiently large positive number r_0, now divide this same angular domain $|\arg z - \theta_0| < \eta - \varepsilon$ by arcs of the circumferences of the circles

$$|z| = r_0, \ r_0\left(1 + \frac{\varepsilon}{4}\right), \ r_0\left(1 + \frac{\varepsilon}{4}\right)^2, \ \cdots.$$

Thus the region $(|z| > r_0) \cap (|\arg z - \theta_0| < \eta - \varepsilon)$ is divided into small quadrangles Ω_{jl} $(j = 1, 2, \cdots, J; l = 1, 2, \cdots)$. For every Ω_{jl}, there are disks Γ_{jl} and Γ'_{jl} such that

$$\Omega_{jl} \subset \Gamma_{jl} \subset \Gamma'_{jl} \subset (|\arg z - \theta_0| < \eta),$$

where the radius of Γ'_{jl} is twice that of Γ_{jl}. For every Γ'_{jol_0}, the number of other Γ'_{jl} having a non-empty intersection with Γ'_{jol_0}, has a fixed upper bound. For example, we can take one hundred as this upper bound.

From the convergence of (3.4.2) and Lemma 3.5', we have

$$\int^\infty \frac{n(r,\theta_0,\eta,f=a_\nu)}{r^{\sigma+1}}dr < \infty, \quad \nu = 1,2,3.$$

But Theorem 3.4 gives

$$n(\Omega_{jl},\ f=a) < C\Big\{ \sum_{\nu=1}^{3} n(\Gamma'_{jl},\ f=a_\nu) + \log \frac{1}{e^{-(r_0(1+\frac{\varepsilon}{4})^l)^{\frac{\sigma}{2}}}} \Big\} \qquad (3.4.3)$$

in every Γ'_{jl}, except for those complex numbers contained in a spherical disk e_{jl} with radius $e^{-(r_0(1+\frac{\varepsilon}{4})^l)^{\frac{\sigma}{2}}}$.

Setting

$$e_j = \overset{J}{\underset{j=1}{\cup}} e_{jl}, \quad \text{and} \quad e = \overset{\infty}{\underset{k=1}{\cap}} \Big(\overset{\infty}{\underset{l=k}{\cup}} e_l \Big),$$

it is clear that

$$\text{mese} = \lim_{k\to\infty} \text{mes}\Big(\overset{\infty}{\underset{l=k}{\cup}} e_l \Big) = 0.$$

For any complex number a outside the set e, there exists a positive integer l_0 such that $a \bar{\in} \cup_{l=l_0}^{\infty} e_l$. Thus (3.4.3) holds for $j = 1,2,\cdots,J$ and a, when $l \geq l_0$. Consequently

$$\sum_{l=l_0}^{L} \sum_{j=1}^{J} n(\Omega_{jl},\ f=a) < C\Big\{ \sum_{l=l_0}^{L} \sum_{j=1}^{J} \sum_{\nu=1}^{3} n\big(\Gamma'_{jl},\ f=a_\nu\big)$$

$$+ J \sum_{l=l_0}^{L} \big(r_0(1+\tfrac{\varepsilon}{4})^l\big)^{\frac{\sigma}{2}} \Big\},$$

so that

$$n(r,\ \theta_0,\ \eta-\varepsilon,\ f=a) < C\Big\{ \sum_{\nu=1}^{3} n(2r,\ \theta_0,\ \eta,\ f=a_\nu) + (2r)^{\frac{\sigma}{2}} \Big\}.$$

Therefore

$$\int^\infty \frac{n(r,\ \theta_0,\ \eta-\varepsilon,\ f=a)}{r^{\sigma+1}}dr < \infty$$

for any complex number a outside the set e. The conclusion of Theorem 3.9 follows from this fact and Theorem 3.5'.   □

The following important result will be deduced from Theorem 3.9.

Theorem 3.10   Let $f(z)$ be meromorphic and of order λ $(0 < \lambda < \infty)$ in the finite plane. If $f(z)$ has no Borel direction of order λ in the angle $\theta_1 < \arg z < \theta_2$, then, for any small positive number α, there are three distinct complex numbers a_ν $(\nu = 1, 2, 3)$ and a positive number τ, $\tau < \lambda$, such that

$$\sum_{\nu=1}^{3} n(r, \ \theta_1 + \alpha, \ \theta_2 - \alpha, \ f = a_\nu) < r^\tau. \tag{3.4.4}$$

Proof.   Since $\arg z = \varphi$ is not a Borel direction of order λ of $f(z)$, for every value φ on $[\theta_1 + \alpha, \ \theta_2 - \alpha]$ there are three distinct complex numbers $\beta_j(\varphi)$, a small positive number $\varepsilon(\varphi)$ and a positive number $\tau(\varphi)$, $\tau(\varphi) < \tau$ such that

$$n(r, \ \varphi - \varepsilon(\varphi), \ \varphi + \varepsilon(\varphi), \ f = \beta_j(\varphi)) < r^{\tau(\varphi)}, \quad j = 1, 2, 3.$$

Thus we have

$$\int^{\infty} n(t, \ \varphi - \varepsilon(\varphi), \ \varphi + \varepsilon(\varphi), \ f = \beta_j(\varphi)) \frac{dt}{t^{\tau_1(\varphi)+1}} < \infty, \quad j = 1, 2, 3$$

for $\tau_1(\varphi)$ with $\tau(\varphi) < \tau_1(\varphi) < \lambda$. It follows from Lemma 3.5' that

$$\sum \{r(\varphi, \ \varepsilon(\varphi), \ f = \beta_j(\varphi))\}^{-\tau_1(\varphi)} < \infty, \quad j = 1, 2, 3.$$

Using Theorem 3.9, the series

$$\sum \left\{ r\left(\varphi, \ \frac{\varepsilon(\varphi)}{2}, \ f = a\right) \right\}^{-\tau_1(\varphi)}$$

converges for any complex number a, except for a set $e(\varphi)$ of a with linear measure zero.

Since $\{(\varphi - \varepsilon(\varphi)/2, \ \varphi + \varepsilon(\varphi)/2) : \theta_1 + \alpha \le \varphi \le \theta_2 - \alpha\}$ is a convering of the closed interval $[\theta_1 + \alpha, \ \theta_2 - \alpha]$, there exists a finite number of intervals $(\varphi_l - \varepsilon_l/2, \ \varphi_l + \varepsilon_l/2)(l = 1, 2, \cdots, L; \ \varepsilon_l = \varepsilon(\varphi_l))$ which constitute a convering of $[\theta_1 + \alpha, \ \theta_2 - \alpha]$.

If the exceptional set with linear measure zero, which corresponds to every φ_l, is denoted by $e_l = e(\varphi_l)$, it is clear that $e = \cup_{l=1}^L e_l$ has measure zero. By setting $\tau_1 = \max_{1 \le l \le L} \tau_1(\varphi_l)$, we can see that $0 < \tau_1 < \lambda$ and

$$\sum \left\{ r\left(\varphi_l, \frac{\varepsilon_l}{2}, f = a\right) \right\}^{-\tau_1} < \infty, \quad l = 1, 2, \cdots, L$$

for any complex number a outside of e. Thus,

$$\sum \left\{ r\left(\frac{\theta_1 + \theta_2}{2}, \frac{\theta_2 - \theta_1}{2} - \alpha, f = a\right) \right\}^{-\tau_1} < \infty.$$

By Lemma 3.5', we obtain

$$\int^\infty n(t, \theta_1 + \alpha, \theta_2 - \alpha, f = a) \frac{dt}{t^{\tau_1 + 1}} < \infty.$$

Therefore, (3.4.4) holds for three distinct complex numbers $a_\nu (\nu = 1, 2, 3)$ not belonging to the set e and a positive number τ with $\tau_1 < \tau < \lambda$. □

3.4.3 A sequence of filling disks determined by a Borel direction.

In §3.3, we have seen that a Borel direction is derived from a sequence of filling disks of order λ. Conversely, the following Rauch's theorem states that a sequence of filling disks can be obtained from a Borel direction.

Theorem 3.11 *Let $f(z)$ be meromorphic and of order $\lambda(0 < \lambda < \infty)$ in the finite plane. If*

$$B : \arg z = \theta_0, \quad 0 \le \theta_0 < 2\pi$$

is a Borel direction of $f(z)$, then there exists a sequence of disks

$$\Gamma_j : |z - z_j| < \varepsilon_j |z_j|, \quad z_j = |z_j| e^{i\theta_0}, \quad j = 1, 2, \cdots,$$

$$\lim_{j \to \infty} |z_j| = \infty, \quad \lim_{j \to \infty} \varepsilon_j = 0$$

such that $f(z)$ takes every complex number at least $|z_j|^{\lambda - \delta_j}$ times in Γ_j, except possibly for those numbers contained in two spherical disks each with radius 2^{-j}, where $\lim_{j \to \infty} \delta_j = 0$.

Proof. Choose a small angular region with B as its bisector,

$$\Omega : \quad |\arg z - \theta_0| < \frac{\eta}{2}.$$

Using the arcs of circumferences $|z| = 2^j (j = 1, 2, \cdots)$, divide Ω into a sequence of small quadrangles

$$\Omega_j : \quad (2^j \leq |z| < 2^{j+1}) \cap \left(|\arg z - \theta_0| < \frac{\eta}{2}\right).$$

Then, divide every Ω_j by $s - 1$ arcs of $|z| = 2^j (1 + \eta)^l (l = 1, 2, \cdots, s - 1$, where s is a positive integer determined by $2^j (1 + \eta)^{s-1} < 2^{j+1} \leq 2^j (1 + \eta)^s)$. We get s smaller quadrangles $\Omega_{jl} (l = 1, 2, \cdots, s)$. Choose concentric disks Γ_{jl} and Γ'_{jl} with radius $C \cdot 2^{j+1} \eta$ and $2C \cdot 2^{j+1} \eta$, respectively, such that

$$\Omega_{jl} \subset \Gamma_{jl} \subset \Gamma'_{jl}.$$

Set

$$n_{jl} = \min\left\{ \sum_{\nu=1}^{3} n(\Gamma'_{jl}, \ f = a_\nu) \right\},$$

where the minimum is taken over all the triples of complex numbers, provided that the mutual spherical distances among these three complex numbers are greater than or equal to 2^{-j}. Using Theorem 3.4, $f(z)$ takes every complex number a at most

$$C\{n_{jl} + j\}$$

times in Ω_{jl}, except for those complex numbers contained in a disk with spherical radius 2^{-j}.

There are s exceptional spherical disks for every fixed j. When j varies, the sum of the radii of all the exceptional spherical disks is equal to

$$\sum_{j=1}^{\infty} \frac{s}{2^j}.$$

Choosing j_0 sufficiently large, we can let

$$\sum_{j=j_0}^{\infty} \frac{s}{2^j} < \frac{1}{2}.$$

Since B is a Borel direction of order λ of $f(z)$, the series

$$\sum \left\{ r_n(\theta_0, \frac{\eta}{2}, f = a) \right\}^{-\tau}$$

converges for any positive number τ less than λ and every complex value a, except for at most two values. If a_0 is not an exceptional value and does not belong to all the exceptional spherical disks, then we have

$$\sum \left\{ r_n \left(\theta_0, \frac{\eta}{2}, f = a \right) \right\}^{-\tau} = \infty. \tag{3.4.5}$$

On the other hand, by setting

$$n_j = \max_{1 \leq l \leq s} n_{jl},$$

we have

$$\sum \left\{ r_n \left(\theta_0, \frac{\eta}{2}, f = a \right) \right\}^{-\tau} < C \left\{ \sum_j \frac{s n_j}{2^{j\tau}} + \sum_j \frac{s j}{2^{j\tau}} \right\}$$

$$< C \left\{ \sum_j \frac{n_j}{\eta 2^{j\tau}} + \sum_j \frac{j}{\eta 2^{j\tau}} \right\}, \quad 0 < \tau < \lambda. \tag{3.4.6}$$

Choose a decreasing sequence (η_l) of positive numbers tending to zero and an increasing sequence of positive numbers (τ_l) tending to λ. As a consequence of $\sum_j j / \eta_1 2^{j\tau_1} < \infty$, (3.4.5) and (3.4.6), we have $\sum_j n_j / \eta_1 2^{j\tau_1} = \infty$. Thus there is a positive integer j_1 such that

$$n_{j_1} > \frac{\eta_1 2^{(j_1+1)\tau_1}}{j_1^2} > 2^{(j_1+1)(2\tau_1 - \lambda)}.$$

Similarly, from $\sum_{j=j_1+1}^{\infty} j / \eta_2 2^{j\tau_2} < \infty$, we have $\sum_{j=j_1+1}^{\infty} n_j / \eta_2 2^{j\tau_2} = \infty$. There is an integer j_2 with $j_2 > j_1$ and

$$n_{j_2} > \frac{\eta_2 2^{(j_2+1)\tau_2}}{j_2^2} > 2^{(j_2+1)(2\tau_2 - \lambda)}.$$

Therefore, we can find a sequence of disks $\Gamma_l : |z - z_l| < C \eta_l |z_l|$, $\arg z_l = \theta_0$, such that $f(z)$ takes every complex value at least $|z_l|^{\lambda - 2(\lambda - \tau_l)}$ times, except possibly for those values contained in two spherical disks each with radius 2^{-j_l}. □

Chapter 4

Value Distribution of Meromorphic Functions Together with Their Derivatives

The present chapter will be devoted to an important topic of the theory of value distribution, namely, the study of meromorphic functions together with their derivatives.

4.1 Comparison Between Growths of $T(r, f)$ and $T(r, f')$

4.1.1 Two lemmas. Let $f(z)$ be meromorphic in the finite plane. It is easy to bound $T(r, f')$ in terms of $T(r, f)$. However a reverse estimate, for example, an inequality of Chuang Chi-tai, is more difficult. In order to introduce this inequality, we need the following two lemmas.

Lemma 4.1 *Let $f(z)$ be meromorphic in the finite plane and $f(0) \neq \infty$. If R and R' ($R < R'$) are two positive numbers, then there exists a real number θ_0 such that*

$$\log^+ \left| f(re^{i\theta_0}) \right| \leq \frac{R' + R}{R' - R} m(R', f) + n(R', f) \log 4 + N(R', f) \quad (4.1.1)$$

for $0 \leq r \leq R$.

Proof. Let $z = re^{i\theta}$, $0 \leq r \leq R$, be an arbitrary point distinct from the poles of $f(z)$. The Poisson-Jensen formula gives

$$\log \left| f(z) \right| \leq \frac{1}{2\pi} \int_0^{2\pi} \log \left| f(R'e^{i\phi}) \right| \frac{R'^2 - r^2}{R'^2 - 2R'r \cos(\varphi - \theta) + r^2} d\varphi$$

$$+ \sum_{|b_\mu| \leq R'} \log \left| \frac{R'^2 - \bar{b}_\mu z}{R'(z - b_\mu)} \right|,$$

where b_μ are the poles of $f(z)$ on $|z| \leq R'$. Thus

$$
\begin{aligned}
\log^+ |f(z)| &\leq \frac{R'+r}{R'-r} m(R', f) + \sum_{|b_\mu| \leq R'} \log \frac{2R'}{|z - b_\mu|} \\
&\leq \frac{R'+r}{R'-r} m(R', f) + \log \frac{(2R')^{n(R',f)}}{\displaystyle\prod_{|b_\mu| \leq R'} |z - b_\mu|}.
\end{aligned}
\tag{4.1.2}
$$

Writing $b_\mu = |b_\mu| e^{i\varphi_\mu}$, we have

$$
|re^{i\theta} - |b_\mu| e^{i\varphi_\mu}| \geq |b_\mu| |\sin(\theta - \varphi_\mu)|,
$$

thus

$$
\prod_{|b_\mu| \leq R'} |z - b_\mu| \geq \left(\prod_{\mu=1}^{n(R',f)} |b_\mu| \right) \left(\prod_{\mu=1}^{n(R',f)} |\sin(\theta - \varphi_\mu)| \right).
\tag{4.1.3}
$$

Since

$$
\int_0^\pi \log \left| \prod_{\mu=1}^{n(R',f)} \sin(\theta - \varphi_\mu) \right| d\theta = n(R', f) \int_0^\pi \log |\sin \theta| d\theta
$$

$$
= -n(R', f)\pi \log 2,
$$

there exists at least one real number θ_0 such that

$$
\left| \prod_{\mu=1}^{n(R',f)} \sin(\theta_0 - \varphi_\mu) \right| > \frac{1}{2^{n(R',f)}}.
\tag{4.1.4}
$$

It is obvious that $z = re^{i\theta_0}$ $(0 \leq r \leq R)$ is not a pole of $f(z)$. Otherwise, one factor on the left-hand side of the above inequality is zero. Substituting (4.1.3) and (4.1.4) into (4.1.2), we have

$$
\begin{aligned}
\log^+ \left| f(re^{i\theta_0}) \right| &\leq \frac{R'+R}{R'-R} m(R', f) + \log 4^{n(R',f)} + \sum \log \frac{R'}{|b_\mu|} \\
&\leq \frac{R'+R}{R'-R} m(R', f) + n(R', f) \log 4 + N(R', f)
\end{aligned}
$$

for $0 \leq r \leq R$. □

Lemma 4.2 *Let $f(z)$ be meromorphic in the finite plane. If R, R' and R'' are three positive numbers and $R < R' < R''$, then there exists a positive number ρ such that $R \le \rho \le R'$ and*

$$\log^+ |f(z)| \le \frac{R'' + R'}{R'' - R'} m(R'', f) + n(R'', f) \log \frac{8eR''}{R' - R} \qquad (4.1.5)$$

on $|z| = \rho$.

Proof. Suppose $b_\mu (\mu = 1, 2, \cdots, n(R'', f))$ are the poles of $f(z)$ on $|z| \le R''$. According to the Boutroux-Cartan theorem, the inequality

$$\prod_{\mu=1}^{n(R'',f)} |z - b_\mu| \ge \left(\frac{R' - R}{4e} \right)^{n(R'',f)} \qquad (4.1.6)$$

holds, except for those points in a group (γ) of disks, where the total sum of their radii does not exceed $(R' - R)/2$. Since there is a circle $|z| = \rho$ in the annulus $R \le |z| \le R'$, not intersecting (γ), (4.1.6) holds on $|z| = \rho$.

From the Poisson-Jensen formula and (4.1.6), we have

$$\log^+ |f(z)| \le \frac{R'' + \rho}{R'' - \rho} m(R'', f) + \sum_{|b_\mu| \le R''} \log \left| \frac{R''^2 - \bar{b}_\mu z}{R''(z - b_\mu)} \right|$$

$$\le \frac{R'' + R'}{R'' - R'} m(R'', f) + n(R'', f) \log \frac{8eR''}{R' - R}$$

for any point z on $|z| = \rho$. $\square$

4.1.2 Comparison between $T(r,f)$ and $T(r,f')$. Now let us prove the Chuang Chi-tai's inequality.

Theorem 4.1 *If $f(z)$ is meromorphic in the finite plane and $f(0) \ne \infty$, then*

$$T(r, f) < C_\tau T(\tau r, f') + \log^+(\tau r) + 4 + \log^+ |f(0)| \qquad (4.1.7)$$

for $\tau > 1$ and $r > 0$.

Proof. Write $\sigma = \tau^{1/3}$, $r_1 = \sigma r$, $r_2 = \sigma r_1$ and $r_3 = \sigma r_2$. According to Lemma 4.1, there is a real number θ_0 such that

$$\log^+ \left| f'(te^{i\theta}) \right| \le \frac{r_2 + r_1}{r_2 - r_1} m(r_2, f') + n(r_2, f') \log 4 + N(r_2, f') \qquad (4.1.8)$$

for $0 \le t \le r_1$. By Lemma 4.2, a number ρ in $[r, r_1]$ can be found with the property

$$\log^+ \left| f'(z) \right| \le \frac{r_2 + r_1}{r_2 - r_1} m(r_2, f') + n(r_2, f') \log \frac{8er_2}{r_1 - r} \qquad (4.1.9)$$

on $|z| = \rho$.

If L is the union of the circumference $|z| = \rho$ and the segment between the origin and $\rho e^{i\theta_0}$, then its length does not exceed $(2\pi + 1)\rho$. Denoting by M the maximum modulus of $f'(z)$ on L, we have

$$|f(z)| \le |f(0)| + M(2\pi + 1)\rho$$

on $|z| = \rho$. Thus

$$\log^+ |f(z)| \le \log^+ |f(0)| + \log^+ M + \log^+ \rho + \log 8\pi.$$

On the other hand, (4.1.8) and (4.1.9) give us

$$\log^+ M \le \frac{r_2 + r_1}{r_2 - r_1} m(r_2, f') + n(r_2, f') \log \frac{8er_2}{r_1 - r} + N(r_2, f')$$

$$\le \left\{ \frac{\log \dfrac{8er_2}{r_1 - r}}{\log \dfrac{r_3}{r_2}} + \frac{r_2 + r_1}{r_2 - r_1} \right\} T(r_3, f')$$

$$= \left\{ \frac{\log \dfrac{8e\sigma^2}{\sigma - 1}}{\log \sigma} + \frac{\sigma + 1}{\sigma - 1} \right\} T(r_3, f')$$

$$= C'_\tau T(r_3, f').$$

Therefore

$$m(\rho, f) < C'_\tau T(r_3, f') + \log^+(\tau r) + 4 + \log^+ |f(0)|,$$

and hence

$$T(r, f) \le T(\rho, f)$$
$$< (C'_\tau + 1)T(r_3, f') + \log^+(\tau r) + 4 + \log^+ |f(0)|. \quad \square$$

Corollary. *If $f(z)$ is meromorphic in the finite plane, then*

$$T(r, f) < O\left\{ T(2r, f') + \log r \right\}, \quad r \to \infty. \qquad (4.1.10)$$

When $f(0) \neq \infty$, (4.1.10) can be obtained immediately from Theorem 4.1. When $f(0) = \infty$, we set

$$g(z) = f(z) - R(z),$$

where $R(z)$ is the principal part of the Laurent expansion of $f(z)$ in the neighborhood of the origin. Thus $g(0) \neq \infty$, and hence

$$T(r, g) < O\{T(2r, g') + \log r\}.$$

Therefore, (4.1.10) follows immediately from

$$T(r, f) = T(r, g) + O(\log r)$$

and

$$T(2r, g') = T(2r, f') + O(\log r).$$

Theorem 4.2   If $f(z)$ is meromorphic in the finite plane, then $f(z)$ and its derivative $f'(z)$ have the same order and lower order.

Proof.   Since

$$T(r, f') = m(r, f') + N(r, f')$$

$$\leq m(r, f) + m\left(r, \frac{f'}{f}\right) + 2N(r, f)$$

$$\leq 2T(r, f) + m\left(r, \frac{f'}{f}\right),$$

the order and the lower order of $f'(z)$ do not exceed respectively, those of $f(z)$. On the other hand, we can see from the Corollary of Theorem 4.1 that the order and the lower order of $f(z)$ are also less than or qual to those of $f'(z)$ respectively.   □

The assertion that $f(z)$ and $f'(z)$ have the same order was made by G. Valiron, but the first proof was given later by J. M. Whittaker [1].

If $f(z)$ is an entire function of finite order, then we have

$$T(r, f') = m(r, f') \leq m(r, f) + m\left(r, \frac{f'}{f}\right)$$

$$= \left(1 + o(1)\right)T(r, f).$$

Thus

$$\varlimsup_{r \to \infty} \frac{T(r, f')}{T(r, f)} \le 1.$$

Hence Nevanlinna conjectured that:

$$\lim_{r \to \infty} \frac{T(r, f')}{T(r, f)} = 1.$$

W. K. Hayman [3] disproved this conjecture in 1965. In order to formulate his result, we need to introduce the concept of lower logarithmic density of a set.

If E is a set in $[1, \infty)$ and $E(r)$ denotes the portion of E in $[1, r]$, the lower logarithmic density of E is defined as

$$\underline{\log \mathrm{dens}} E = \varliminf_{r \to \infty} \frac{1}{\log r} \int_{E(r)} \frac{dt}{t}.$$

Hayman's result can now be formulated as follows:

Given two positive numbers λ and K, there is an entire function $f(z)$ of order λ such that

$$T(r, f) > KT(r, f')$$

on a set E with positive lower logarithmic density.

Furthermore S. Toppila [1] constructed an entire function $f(z)$ of finite order such that

$$\frac{T(r, f)}{T(r, f')} \ge 1 + \frac{7}{10^7}$$

for all sufficiently large values of r.

Recently, Hayman and Miles [1] proved the following theorem.

Theorem 4.2′   *Given a transcendental meromorphic function $f(z)$ and a constant $K > 1$. Then there exists a set $M(K)$ whose upper logarithmic density is at most*

$$\delta(K) = \min \left\{ (2e^{K-1} - 1)^{-1}, (1 + e(K - 1)) \exp(e(1 - K)) \right\}$$

such that for every positive integer k,

$$\varlimsup_{\substack{r \to \infty \\ r \notin M(K)}} \frac{T(r, f)}{T(r, f^{(k)})} \le 3eK.$$

If f is entire we can replace $3eK$ by $2eK$ in the last inequality.

The upper logarithmic density of a set has a definition similar to that of the lower logarithmic density, but lower limit is now replaced by upper limit.

Thus as $r \to \infty$ on a set of positive lower logarithmic density, we have for every meromorphic function $f(z)$ and every $\varepsilon > 0$

$$\frac{1}{2} - \varepsilon < \frac{T(r, f)}{T(r, f')} < 3e + \varepsilon,$$

and for every entire function

$$1 - \varepsilon < \frac{T(r, f)}{T(r, f')} < 2e + \varepsilon.$$

The constants $1/2$ and 1 on the left-hand side are sharp. The quantities $2e$ and $3e$ on the right-hand side are probably not sharp, but cannot at any rate be replaced by 1.

4.2 Modular Distribution of Meromorphic Functions Together with Their Derivatives

4.2.1 Generalization of the Nevanlinna fundamental lemma on logarithmic derivative.   In order to discuss the value distribution of meromorphic functions together with their derivatives, we need the following useful lemma, which was given by Hiong King-lai [2] as a generalization of the Nevanlinna lemma on the logarithmic derivative.

Lemma 4.3 *Let $f(z)$ be meromorphic in $|z| < R(\le \infty)$. If $f(0) \ne 0, \infty$, then for every positive integer k,*

$$m\left(r, \frac{f^{(k)}}{f}\right) < C_k\left\{1 + \log^+ \log^+ \frac{1}{|f(0)|} + \log^+ \frac{1}{r} + \log^+ \frac{1}{\rho - r}\right. \tag{4.2.1}$$
$$\left. + \log^+ \rho + \log^+ T(\rho, f)\right\},$$

where $0 < r < \rho < R$ and C_k is a constant depending only on k[1].

---

1) Henceforth, we shall denote by C_k a constant depending only on k, but one which may vary with each occurrence.

Proof. When $k = 1$, (4.2.1) is obvious from the Nevanlinna lemma (Lemma 1.3). Suppose (4.2.1) is true for $k = 1, 2, \cdots, l - 1$. Let us prove that it is also true for $k = l$.

Let $z = re^{i\theta}$ be a point other than a zero or a pole of $f(z)$. Then the Poisson-Jensen formula gives us

$$\frac{f'(z)}{f(z)} = \frac{1}{2\pi} \int_0^{2\pi} \log \left| f(\rho e^{i\varphi}) \right| \frac{2\rho e^{i\varphi}}{(\rho e^{i\varphi} - z)^2} d\varphi$$

$$+ \sum_{|a_\mu| \le \rho} \left\{ \frac{1}{z - a_\mu} + \frac{\bar{a}_\mu}{\rho^2 - \bar{a}_\mu z} \right\}$$

$$- \sum_{|b_\nu| \le \rho} \left\{ \frac{1}{z - b_\nu} + \frac{\bar{b}_\nu}{\rho^2 - \bar{b}_\nu z} \right\},$$

where a_μ and b_ν are respectively the zeros and poles of $f(z)$ in $|z| \le \rho$ with their multiplicities counted.

Thus

$$\frac{d^{l-1}}{dz^{l-1}} \left(\frac{f'}{f} \right) = \frac{l!}{2\pi} \int_0^{2\pi} \log \left| f(\rho e^{i\varphi}) \right| \frac{2\rho e^{i\varphi}}{(\rho e^{i\varphi} - z)^{l+1}} d\varphi$$

$$+ (l - 1)! \sum_{|a_\mu| \le \rho} \left\{ \frac{(-1)^{l-1}}{(z - a_\mu)^l} + \frac{\bar{a}_\mu^l}{(\rho^2 - \bar{a}_\mu z)^l} \right\}$$

$$- (l - 1)! \sum_{|b_\nu| \le \rho} \left\{ \frac{(-1)^{l-1}}{(z - b_\nu)^l} + \frac{\bar{b}_\nu^l}{(\rho^2 - \bar{b}_\nu z)^l} \right\}.$$

Since

$$\left| \frac{(-1)^{l-1}}{(z - a_\mu)^l} + \frac{\bar{a}_\mu^l}{(\rho^2 - \bar{a}_\mu z)^l} \right|$$

$$\le \left| \frac{\rho^2 - \bar{a}_\mu z}{\rho(z - a_\mu)} \right|^l \left\{ \frac{\rho^l}{|\rho^2 - \bar{a}_\mu z|^l} + \frac{\rho^{2l} |z - a_\mu|^l}{|\rho^2 - \bar{a}_\mu z|^{2l}} \right\}$$

$$\le \left| \frac{\rho^2 - \bar{a}_\mu z}{\rho(z - a_\mu)} \right|^l \left\{ \frac{1}{(\rho - r)^l} + \frac{(2\rho)^l}{(\rho - r)^{2l}} \right\}$$

$$\le \left| \frac{\rho^2 - \bar{a}_\mu z}{\rho(z - a_\mu)} \right|^l \cdot \frac{2^{l+1}\rho^l}{(\rho - r)^{2l}}$$

and

$$\left| \frac{(-1)^{l-1}}{(z - b_\nu)^l} + \frac{\bar{b}_\nu^l}{(\rho^2 - \bar{b}_\nu z)^l} \right| \le \left| \frac{\rho^2 - \bar{b}_\nu z}{\rho(z - b_\nu)} \right|^l \cdot \frac{2^{l+1}\rho^l}{(\rho - r)^{2l}},$$

we obtain

$$\left|\frac{d^{l-1}}{dz^{l-1}}\left(\frac{f'}{f}\right)\right| \le l!\frac{2\rho}{(\rho-r)^{l+1}}\left\{2T(r,f)+\log^+\frac{1}{|f(0)|}\right\}$$

$$+\frac{2^{l+1}\rho^l}{(\rho-r)^{2l}}\left\{\sum_{|a_\mu|\le\rho}\left|\frac{\rho^2-\bar{a}_\mu z}{\rho(z-a_\mu)}\right|^l\right.$$

$$\left.+\sum_{|b_\nu|\le\rho}\left|\frac{\rho^2-\bar{b}_\nu z}{\rho(z-b_\nu)}\right|^l\right\}. \tag{4.2.2}$$

Note

$$\frac{d^{l-1}}{dz^{l-1}}\left(\frac{f'}{f}\right) = \frac{f^{(l)}}{f} + P_l\left(\frac{f'}{f},\frac{f''}{f},\cdots,\frac{f^{(l-1)}}{f}\right),$$

where P_l is a polynomial of degree l, we have

$$\log^+\left|\frac{f^{(l)}}{f}\right| \le C_l\left\{\sum_{k=1}^{l-1}\log^+\left|\frac{f^{(k)}}{f}\right|+1\right\} + \log^+\left|\frac{d^{l-1}}{dz^{l-1}}\left(\frac{f'}{f}\right)\right|.$$

This fact, together with (4.2.2) and the hypothesis of induction, yields

$$m\left(r,\frac{f^{(l)}}{f}\right) < C_l\left\{1+\log^+\log^+\frac{1}{|f(0)|}+\log^+\frac{1}{r}\right.$$

$$\left.+\log^+\frac{1}{\rho-r}+\log^+\rho+\log^+T(\rho,f)\right\} \tag{4.2.3}$$

$$+m\left(r,\sum_{|a_\mu|\le\rho}\left|\frac{\rho^2-\bar{a}_\mu z}{\rho(z-a_\mu)}\right|^l+\sum_{|b_\nu|\le\rho}\left|\frac{\rho^2-\bar{b}_\nu z}{\rho(z-b_\nu)}\right|^l\right).$$

To estimate the last term of (4.2.3), we note that the inequality

$$m\left(r,\sum_{|a_\mu|\le\rho}\left|\frac{\rho^2-\bar{a}_\mu z}{\rho(z-a_\mu)}\right|^l+\sum_{|b_\nu|\le\rho}\left|\frac{\rho^2-\bar{b}_\nu z}{\rho(z-b_\nu)}\right|^l\right)$$

$$\le l\sum_{|a_\mu|\le\rho}m\left(r,\frac{\rho^2-\bar{a}_\mu z}{\rho(z-a_\mu)}\right)+l\sum_{|b_\nu|\le\rho}m\left(r,\frac{\rho^2-\bar{b}_\nu z}{\rho(z-b_\nu)}\right) \tag{4.2.4}$$

$$+\log\left(n\left(\rho,\frac{1}{f}\right)+n(\rho,f)\right)$$

and the estimate

$$\sum_{|a_\mu|\le\rho} m\left(r,\frac{\rho^2-\bar{a}_\mu z}{\rho(z-a_\mu)}\right) + \sum_{|b_\nu|\le\rho} m\left(r,\frac{\rho^2-\bar{b}_\nu z}{\rho(z-b_\nu)}\right)$$

$$= N\left(\rho,\frac{1}{f}\right) + N(\rho,f) - N\left(r,\frac{1}{f}\right) - N(r,f)$$

$$\le \frac{\rho'(\rho-r)}{r(\rho'-r)}\left\{2T(\rho',f)+\log^+\frac{1}{|f(0)|}\right\}$$

both appeared in the proof of Lemma 1.3 of Chapter 1.

Choosing

$$\rho = r + \frac{r(\rho'-r)}{2\rho'\left\{T(\rho',f)+\log^+\frac{1}{|f(0)|}+1\right\}}, \tag{4.2.5}$$

one can complete the proof as on p.p. 19–21. $\square$

4.2.2 An inequality of Milloux .   In 1940, H. Milloux obtained the first result involving the modular distribution of meromorphic functions together with their derivatives as follows.

Theorem 4.3   Let $f(z)$ be meromorphic in $|z| < R$ $(\le \infty)$ with $f(0) \ne 0, \infty$, $f^{(k)}(0) \ne 1$ and $f^{(k+1)}(0) \ne 0$, where k is a positive integer. Then

$$T(r,f) < \overline{N}(r,f) + N\left(r,\frac{1}{f}\right) + N\left(r,\frac{1}{f^{(k)}-1}\right)$$

$$-N\left(r,\frac{1}{f^{(k+1)}}\right) + S(r,f), \tag{4.2.6}$$

where

$$S(r,f) = m\left(r,\frac{f^{(k)}}{f}\right) + m\left(r,\frac{f^{(k+1)}}{f}\right) + m\left(r,\frac{f^{(k+1)}}{f^{(k)}-1}\right)$$

$$+\log\left|\frac{f(0)(f^{(k)}(0)-1)}{f^{(k+1)}(0)}\right| + \log 2. \tag{4.2.7}$$

Proof.   First of all, let us note that the identity

$$\frac{1}{f} \equiv \frac{f^{(k)}}{f} - \frac{f^{(k)}-1}{f^{(k+1)}}\cdot\frac{f^{(k+1)}}{f}$$

implies

$$m\left(r, \frac{1}{f}\right) \leq m\left(r, \frac{f^{(k)}}{f}\right) + m\left(r, \frac{f^{(k)} - 1}{f^{(k+1)}}\right) + m\left(r, \frac{f^{(k+1)}}{f}\right) + \log 2.$$

Next, the Jensen-Nevanlinna formula (1.2.5) shows that

$$m\left(r, \frac{1}{f}\right) = T(r, f) - N\left(r, \frac{1}{f}\right) + \log \frac{1}{|f(0)|}$$

and

$$m\left(r, \frac{f^{(k)} - 1}{f^{(k+1)}}\right) = m\left(r, \frac{f^{(k+1)}}{f^{(k)} - 1}\right) + \log \left|\frac{f^{(k)}(0) - 1}{f^{(k+1)}(0)}\right|$$

$$+ \left\{N\left(r, \frac{f^{(k+1)}}{f^{(k)} - 1}\right) - N\left(r, \frac{f^{(k)} - 1}{f^{(k+1)}}\right)\right\}.$$

Thus

$$T(r, f) \leq N\left(r, \frac{1}{f}\right) + S(r, f)$$

$$+ \left\{N\left(r, \frac{f^{(k+1)}}{f^{(k)} - 1}\right) - N\left(r, \frac{f^{(k)} - 1}{f^{(k+1)}}\right)\right\},$$

(4.2.8)

where $S(r, f)$ is given by (4.2.7).

Moreover Lemma 1.2 yields

$$N\left(r, \frac{f^{(k+1)}}{f^{(k)} - 1}\right) - N\left(r, \frac{f^{(k)} - 1}{f^{(k+1)}}\right)$$

$$= N\left(r, \frac{1}{f^{(k)} - 1}\right) + N(r, f^{(k+1)}) - N(r, f^{(k)}) - N\left(r, \frac{1}{f^{(k+1)}}\right).$$

(4.2.9)

Since $f(z)$, $f^{(k)}(z)$ and $f^{(k+1)}(z)$ continue to have the same poles but with different multiplicities, we see that a pole z_0 of multiplicity $j (\geq 1)$ for $f(z)$ is also a pole of multiplicity $j+k$ for $f^{(k)}(z)$ and that of multiplicity $j+k+1$ for $f^{(k+1)}(z)$. By this analysis, we have

$$N(r, f^{(k+1)}) - N(r, f^{(k)}) = \overline{N}(r, f),$$

(4.2.10)

and (4.2.6) therefore follows from substituting (4.2.9) and (4.2.10) into (4.2.8). This completes the proof. $\square$

Now let us use Lemma 4.3 to discuss the error term of Milloux's inequality.

Let $f(z)$ be meromorphic in the finite plane. When the order of $f(z)$ is finite, setting $\rho = 2r$ we have

$$\log^+ T(\rho, f) = O(\log r), \quad r \to \infty.$$

On combining this with Lemma 4.3, we deduce that

$$m\left(r, \frac{f^{(k)}}{f}\right) = O(\log r).$$

Next, from

$$T(\rho, f^{(k)}) = N(\rho, f^{(k)}) + m(\rho, f^{(k)})$$

$$\leq N(\rho, f) + k\overline{N}(\rho, f) + m(\rho, f) + m\left(\rho, \frac{f^{(k)}}{f}\right)$$

$$\leq (k+1)T(\rho, f) + O(\log \rho),$$

we obtain

$$\log^+ T(\rho, f^{(k)}) = O(\log r),$$

and so we have

$$m\left(r, \frac{f^{(k+1)}}{f^{(k)} - 1}\right) = O(\log r).$$

Therefore the error term of Milloux's inequality is of the form

$$S(r, f) = O(\log r)$$

when the order of $f(z)$ is finite.

When the order of $f(z)$ is infinite, taking $\rho = r + 1/T(r, f)$ we have by Lemma 4.3 and Lemma 1.4,

$$m\left(r, \frac{f^{(l)}}{f}\right) = O\left\{\log(rT(r, f))\right\}, \quad l = k, k+1$$

except on a set of r with finite linear measure. Then nothing remains but to estimate $m(r, f^{(k+1)}/(f^{(k)} - 1))$ in the error term $S(r, f)$ expressed by

(4.2.7). First, by the Nevanlinna Lemma for r and $\rho' = (r+\rho)/2$, we have

$$m\left(r, \frac{f^{(k+1)}}{f^{(k)}-1}\right) < 10 + 4\log^+\log^+ \frac{1}{|f^{(k)}(0)-1|} + 2\log^+ \frac{1}{r}$$

$$+ 4\log^+ \rho' + 4\log^+ T\left(\rho', f^{(k)}-1\right) + 3\log^+ \frac{1}{\rho'-r}$$

$$< 10 + 7\log 2 + 4\log^+\log^+ \frac{1}{|f^{(k)}(0)-1|} + 2\log^+ \frac{1}{r}$$

$$+ 3\log^+ \frac{1}{\rho-r} + 4\log^+ \rho + 4\log^+ T(\rho', f^{(k)}).$$

$$(4.2.11)$$

Next,

$$T(\rho', f^{(k)}) \le (k+1)T(\rho', f) + m\left(\rho', \frac{f^{(k)}}{f}\right)$$

implies

$$4\log^+ T(\rho', f^{(k)})$$

$$\le 4\log(k+1) + 4\log^+ T(\rho', f) + 4\log^+ m\left(\rho', \frac{f^{(k)}}{f}\right) + 4\log 2$$

$$(4.2.12)$$

$$\le 4\log(k+1) + 4\log^+ T(\rho, f) + m\left(\rho', \frac{f^{(k)}}{f}\right)$$

$$+ 4(\log 4 - 1) + 4\log 2.$$

Finally, Lemma 4.3 yields

$$m\left(\rho', \frac{f^{(k)}}{f}\right) < C_k\left\{1 + \log^+\log^+ \frac{1}{|f(0)|} + \log^+ \frac{1}{\rho'}\right.$$

$$+ \log^+ \frac{1}{\rho-\rho'} + \log^+ \rho + \log^+ T(\rho, f)\right\}$$

$$(4.2.13)$$

$$< C_k\left\{1 + \log^+\log^+ \frac{1}{|f(0)|} + \log^+ \frac{1}{r}\right.$$

$$+ \log^+ \frac{2}{\rho-r} + \log^+ \rho + \log^+ T(\rho, f)\right\}.$$

Therefore (4.2.11), (4.2.12) and (4.2.13) taken together give

$$m\left(r, \frac{f^{(k+1)}}{f^{(k)}-1}\right) = O\left\{\log(rT(r, f))\right\},$$

except on a set of r with finite linear measure. It follows that $S(r, f)$ has the same property.

The following corollaries can be deduced immediately from the Milloux inequality and the property of the error term.

Corollary 1.   *If $f(z)$ is a transcendental entire function, then the following two cases cannot occur simultaneously:*
 (i) $f(z) - a$ *has a finite number of zeros,*
 (ii) $f^{(k)}(z) - b$ *has a finite number of zeros for a positive integer k, where a and $b(\neq 0)$ are two finite complex values.*

In fact, if (i) and (ii) occur simultaneously, then applying the Milloux inequality to

$$g(z) = \frac{f(z) - a}{b},$$

we obtain

$$\varliminf_{r \to \infty} \frac{T(r, g)}{\log r} < \infty.$$

By Lemma 1.5, $g(z)$ must be rational, so $f(z)$ is also rational. This contradicts the hypothesis.

Corollary 2.   *Let $f(z)$ be an entire function of finite and positive order. Then the following two cases cannot occur simultaneously:*
 (i) $f(z)$ *has a finite Borel exceptional value,*
 (ii) $f^{(k)}(z)$ *has a finite and non-zero Borel exceptional value for a positive integer k.*

4.2.3 An inequality of Hiong King-lai .   Now let us derive an inequality of Hiong King-lai, similar to Milloux's inequality, but which does not involve the counting function of poles.

Theorem 4.4   *Let $f(z)$ be meromorphic in the finite plane and k a positive integer. Then for any three finite complex numbers a, b and c, where $b \neq 0$, $c \neq 0$ and $b \neq c$, the following inequality holds:*

$$T(r, f) < N\left(r, \frac{1}{f - a}\right) + N\left(r, \frac{1}{f^{(k)} - b}\right)$$

$$+ N\left(r, \frac{1}{f^{(k)} - c}\right) - N\left(r, \frac{1}{f^{(k+1)}}\right) + Q(r, f),$$

(4.2.14)

where

$$Q(r, f) = O\{\log(rT(r, f))\}, \tag{4.2.15}$$

except on a set of r with finite linear measure when the order of f is infinite.

Proof. Combining the inequality

$$m\left(r, \frac{1}{f-a}\right) \leq m\left(r, \frac{1}{f^{(k)}}\right) + m\left(r, \frac{f^{(k)}}{f-a}\right)$$

together with

$$m\left(r, \frac{1}{f-a}\right) = T(r, f) - N\left(r, \frac{1}{f-a}\right) + O(1)$$

and

$$m\left(r, \frac{1}{f^{(k)}}\right) = T(r, f^{(k)}) - N\left(r, \frac{1}{f^{(k)}}\right) + O(1),$$

we obtain

$$T(r, f) < N\left(r, \frac{1}{f-a}\right) + T(r, f^{(k)})$$

$$-N\left(r, \frac{1}{f^{(k)}}\right) + m\left(r, \frac{f^{(k)}}{f-a}\right) + O(1). \tag{4.2.16}$$

On the other hand, applying the Nevanlinna second fundamental theorem to $f^{(k)}(z)$ yields

$$T(r, f^{(k)}) < N\left(r, \frac{1}{f^{(k)}}\right) + N\left(r, \frac{1}{f^{(k)} - b}\right) + N\left(r, \frac{1}{f^{(k)} - c}\right)$$

$$-N\left(r, \frac{1}{f^{(k+1)}}\right) + Q_1(r, f^{(k)}). \tag{4.2.17}$$

When the order of $f(z)$ is finite, so is that of $f^{(k)}(z)$ by Theorem 4.2, and we have $Q_1(r, f^{(k)}) = O(\log r)$. When the order of $f(z)$ is infinite, we have

$$Q_1(r, f^{(k)}) = O\{\log(rT(r, f^{(k)}))\} = O\{\log(rT(r, f))\},$$

except on a set of r with finite linear measure.

Comparing (4.2.16) with (4.2.17), we derive the conclusion of Theorem 4.4. □

4.3 An Inequality of Hayman

4.3.1 A lemma of Hayman.   In the second fundamental theorem of Nevanlinna (Theorem 1.3), three counting functions of distinct values are needed to bound the characteristic function $T(r, f)$. We have seen in §4.2 that one or two of these three terms can be replaced by the counting functions of the derivatives of f. In 1959, Hayman [1] obtained the suprising result that $T(r, f)$ can be bounded by only two counting functions, one for $f(z)$ and another for $f^{(k)}(z)$. Clearly this would be impossible without involving the derivatives.

The following lemma plays a key role in the proof of the Hayman inequality.

Lemma 4.4   *Let $f(z)$ be meromorphic and transcendental in $|z| < R(\leq \infty)$, and let*

$$g(z) = \frac{(f^{(k+1)}(z))^{k+1}}{(1 - f^{(k)}(z))^{k+2}}, \tag{4.3.1}$$

where k is a positive integer. Then

$$kN_{1)}(r, f) \leq \overline{N}_{(2}(r, f) + \overline{N}\left(r, \frac{1}{f^{(k)} - 1}\right)$$

$$+ N_0\left(r, \frac{1}{f^{(k+1)}}\right) + m\left(r, \frac{g'}{g}\right) + \log\left|\frac{g(0)}{g'(0)}\right| \tag{4.3.2}$$

for $0 < r < R$ where $N_{1)}(r, f)$ denotes the counting function of simple poles, $\overline{N}_{(2}(r, f)$ the reduced counting function of multiple poles and $N_0(r, 1/f^{(k+1)})$ corresponds only to the zeros of $f^{(k+1)}(z)$ and not to those of $f^{(k)}(z) - 1$.

Proof.   If z_0 is a simple pole of $f(z)$, then

$$f(z) = \frac{a}{z - z_0} + O(1) \tag{4.3.3}$$

with $a \neq 0$. Differentiating both sides k times, we deduce that

$$1 - f^{(k)}(z) = \frac{(-1)^{k+1} a k!}{(z - z_0)^{k+1}} + O(1)$$

$$= \frac{(-1)^{k+1} a k!}{(z - z_0)^{k+1}}\left\{1 + O\left((z - z_0)^{k+1}\right)\right\}.$$

Similarly, differentiating both sides $k + 1$ times yields

$$f^{(k+1)}(z) = \frac{(-1)^{k+1} a(k+1)!}{(z - z_0)^{k+2}} \left\{ 1 + O\left((z - z_0)^{k+2} \right) \right\}.$$

Thus

$$g(z) = \frac{(-1)^{k+1}(k+1)^{k+1}}{ak!} \left\{ 1 + O\left((z - z_0)^{k+1} \right) \right\};$$

hence $g(z_0) \neq 0, \infty$, and $g'(z)$ has a zero of order at least k at z_0. This analysis gives us

$$k N_{1)}(r, f) \leq N_0\left(r, \frac{1}{g'} \right), \tag{4.3.4}$$

where only the zeros of $g'(z)$, and not of $g(z)$, are involved in $N_0(r, 1/g')$.

Now applying the Jensen formula to $g'(z)/g(z)$ shows that

$$N\left(r, \frac{g}{g'} \right) - N\left(r, \frac{g'}{g} \right) = m\left(r, \frac{g'}{g} \right) - m\left(r, \frac{g}{g'} \right) - \log\left| \frac{g'(0)}{g(0)} \right|. \tag{4.3.5}$$

By Lemma 1.2, the left-hand side of (4.3.5) is

$$N(r, g) - N(r, g') + N\left(r, \frac{1}{g'} \right) - N\left(r, \frac{1}{g} \right)$$

$$= N\left(r, \frac{1}{g'} \right) - \overline{N}(r, g) - N\left(r, \frac{1}{g} \right) \tag{4.3.6}$$

$$= N_0\left(r, \frac{1}{g'} \right) - \overline{N}(r, g) - \overline{N}\left(r, \frac{1}{g} \right).$$

Comparing (4.3.5) with (4.3.6), we have

$$N_0\left(r, \frac{1}{g'} \right) \leq \overline{N}(r, g) + \overline{N}\left(r, \frac{1}{g} \right) + m\left(r, \frac{g'}{g} \right) + \log\left| \frac{g(0)}{g'(0)} \right|. \tag{4.3.7}$$

It is clear from (4.3.1) that the zeros and poles of $g(z)$ can occur only at multiple poles of $f(z)$, the zeros of $f^{(k)}(z) - 1$, or the zeros of $f^{(k+1)}(z)$ other than those of $f^{(k)}(z) - 1$. Thus

$$\overline{N}(r, g) + \overline{N}\left(r, \frac{1}{g} \right) \leq \overline{N}_{(2}(r, f) + \overline{N}\left(r, \frac{1}{f^{(k)} - 1} \right) + N_0\left(r, \frac{1}{f^{(k+1)}} \right). \tag{4.3.8}$$

Therefore combining (4.3.4), (4.3.7) and (4.3.8) gives the conclusion of the Lemma. $\square$

4.3.2 The Hayman inequality.   Now let us prove the Hayman inequality.

Theorem 4.5   Let $f(z)$ be meromorphic and transcendental in $|z| < R(\leq \infty)$ and let k be a positive integer. If $f(0) \neq 0, \infty$, $f^{(k)}(0) \neq 1$, $f^{(k+1)}(0) \neq 0$ and

$$(k+1)f^{(k+2)}(0)(f^{(k)}(0)-1) - (k+2)f^{(k+1)}(0)^2 \neq 0,$$

then

$$T(rf) < \left(2+\frac{1}{k}\right)N\left(r,\frac{1}{f}\right) + \left(2+\frac{2}{k}\right)\overline{N}\left(r,\frac{1}{f^{(k)}-1}\right) + S^*(r,f) \quad (4.3.9)$$

for $0 < r < R$, where

$$S^*(r,f) = \left(2+\frac{2}{k}\right)m\left(r,\frac{f^{(k+1)}}{f^{(k)}-1}\right) + \left(2+\frac{1}{k}\right)\left\{m\left(r,\frac{f^{(k+1)}}{f}\right) + m\left(r,\frac{f^{(k)}}{f}\right)\right\}$$

$$+\frac{1}{k}m\left(r,\frac{f^{(k+2)}}{f^{(k+1)}}\right) + 4 + \left(2+\frac{1}{k}\right)\log\left|\frac{f(0)(f^{(k)}(0)-1)}{f^{(k+1)}(0)}\right|$$

$$+\frac{1}{k}\log\left|\frac{f^{(k+1)}(0)(f^{(k)}(0)-1)}{(k+1)f^{(k+2)}(0)(f^{(k)}(0)-1) - (k+2)f^{(k+1)}(0)^2}\right|.$$

$$(4.3.10)$$

Proof.   Since every multiple pole of $f(z)$ is counted at least twice in $N(r,f)$ and only once in $\overline{N}(r,f)$, we have

$$\overline{N}_{(2}(r,f) \leq N(r,f) - \overline{N}(r,f) \leq T(r,f) - \overline{N}(r,f)$$

$$\leq N\left(r,\frac{1}{f}\right) + \overline{N}\left(r,\frac{1}{f^{(k)}-1}\right) - N_0\left(r,\frac{1}{f^{(k+1)}}\right) + S(r,f),$$

where

$$S(r,f) = m\left(r,\frac{f^{(k)}}{f}\right) + m\left(r,\frac{f^{(k+1)}}{f}\right) + m\left(r,\frac{f^{(k+1)}}{f^{(k)}-1}\right)$$

$$+\log\left|\frac{f(0)(f^{(k)}(0)-1)}{f^{(k+1)}(0)}\right| + \log 2.$$

Now using Lemma 4.4 yields

$$\overline{N}(r, f) = N_{1)}(r, f) + \overline{N}_{(2}(r, f)$$

$$\leq \left(1 + \frac{1}{k}\right)\overline{N}_{(2}(r, f) + \frac{1}{k}\left\{\overline{N}\left(r, \frac{1}{f^{(k)} - 1}\right) + N_0\left(r, \frac{1}{f^{(k+1)}}\right)\right.$$

$$\left. + m\left(r, \frac{g'}{g}\right) + \log\left|\frac{g(0)}{g'(0)}\right|\right\}$$

$$\leq \left(1 + \frac{1}{k}\right)\left\{N\left(r, \frac{1}{f}\right) + \overline{N}\left(r, \frac{1}{f^{(k)} - 1}\right) - N_0\left(r, \frac{1}{f^{(k+1)}}\right) + S(r, f)\right\}$$

$$+ \frac{1}{k}\left\{\overline{N}\left(r, \frac{1}{f^{(k)} - 1}\right) + N_0\left(r, \frac{1}{f^{(k+1)}}\right) + m\left(r, \frac{g'}{g}\right) + \log\left|\frac{g(0)}{g'(0)}\right|\right\},$$

where $g(z)$ is given by (4.3.1).

Substituting the above inequality into the Milloux inequality (4.2.9), we deduce that

$$T(r, f) < \left(2 + \frac{1}{k}\right)N\left(r, \frac{1}{f}\right) + \left(2 + \frac{2}{k}\right)\overline{N}\left(r, \frac{1}{f^{(k)} - 1}\right)$$

$$- 2N_0\left(r, \frac{1}{f^{(k+1)}}\right) + S_k(r, f)$$

where

$$S_k(r, f) = \left(2 + \frac{1}{k}\right)S(r, f) + \frac{1}{k}\left\{m\left(r, \frac{g'}{g}\right) + \log\left|\frac{g(0)}{g'(0)}\right|\right\}.$$

From expression (4.3.1) of $g(z)$ in Lemma 4.4, we can see that

$$\frac{g'(z)}{g(z)} = \frac{(k+1)f^{(k+2)}(z)}{f^{(k+1)}(z)} - \frac{(k+2)f^{(k+1)}(z)}{f^{(k)}(z) - 1},$$

so that

$$\log\left|\frac{g(0)}{g'(0)}\right| = \log\left|\frac{f^{(k+1)}(0)(f^{(k)}(0) - 1)}{(k+1)f^{(k+2)}(0)(f^{(k)}(0) - 1) - (k+2)f^{(k+1)}(0)^2}\right|.$$

On the other hand, since

$$\left|\frac{g'(z)}{g(z)}\right| \leq (k+2)\left(\left|\frac{f^{(k+2)}(z)}{f^{(k+1)}(z)}\right| + \left|\frac{f^{(k+1)}(z)}{f^{(k)}(z) - 1}\right|\right),$$

we have

$$m\left(r, \frac{g'}{g}\right) \leq m\left(r, \frac{f^{(k+2)}}{f^{(k+1)}}\right) + m\left(r, \frac{f^{(k+1)}}{f^{(k)} - 1}\right)$$

$$+ \log(k + 2) + \log 2.$$

Thus

$$S_k(r,f) \le \left(2+\frac{2}{k}\right)m\left(r,\frac{f^{(k+2)}}{f^{(k)}-1}\right) + \frac{1}{k}m\left(r,\frac{f^{(k+2)}}{f^{(k+1)}}\right)$$

$$+\left(2+\frac{1}{k}\right)m\left(r,\frac{f^{(k)}}{f}\right) + \left(2+\frac{1}{k}\right)m\left(r,\frac{f^{(k+1)}}{f}\right)$$

$$+\left(2+\frac{1}{k}\right)\log\left|\frac{f(0)(f^{(k)}(0)-1)}{f^{(k+1)}(0)}\right|$$

$$+\frac{1}{k}\log\left|\frac{f^{(k+1)}(0)(f^{(k)}(0)-1)}{(k+1)f^{(k+2)}(0)(f^{(k)}(0)-1)-(k+2)f^{(k+1)}(0)^2}\right|$$

$$+\frac{1}{k}\log(2k+4) + \left(2+\frac{1}{k}\right)\log 2.$$

Noting that

$$\frac{1}{k}\log(2k+4) + \left(2+\frac{1}{k}\right)\log 2 \le \log 6 + 3\log 2 < 4,$$

the proof is finally complete.   □

Corollary.   Let $f(z)$ be meromorphic and transcendental in the finite plane. Then either $f(z)$ takes every finite complex value infinitely many times, or for every positive integer k, $f^{(k)}(z)$ takes every finite non-zero complex value infinitely many times.

4.4 A General Criterion for Normality

4.4.1 Preliminary lemma.   Corresponding to his suprising inequality, Hayman [4] made the following important conjecture. Suppose $\mathcal{F}$ is a family of meromorphic functions in a domain D and k is a positive integer. If $f(z) \ne 0$ and $f^{(k)}(z) \ne 1$ in D for all $f(z)$ in $\mathcal{F}$, then $\mathcal{F}$ is normal in D. Gu Yongxing [2] later succeeded in proving this. Let us give a simpler proof here. In order to do so, we first establish the following lemma.

Lemma 4.5   Let $f(z)$ be meromorphic in $|z| < R(0 < R \le \infty)$, where $f(z) \ne 0$ and $f^{(k)}(z) \ne 1$. If $f(0) \ne 0, \infty$, $f^{(k)}(0) \ne 1$, $f^{(k+1)}(0) \ne 0$

and
$$(k+1)f^{(k+2)}(0)\{f^{(k)}(0) - 1\} - (k+2)f^{(k+1)}(0)^2 \neq 0,$$

then we have
$$\log M\left(r, \frac{1}{f}\right) < C_k \frac{R}{R-r}\left\{1 + B_k + \log \frac{2R}{R-r}\right\}, \qquad (4.4.1)$$

$$B_k = \log^+ R + \log^+ \frac{1}{R} + \log^+ |f(0)| + \log^+ |f^{(k)}(0)| + \log^+ \frac{1}{|f^{(k+1)}(0)|}$$

$$+ \log^+ \frac{1}{|(k+1)f^{(k+2)}(0)\{f^{(k+1)}(0) - 1\} - (k+2)\{f^{(k+1)}(0)\}^2|},$$

$$(4.4.2)$$

whenever $0 < r < R$.

Proof.   According to the hypothesis and (4.3.9), we have
$$T(r, f) < S^*(r, f),$$

where $S^*(r, f)$ is as given in (4.3.10).

Let us estimate the terms of (4.3.10). The Nevanlinna lemma (Lemma 1.3) gives us

$$m\left(r, \frac{f^{(k+1)}}{f^{(k)} - 1}\right) < C\left\{ 1 + \log^+ \rho' + \log^+ \frac{1}{r} + \log^+ \frac{1}{\rho' - r}\right.$$
$$\left. + \log^+ \log^+ \frac{1}{|f^{(k)}(0) - 1|} + \log^+ T(\rho', f^{(k)})\right\} \qquad (4.4.3)$$

and

$$m\left(r, \frac{f^{(k+2)}}{f^{(k+1)}}\right) < C\left\{ 1 + \log^+ \rho' + \log^+ \frac{1}{r} + \log^+ \frac{1}{\rho' - r}\right.$$
$$\left. + \log^+ \log^+ \frac{1}{|f^{(k+1)}(0)|} + \log^+ T(\rho', f^{(k+1)})\right\}, \qquad (4.4.4)$$

where ρ and ρ' are chosen such that $0 < r < \rho' = (r + \rho)/2 < \rho < R$. (The usual estimate gives (4.4.3) with $\log^+ T(\rho', f^{(k)} - 1)$ as its last term, and $|T(\rho', f^{(k)} - 1) - T(\rho', f^{(k)})| \leq \log 2$.)

For the terms $\log^+ T(\rho', f^{(j)})(j = k, k+1)$ in (4.4.3) and (4.4.4) we have

$$\log^+ T(\rho', f^{(j)}) \leq \log^+ \left\{(j + 1)T(\rho', f) + m\left(\rho', \frac{f^{(j)}}{f}\right)\right\}$$

$$< \log^+ T(\rho', f) + m\left(\rho', \frac{f^{(j)}}{f}\right) + C,$$

and hence

$$T(r, f) < C_k \left\{ 1 + \log^+ \rho' + \log^+ \frac{1}{r} + \log^+ \frac{1}{\rho' - r} + \log^+ \log^+ \frac{1}{|f^{(k)}(0) - 1|} \right.$$

$$\left. + \log^+ T(\rho', f) + \log^+ \log^+ \frac{1}{|f^{(k+1)}(0)|} \right\}$$

$$+ \left(2 + \frac{2}{k}\right) \log |f^{(k)}(0) - 1| + 2 \log \frac{1}{|f^{(k+1)}(0)|}$$

$$+ \frac{1}{k} \log \frac{1}{|(k+1)f^{(k+2)}(0)(f^{(k)}(0) - 1) - (k+2)f^{(k+1)}(0)^2|}$$

$$+ \left(2 + \frac{1}{k}\right) \log |f(0)| + C \left\{ m\left(\rho', \frac{f^{(k)}}{f}\right) + m\left(\rho', \frac{f^{(k+1)}}{f}\right) \right\}$$

$$(4.4.5)$$

in view of (4.3.9), (4.3.10), (4.4.3) and (4.4.4).

Applying Lemma 4.3 with r, ρ equal to ρ' and ρ respectively to the last term of (4.4.5) inside { }, and noting the relationships among r, ρ, ρ' and R, we have

$$T\left(r, \frac{1}{f}\right) < C_k \left\{ 1 + \log^+ R + \log^+ \frac{1}{R} + \log^+ \frac{1}{\rho - r} + \log^+ \log^+ \frac{1}{|f(0)|} \right.$$

$$+ \log^+ \log^+ \frac{1}{|f^{(k)}(0) - 1|} + \log^+ \log^+ \frac{1}{|f^{(k+1)}(0)|}$$

$$\left. + \log^+ T(\rho, f) \right\} + \left(2 + \frac{1}{k}\right) \log |f(0)|$$

$$+ \left(2 + \frac{2}{k}\right) \log |f^{(k)}(0) - 1| + 2 \log^+ \frac{1}{|f^{(k+1)}(0)|}$$

$$+ \frac{1}{k} \log \frac{1}{|(k+1)f^{(k+2)}(0)(f^{(k)}(0) - 1) - (k+2)f^{(k+1)}(0)^2|},$$

$$(4.4.6)$$

whenever $R/2 < r < \rho < R$. Moreover Lemma 2.3 implies

$$\beta \log x + C_k \log^+ \log^+ \frac{1}{x} < \beta \log^+ x + C_k$$

for $\beta > 0$, $0 < x < \infty$. Applying this successively with $x = |f(0)|$ and $x = |f^{(k)}(0) - 1|$, and noting that

$$\log^+ T(\rho, f) = \log^+ \left\{ T\left(\rho, \frac{1}{f}\right) + \log |f(0)| \right\}$$

$$\leq \log^+ T\left(\rho, \frac{1}{f}\right) + \log^+ |f(0)| + 1,$$

we see from (4.4.6) that whenever $R/2 < r < \rho < R$

$$T\left(r, \frac{1}{f}\right) < C_k\left\{1 + B_k + \log^+ \frac{1}{\rho - r} + \log^+ T\left(\rho, \frac{1}{f}\right)\right\},$$

where B_k is given by (4.4.2). Hence

$$T\left(r, \frac{1}{f}\right) < C_k\left\{1 + B_k + \log \frac{2R}{R - r}\right\}, \quad \frac{R}{2} < r < R, \qquad (4.4.7)$$

by applying Lemma 2.4 to $T(r, 1/f)$. The conclusion of the Lemma follows immediately in view of the fact that

$$\log M\left(r, \frac{1}{f}\right) \le \frac{R + 3r}{R - r} T\left(\frac{r + R}{2}, \frac{1}{f}\right)$$

for any r with $0 < r < R$. This completes the proof. □

4.4.2 A general criterion for normality.   Now let us prove the following theorem, from which Gu's criterion follows immediately.

Theorem 4.6 (Yang Lo [5])   *If k is a positive integer, $f(z)$ is meromorphic in $|z| < 1$ and $f(z) \ne 0$, $f^{(k)}(z) \ne 1$ there, then inside $|z| < 1/32$, either $|f(z)| < 1$ for all z or $|f(z)| > C_k$ for all z.*

For brevity, we shall express the last assertion as: either $|f(z)| < 1$ or $|f(z)| > C_k$ uniformly in $|z| < 1/32$.

Proof.   The conclusion holds for $C_k = 1$ unless there are points z' and z'' such that $|f(z')| \ge 1$, $|f(z'')| \le 1$, $|z'| < 1/32$ and $|z''| < 1/32$. Then because of the continuity of f, there is a point z_1 such that

$$|f(z_1)| = 1, \quad |z_1| < \frac{1}{32}. \qquad (4.4.8)$$

Let us show that (4.4.8) implies $|f(z)| > C_k$ uniformly in $|z| < 1/32$. There are two mutually exclusive cases.

   Case 1. The inequality

$$\sum_{j=0}^{k+1} |f^{(j)}(z)| \ge \frac{1}{4}$$

holds uniformly in $|z| < 1/8$. In this case we have

$$\frac{1}{|f|} \le 4 \sum_{j=0}^{k+1} \left| \frac{f^{(j)}}{f} \right|, \quad |z| < \frac{1}{8},$$

and so if $m(r, z_1, f)$ and $T(r, z_1, f)$ denote $m(r, f(z + z_1))$ and $T(r, f(z + z_1))$ respectively, we have

$$m\left(r, z_1, \frac{1}{f}\right) \le \sum_{j=0}^{k+1} m\left(r, z_1, \frac{f^{(j)}}{j}\right) + \log 4(k+2), \quad 0 < r < \frac{3}{32}. \quad (4.4.9)$$

According to Lemma 4.3 applied to $f(z + z_1)$, (4.4.9) implies

$$T\left(r, z_1, \frac{1}{f}\right) = m\left(r, z_1, \frac{1}{f}\right) \le C_k \left\{ 1 + \log^+ \frac{1}{\rho - r} + \log^+ T(\rho, z_1, f) \right\},$$

for $1/32 < r < \rho < 3/32$, since $N(r, z_1, 1/f) = 0$. Next, since $|f(z_1)| = 1$, Jensen's formula implies that the last term on the right side of the above inequality can be replaced by $\log^+ T(\rho, z_1, 1/f)$. Finally, we apply Lemma 2.4 to $T(r, z_1, 1/f)$ in $[1/32, 2/32]$ to obtain

$$T\left(r, z_1, \frac{1}{f}\right) < C_k \left\{ 1 + \log \frac{2}{\frac{3}{32} - r} \right\}.$$

Thus

$$T\left(\frac{5}{64}, z_1, \frac{1}{f}\right) < C_k,$$

$$\log M\left(\frac{1}{32}, \frac{1}{f}\right) \le \log M\left(\frac{1}{16}, z_1, \frac{1}{f}\right) \le 9T\left(\frac{5}{64}, z_1, \frac{1}{f}\right) < C_k.$$

Case 2. There is a point z_2 such that

$$\sum_{j=0}^{k+1} \left| f^{(j)}(z_2) \right| < \frac{1}{4}, \quad |z_2| < \frac{1}{8}. \quad (4.4.10)$$

We assert that there exists a point z_0 on the segment $\overline{z_2 z_1}$ such that

$$\left| f^{(k+2)}(z_0) \right| \ge 1, \quad \frac{1}{12} < \left| f^{(k+1)}(z_0) \right| < \frac{1}{2},$$

$$\left| f^{(k)}(z_0) \right| < \frac{1}{2}, \quad |f(z_0)| < \frac{1}{2}. \quad (4.4.11)$$

In fact, if $|f^{(k+1)}(z)| < 1/4$ on $\overline{z_1 z_2}$, inequality (4.4.10) leads to

$$\left|f^{(k)}(z)\right| \leq \left|f^{(k)}(z_2)\right| + \left|\int_{z_2 z} f^{(k+1)}(t)dt\right| < \frac{1}{4} + \frac{1}{4}|z_2 - z| < \frac{1}{3},$$

and hence successively to

$$|f^{(j)}(z)| < \frac{1}{3}, \quad j = k-1, k-2, \cdots, 1, 0.$$

The last inequality contradicts the fact that $|f(z_1)| = 1$. Thus there is a point z_3 on $\overline{z_2 z_1}$ such that

$$|f^{(k+1)}(z_3)| = \frac{1}{4} \text{ and } |f^{(k+1)}(z)| < \frac{1}{4} \text{ on } \overline{z_2 z_3}.$$

Noting that

$$\left|f^{(k)}(z_3)\right| \leq \left|f^{(k)}(z_2)\right| + \left|\int_{\overline{z_2 z_3}} f^{(k+1)}(t)dt\right| < \frac{1}{3},$$

we have by a similar argument

$$|f^{(j)}(z_3)| < \frac{1}{3}, \quad j = k-1, \cdots, 1, 0.$$

If $|f^{(k+2)}(z_3)| \geq 1$, we may take z_0 in (4.4.11) to be z_3, and our assertion follows.

If $|f^{(k+2)}(z_3)| < 1$, note that $|f^{(k+2)}(z)| < 1$ on $\overline{z_3 z_1}$ implies

$$|f^{(k+1)}(z)| < \frac{1}{4} + \frac{1}{8} + \frac{1}{32} < \frac{1}{2}$$

on $\overline{z_2 z_1}$; hence

$$|f^{(k)}(z)| \leq |f^{(k)}(z_2)| + |z_2 - z_1| \cdot \max_{t \in \overline{z_2 z}} |f^{(k+1)}(t)| < \frac{1}{3}.$$

We then obtain $|f^{(j)}(z)| < 1/3$ on $\overline{z_2 z_1}$ for $j = 0, 1, \cdots, k$, which contradicts the fact that $|f(z_1)| = 1$. Therefore a point z_4 on $\overline{z_3 z_1}$ can be found such that $|f^{(k+2)}(z_4)| = 1$ and $|f^{(k+2)}(z)| < 1$ on $\overline{z_3 z_4}$. Since

$$|z_3 - z_4| < |z_1 - z_2| \leq \frac{1}{32} + \frac{1}{8} = \frac{5}{32},$$

we have for every point of $\overline{z_3 z_4}$,

$$|f^{(k+1)}(z)| \geq |f^{(k+1)}(z_3)| - |z_3 - z_4| \left(\max_{t \in \overline{z_3 z_4}} |f^{(k+2)}(t)|\right) > \frac{1}{12}$$

and

$$|f^{(k+1)}(z)| \leq |f^{(k+1)}(z_3)| + |z_3 - z_4| \Big(\max_{t \in \overline{z_3 z_4}} |f^{(k+2)}(t)| \Big) < \frac{1}{2}.$$

Thus

$$|f^{(k)}(z_4)| \leq |f^{(k)}(z_3)| + |z_3 - z_4| \Big(\max_{t \in \overline{z_3 z_4}} |f^{(k+1)}(t)| \Big) < \frac{1}{2},$$

and we can deduce similarly that

$$|f^{(j)}(z_4)| < \frac{1}{2}, \quad j = 0, 1, \cdots, k-1.$$

Therefore, when $|f^{(k+2)}(z_3)| < 1$, we may choose $z_0 = z_4$ in (4.4.11), and our assertion on (4.4.11) has been established in both cases.

We now apply Lemma 4.5 to $f(z)$ in $|z - z_0| < 7/8$. The only condition which needs checking follows from (4.4.11):

$$\Big| (k+1) \; f^{(k+2)}(z_0) \big\{ f^{(k)}(z_0) - 1 \big\} - (k+2) f^{(k+1)}(z_0)^2 \Big|$$

$$> \frac{k+1}{2} - \frac{k+2}{4} \geq \frac{1}{4}.$$

From Lemma 4.5 we see that

$$\log M \Big(\frac{1}{2}, z_0, \frac{1}{f} \Big) < C_k,$$

and hence

$$\log M \Big(\frac{1}{32}, \frac{1}{f} \Big) < \log M \Big(\frac{1}{2}, z_0, \frac{1}{f} \Big) < C_k. \quad \square$$

Theorem 4.7   Let $\mathcal{F}$ be a family of functions meromorphic in a region D such that for every function $f(z)$ of $\mathcal{F}$, $f(z) \neq a$ and $f^{(k)}(z) \neq b$ in D, where k is a positive integer, and a and $b(\neq 0)$ are two finite complex numbers. Then $\mathcal{F}$ is normal in D.

Theorem 4.7 follows immediately from Theorem 4.6 and implies the Miranda criterion for normality, which has the same formulation as Theorem 4.7, except that it is usually given only for a family of holomorphic functions. Moreover, Theorem 4.7 can be extended to the following result.

Theorem 4.7′   *Let $\mathcal{F}$ be a family of meromorphic functions in a region D and k a positive integer. Suppose also that $\varphi(z)$ and $\psi(z)$ are holomorphic in D and $\varphi^{(k)}(z) \not\equiv \psi(z)$. If, for every function $f(z)$ of $\mathcal{F}$, $f(z) \neq \varphi(z)$ and $f^{(k)}(z) \neq \psi(z)$, then $\mathcal{F}$ is normal in D.*

In particular, we have the following corollary (see Yang Lo [10]).

Corollary   *If, for every function $f(z)$ of $\mathcal{F}$, neither $f(z)$ nor $f^{(k)}(z)$ has a fixed point in D, then $\mathcal{F}$ is normal there.*

4.4.3 Hayman directions. As another application of Theorem 4.6, we derive a result on the existence of a singular direction (Yang Lo [5]).

Theorem 4.8   *Let $f(z)$ be a meromorphic function in the finite plane. If*

$$\varlimsup_{r\to\infty} \frac{T(r, f)}{(\log r)^3} = \infty, \tag{4.4.12}$$

then there is a number θ_0 such that $0 \leq \theta_0 < 2\pi$, and for every positive number ε and every positive integer k, either $f(z)$ assumes every finite complex value infinitely often or $f^{(k)}(z)$ assumes every finite non-zero complex value infinitely often in the angular region $|\arg z - \theta_0| < \varepsilon$.

Proof.   According to Theorem 3.6, if $f(z)$ satisfies (4.4.12), then there exists a sequence of disks

$$\Gamma_j : |z - z_j| < \varepsilon_j |z_j|, \quad \lim_{j\to\infty} |z_j| = \infty, \quad \lim_{j\to\infty} \varepsilon_j = 0,$$

such that $f(z)$ takes every complex value n_j times in Γ_j, with the exception of those values contained in two spherical disks each with radius e^{-n_j}, provided that $\lim_{j\to\infty} n_j / \log |z_j| = \infty$.

Denote by θ_0 an accumulation point of $\{\arg z_j : j = 1, 2, \cdots\}$. There is no loss of generality in assuming that $\arg z_j \to \theta_0$ $(j \to \infty)$. We shall prove that the ray $\arg z = \theta_0$ has the desired property of Theorem 4.8.

In fact, if it is not true, then there exist a positive number ε, a positive integer k and two finite complex values a, b with $b \neq 0$, such that $f(z) \neq a$ and $f^{(k)}(z) \neq b$ in the angle $|\arg z - \theta_0| < \varepsilon$.

When j is sufficiently large, the disks

$$\Gamma_j' : |z - z_j| < 32\varepsilon_j |z_j|$$

are contained in $|\arg z - \theta_0| < \varepsilon$. For every fixed j, the function

$$g_j(t) = \frac{f(z_j + 32\varepsilon_j|z_j|t) - a}{b(32\varepsilon_j|z_j|)^k}$$

is meromorphic in $|t| < 1$ with $g_j(t) \neq 0$ and $g_j^{(k)}(t) \neq 1$. Theorem 4.6 then implies that either $|g_j(t)| < 1$ or $|g_j(t)| > C$ in $|t| < 1/32$.

(1) Suppose $|g_j(t)| < 1$ uniformly in $|t| < 1/32$, that is,

$$|f(z)| < |a| + |b|(32\varepsilon_j|z_j|)^k < |z_j|^{k+1}$$

for all z in Γ_j when j is sufficiently large.

Since the spherical distance between $|z_j|^{k+1}$ and ∞ is

$$\frac{1}{(1 + |z_j|^{2(k+1)})^{\frac{1}{2}}} > \frac{1}{2|z_j|^{k+1}},$$

the image of Γ_j under $w = f(z)$ lies outside the region G of these points w', such that spherical distance $|w', \infty|$ is less than $(2|z_j|^{k+1})^{-1}$. On the other hand, the image of Γ_j under $w = f(z)$ covers $|w| < \infty$, apart from two spherical disks with radius e^{-n_j}, where $\lim_{j \to \infty} n_j / \log |z_j| = \infty$. Setting $n_j = m_j \log |z_j|$, we have $\lim_{j \to \infty} m_j = \infty$. Thus the values which are not taken by $f(z)$ in Γ_j can be contained in two spherical disks each with radius

$$e^{-n_j} = e^{-m_j \log |z_j|} = \frac{1}{|z_j|^{m_j}}.$$

Clearly these two disks can not contain the spherical disk $|w, \infty| < \dfrac{1}{2|z_j|^{k+1}}$ and so we derive a contradiction.

(2) Suppose $|g_j(t)| > C$ uniformly in $|t| < 1/32$.

Now we can assume that $\varepsilon_j|z_j| > 1$ $(j \to \infty)$, for otherwise we can choose $\varepsilon_j' = \max(\varepsilon_j, 2/|z_j|)$ and replace the disks Γ_j by the larger disks $|z - z_j| < \varepsilon_j'|z_j|$, which satisfy the same conditions. Thus in Γ_j, we have

$$|f(z) - a| > C|b|(32\varepsilon_j|z_j|)^k > (32)^k|b|C,$$

and the image of Γ_j under $w = f(z)$ is entirely disjoint from the fixed disk $|w - a| \le C$. Except for large j this disk is not contained in any two spherical disks of each radius e^{-n_j}. Thus we have a contradiction and Theorem 4.8 is proved. $\square$

It is convenient to call this type of direction a Hayman direction.

Since the condition which ensures a meromorphic function having a Julia direction is

$$\varlimsup_{r \to \infty} \frac{T(r, f)}{(\log r)^2} = \infty, \tag{4.4.13}$$

Drasin[1] raised the following question in 1984.

Problem. Is Theorem 4.8 still true, if condition (4.4.12) is replaced by (4.4.13)?

The answer would appear to be positive, since H. H. Chen [2] recently proved the following result.

Theorem 4.8′   Let $f(z)$ be a meromorphic function in the finite plane. If the condition (4.4.12) is satisfied, then there exists a direction (H') : arg $z = \theta_0$ such that for any positive number ε, an arbitrary positive integer k and any two finite complex values a and b with $b \neq 0$, we have

$$\varlimsup_{r \to \infty} \frac{n(r, \theta_0, \varepsilon, f = a) + n(r, \theta_0, \varepsilon, f^{(k)} = b)}{(\log r)^2} = \infty.$$

4.5 Total Deficiency of Meromorphic Derivatives

4.5.1 A lemma of Frank and Weissenborn.   Let $f(z)$ be a transcendental meromorphic function in the finite plane and k a positive integer. Now we discuss the precise estimate of the total deficiency for $f^{(k)}(z)$. In order to do this, the following lemma due to Frank and Weissenborn [1] is needed.

Lemma 4.6   If ε is an arbitrary positive number, then we have

$$k\overline{N}(r, f) < (1 + \varepsilon)N\left(r, \frac{1}{f^{(k+1)}}\right) + (1 + \varepsilon)N_1(r, f) + S(r, f), \tag{4.5.1}$$

where $N_1(r, f) = N(r, f) - \overline{N}(r, f)$ and $S(r, f) = O(\log(rT(r, f)))$, except possibly on a set of r with finite linear measure, when the order of $f(z)$ is infinite.

---

1) See Barth, K. F., Brannan, D. A. and Hayman, W. K. [1].

Proof.   For a given $\varepsilon > 0$, we choose an integer $l > k/\varepsilon$ and consider for all $z \in \mathbb{C}$

$$W(z) = W(1, z, z^2, \cdots, z^{k+l}, f(z), zf(z), \cdots, z^l f(z))$$

where $W(f_1, \cdots, f_m)$ denotes the Wronskian of the functions $f_1, \cdots, f_m$. Since f is transcendental, $W(z)$ is not identical to zero.

Noting that

$$\frac{W(z)}{(f^{(k+1)})^{l+1}} = \frac{W(1, z, z^2, \cdots, z^{k+l}) \cdot W(f^{(k+l+1)}, \cdots, (z^l f)^{(k+l+1)})}{(f^{(k+1)})^{l+1}}$$

$$= W(1, z, z^2, \cdots, z^{k+l}) W\left(\frac{f^{(k+l+1)}}{f^{(k+1)}}, \cdots, \frac{(z^l f)^{(k+l+1)}}{f^{(k+1)}}\right),$$

we have

$$m(r, A) = S(r, f), \tag{4.5.2}$$

where $A(z) = W(z)/(f^{(k+1)})^{l+1}$. From the first fundamental theorem, we obtain

$$N\left(r, \frac{1}{A}\right) \leq T(r, A) + O(1) \leq N(r, A) + S(r, f). \tag{4.5.3}$$

We now estimate the number of zeros and poles of A in a disk $|z| \leq r$. A simple property of the Wronskian (see (1.5.12)) gives us

$$W = f^{k+2l+2} W\left(\frac{1}{f}, \frac{z}{f}, \cdots, \frac{z^{k+l}}{f}, 1, z, \cdots, z^l\right).$$

If z_0 is a pole of f of order p, then

$$W(z) = O\left((z - z_0)^{-p(k+2l+2)}\right), \quad \text{for } z \to z_0.$$

Therefore,

$$A(z) = O\left((z - z_0)^{(l+1)(k+p+1)-p(k+2l+2)}\right)$$
$$= O\left((z - z_0)^{lk-(k+l+1)(p-1)}\right). \tag{4.5.4}$$

Let $\overline{N}_p^0(r)$, $\overline{N}_p^\infty(r)$ and $\overline{N}_p^*(r)$ be the counting functions for those poles of f of order p, where A has a zero, pole or finite non-zero value respectively, each pole being counted without regard to multiplicity. From (4.5.3) and

(4.5.4), we get

$$\sum_{\kappa=1}^{\infty} \left(lk - (k+l+1)(p-1) \right) \overline{N}_p^0(r) \leq N\left(r, \frac{1}{A}\right)$$

$$\leq N(r, A) + S(r, f)$$

$$\leq \sum_{p=1}^{\infty} \left((k+l+1)(p-1) - lk \right) \overline{N}_p^{\infty}(r)$$

$$+ (l+1)N\left(r, \frac{1}{f^{(k+1)}}\right) + S(r, f). \tag{4.5.5}$$

If a pole of f makes a positive contribution to $N_p^*(r)$, then by (4.5.4) we have

$$lk - (k+l+1)(p-1) \leq 0,$$

$$lk\overline{N}_p^*(r) \leq (k+l+1)(p-1)\overline{N}_p^*(r).$$

Adding the last inequality for $p = 1, 2, \cdots$, to (4.5.5), we obtain

$$lk \sum_{p=1}^{\infty} \overline{N}_p(r) \leq (k+l+1) \sum_{p=1}^{\infty} (p-1)\overline{N}_p(r)$$

$$+ (l+1)N\left(r, \frac{1}{f^{(k+1)}}\right) + S(r, f). \tag{4.5.6}$$

Here we set $\overline{N}_p(r) = \overline{N}_p^0(r) + \overline{N}_p^{\infty}(r) + \overline{N}p^*(r)$, which counts the poles of order p of f. The sum on the left-hand side of (4.5.6) is exactly $\overline{N}(r, f)$, while the sum of the right-hand side is equal to $N_1(r, f)$. We finally obtain

$$k\overline{N}(r, f) \leq \left(1 + \frac{k+1}{l}\right) N_1(r, f) + \left(1 + \frac{1}{l}\right) N\left(r, \frac{1}{f^{(k+1)}}\right) + S(r, f),$$

as asserted. □

Remark *In Frank and Weissenborn [1], Lemma 4.6 is the key point in the proof of their result. (i.e. Theorem 1.12″ of the present book) The Steinmetz's elegant proof of Theorem 1.13 is inspired mainly by the use of the Wronskian in Lemma 4.6.*

Lemma 4.7

$$N\left(r, \frac{1}{f^{(k+1)}}\right) > (k+1)\overline{N}(r, f) - N(r, f)$$

$$- \varepsilon T(r, f^{(k)}) - S(r, f^{(k)}), \tag{4.5.7}$$

where $S(r, f^{(k)}) = O\left(\log(rT(r, f^{(k)}))\right)$, except possibly on a set of r with finite linear measure, when the order of $f(z)$ is infinite.

In fact, according to Lemma 4.6, we have

$$k\overline{N}(r, f) < N\left(r, \frac{1}{f^{(k+1)}}\right) + \left(N(r, f) - \overline{N}(r, f)\right) + \frac{\varepsilon}{3}N\left(r, \frac{1}{f^{(k+1)}}\right)$$

$$+ \frac{\varepsilon}{3}(N(r, f) - \overline{N}(r, f)) + m\left(r, \frac{W}{(f^{(k+1)})^{l+1}}\right).$$

Noting that

$$N\left(r, \frac{1}{f^{(k+1)}}\right) \leq T(r, f^{(k+1)}) + O(1)$$

$$\leq 2T(r, f^{(k)}) + m\left(r, \frac{f^{(k+1)}}{f^{(k)}}\right) + O(1),$$

$$N(r, f) \leq T(r, f^{(k)}),$$

and utilizing (4.5.2), inequality (4.5.7) follows immediately.

4.5.2 Precise estimate of the total deficiency of meromorphic derivatives.   Now we give the precise estimate of the total deficiency of meromorphic derivatives (see Yang Lo [14]).

Theorem 4.9   Let $f(z)$ be a transcendental meromorphic function in the finite plane and k a positive integer. Then we have

$$\sum_{a \in \mathbb{C}} \delta(a, f^{(k)}) \leq \frac{2k + 2}{2k + 1}. \tag{4.5.8}$$

The summation of the left-hand side is taken over all the finite complex numbers.

Proof. If $a_j(j = 1, 2, \cdots, q)$ are q distinct finite complex numbers, then we have

$$\sum_{j=1}^{q} m\left(r, \frac{1}{f^{(k)} - a_j}\right) \leq m\left(r, \frac{1}{f^{(k+1)}}\right) + S(r, f^{(k)}). \tag{4.5.9}$$

Denote by e the union of exceptional sets corresponding to inequalities (4.5.7) and (4.5.9). Then e has finite linear measure.

We consider two cases which are mutually exclusive.

(1) $\quad \overline{\lim_{\substack{r \to \infty \\ r \tilde{e} e}}} \dfrac{\overline{N}(r, f)}{T(r, f^{(k)})} < \dfrac{1}{2k + 1}.$

In this case, we have

$$\sum_{j=1}^{q} m\left(r, \frac{1}{f^{(k)} - a_j}\right) \leq T(r, f^{(k+1)}) - N\left(r, \frac{1}{f^{(k+1)}}\right) + S(r, f^{(k)})$$

$$\leq T(r, f^{(k)}) + \overline{N}(r, f) - N\left(r, \frac{1}{f^{(k+1)}}\right) + S(r, f^{(k)}).$$
$$(4.5.10)$$

Thus

$$\sum_{j=1}^{q} \delta(a_j, f^{(k)}) \leq \overline{\lim_{\substack{r \to \infty \\ r \tilde{e} e}}} \frac{T(r, f^{(k)}) + \overline{N}(r, f)}{T(r, f^{(k)})} \leq 1 + \frac{1}{2k + 1}.$$

(2) $\quad \overline{\lim_{\substack{r \to \infty \\ r \tilde{e} e}}} \dfrac{\overline{N}(r, f)}{T(r, f^{(k)})} \geq \dfrac{1}{2k + 1}.$

Combining (4.5.7) and (4.5.10), we obtain

$$\sum_{j=1}^{q} m\left(r, \frac{1}{f^{(k)} - a_j}\right)$$

$$\leq T(r, f^{(k)}) - k\overline{N}(r, f) + N(r, f) + \varepsilon T(r, f^{(k)}) + S(r, f^{(k)})$$

$$\leq 2T(r, f^{(k)}) - 2k\overline{N}(r, f) + \varepsilon T(r, f^{(k)}) + S(r, f^{(k)}).$$

Therefore,

$$\sum_{j=1}^{q} \delta(a_j, f^{(k)}) \leq \overline{\lim_{\substack{r \to \infty \\ r \tilde{e} e}}} \left\{2 - 2k\frac{\overline{N}(r, f)}{T(r, f^{(k)})}\right\} + \varepsilon + \overline{\lim_{\substack{r \to \infty \\ r \tilde{e} e}}} \frac{S(r, f^{(k)})}{T(r, f^{(k)})}$$

$$\leq \frac{2k + 2}{2k + 1} + \varepsilon.$$

Letting ε be arbitrarily small and q be arbitrarily large, the proof of Theorem 4.9 is complete. $\quad \square$

Although Theorem 4.8 gives a very good estimate for $\sum_{a \in \mathfrak{C}} \delta(a, f^{(k)})$, it does not include $\delta(\infty, f^{(k)})$. For this reason, we prove another estimate (see Yang Lo [14]).

Theorem 4.10 *Let $f(z)$ be a transcendental meromorphic function of finite order in the finite plane and k a positive integer. Then we have*

$$\sum_{a \in \overline{\mathbb{C}}} \delta(a, f^{(k)}) \leq 2 - \frac{2k(1 - \Theta(\infty, f))}{1 + k(1 - \Theta(\infty, f))},$$

where $\Theta(\infty, f)$ is the ramification index of ∞ with respect to f, defined by

$$\Theta(\infty, f) = 1 - \varlimsup_{r \to \infty} \frac{\overline{N}(r, f)}{T(r, f)}.$$

Proof.   Similar to case (2) of the proof of Theorem 4.9, we have

$$m(r, f^{(k)}) + \sum_{j=1}^{q} m\left(r, \frac{1}{f^{(k)} - a_j}\right)$$

$$\leq 2T(r, f^{(k)}) - 2k\overline{N}(r, f) + \varepsilon T(r, f^{(k)}) + S(r, f^{(k)}).$$

Thus

$$\delta(\infty, f^{(k)}) + \sum_{j=1}^{q} \delta(a_j, f^{(k)})$$

$$\leq \varliminf_{r \to \infty} \left\{ 2 - \frac{2k\overline{N}(r, f)}{T(r, f^{(k)})} + \varepsilon + \frac{S(r, f^{(k)})}{T(r, f^{(k)})} \right\}$$

$$\leq \varliminf_{r \to \infty} \left\{ 2 - \frac{2k\overline{N}(r, f)}{T(r, f^{(k)})} \right\} + \varlimsup_{r \to \infty} \left\{ \varepsilon + \frac{S(r, f^{(k)})}{T(r, f^{(k)})} \right\} \tag{4.5.11}$$

$$\leq 2 - 2k \varlimsup_{r \to \infty} \frac{\overline{N}(r, f)}{T(r, f^{(k)})} + \varepsilon.$$

Since

$$\frac{\overline{N}(r, f)}{T(r, f^{(k)})} \geq \frac{\overline{N}(r, f)}{T(r, f) + k\overline{N}(r, f) + m\left(r, \dfrac{f^{(k)}}{f}\right)},$$

we have

$$\varlimsup_{r\to\infty} \frac{\overline{N}(r,f)}{T(r,f^{(k)})} \geq \varlimsup_{r\to\infty} \frac{\overline{N}(r,f)}{T(r,f) + k\overline{N}(r,f) + m\left(r,\frac{f^{(k)}}{f}\right)}$$

$$\geq \frac{\varlimsup_{r\to\infty} \dfrac{\overline{N}(r,f)}{T(r,f)}}{\varlimsup_{r\to\infty}\left\{1 + k\dfrac{\overline{N}(r,f)}{T(r,f)} + \dfrac{m(r,\frac{f^{(k)}}{f})}{T(r,f)}\right\}}$$

$$= \frac{1 - \Theta(\infty,f)}{1 + k(1 - \Theta(\infty,f))}. \qquad (4.5.12)$$

Combining (4.5.11) and (4.5.12), letting ε tend to zero and q to infinity, we obtain finally

$$\sum_{a\in\overline{\mathbb{C}}} \delta(a,f^{(k)}) \leq 2 - \frac{2k(1 - \Theta(\infty,f))}{1 + k(1 - \Theta(\infty,f))}. \qquad \square$$

The following corollaries are immediate consequences of Theorem 4.10.

Corollary 1. *Suppose $f(z)$ is transcendental meromorphic and of finite order in the finite plane. If $\Theta(\infty,f) < 1$, then we have*

$$\lim_{k\to\infty}\left\{\sum_{a\in\overline{\mathbb{C}}} \delta(a,f^{(k)})\right\} = 0.$$

Corollary 2. *Let $f(z)$ be transcendental meromorphic and of finite order. If $\Theta(\infty,f) = 0$, then for any positive integer k, we have*

$$\sum_{a\in\overline{\mathbb{C}}} \delta(a,f^{(k)}) \leq \frac{2}{k+1}.$$

Corollary 3. *Let $f(z)$ be transcendental meromorphic and of finite order. If there exists a positive integer k_0 such that $\sum_{a\in\overline{\mathbb{C}}} \delta(a,f^{(k_0)}) = 2$, then we have $\Theta(\infty,f) = 1$.*

4.5.3 Problems of Drasin.   It is natural to discuss the precise estimate of the total deficiency of both the function itself and its derivative. On this subject, Drasin [2] posed the following questions.

Problem D_1. Let $f(z)$ be meromorphic and of finite order in the finite plane. If $\sum_{a\in\bar{\mathbb{C}}} \delta(a, f) = 2$ and $\delta(\infty, f) = 0$, must we have

$$\sum_{b\in\bar{\mathbb{C}}} \delta(b, f') = \delta(0, f') = 1? \tag{4.5.13}$$

Problem D_2. Let $f(z)$ be meromorphic in the finite plane with $\delta(\infty, f) = 0$. Can we have

$$\sum_{a\in\bar{\mathbb{C}}} \delta(a, f) + \sum_{b\in\bar{\mathbb{C}}} \delta(b, f') = 4? \tag{4.5.14}$$

If not, what is the best bound?

The author recently proved the following theorem:

Theorem 4.11   *Let $f(z)$ be a transcendental meromorphic function of finite order in the finite plane and k a positive integer. Then we have*

$$\sum_{a\in\mathbb{C}} \delta(a, f) + \sum_{b\in\bar{\mathbb{C}}} \delta(b, f^{(k)}) \leq 3. \tag{4.5.15}$$

The equality of (4.5.15) holds if and only if

(1) $\Theta(\infty, f) = 1,\ \sum_{a\in\mathbb{C}} \delta(a, f) = 1,\ \sum_{b\in\bar{\mathbb{C}}} \delta(b, f^{(k)}) = 2;$

or

(2) $k = 1, \Theta(\infty, f) = 0,\ \sum_{a\in\mathbb{C}} \delta(a, f) = 2,\ \sum_{b\in\bar{\mathbb{C}}} \delta(b, f') = 1.$

Proof. By Theorem 4.10, we have

$$\sum_{b\in\bar{\mathbb{C}}} \delta(b, f^{(k)}) \leq 2 - \frac{2k(1 - \Theta(\infty, f))}{1 + k(1 - \Theta(\infty, f))}.$$

On the other hand,

$$\sum_{a\in\mathbb{C}} \delta(a, f) \leq \sum_{a\in\mathbb{C}} \Theta(a, f) \leq 2 - \Theta(\infty, f).$$

Setting $u = 1 - \Theta(\infty, f)$, we have

$$\sum_{a \in \mathbb{C}} \delta(a, f) + \sum_{b \in \bar{\mathbb{C}}} \delta(b, f^{(k)}) \leq 3 + u - \frac{2ku}{1 + ku}. \qquad (4.5.16)$$

When $0 \leq u \leq 1$, it is easy to see that $u \leq 2ku/(1 + ku)$. Moreover, the equality in (4.5.16) holds if and only if either $u = 0$ or $u = 1$ and $k = 1$. The conclusion of Theorem 4.11 follows immediately. $\square$

Theorem 4.11 gives a positive answer to Problem D_1. In fact, let $f(z)$ be a transcendental meromorphic function of finite order in the finite plane. If $a_j (j = 1, 2, \cdots, q)$ are q distinct finite complex numbers, then we have

$$\sum_{j=1}^{q} m\left(r, \frac{1}{f - a_j}\right) \leq m\left(r, \frac{1}{f'}\right) + O(\log r).$$

Connecting this inequality with the estimate

$$T(r, f') \leq 2T(r, f) + O(\log r),$$

we obtain

$$\sum_{a \in \mathbb{C}} \delta(a, f) \leq 2\delta(0, f'). \qquad (4.5.17)$$

Thus (4.5.13) of Problem D_1 can be deduced immediately from (4.5.15) and (4.5.17). Theorem 4.11 also gives a negative answer to Problem D_2 and yields the best bound of 3 for the left-hand side of (4.5.14), when the order of $f(z)$ is finite.[1]

4.5.4 Conjectures of Frank, Goldberg and Mues.   For the estimate of the total deficiency of the derivative $f^{(k)}(z)$, it was Hayman [1] who first pointed out that the inequality

$$\sum_{a \in \mathbb{C}} \delta(a, f^{(k)}) \leq \frac{k + 2}{k + 1}$$

---

[1] Recently the author and Wang Yaofei (Yang Lo & Wang Yaofei, Drasin problems and Mues conjecture, preprint) noted that the condition of finite order can be omitted in Theorem 4.11. Problem D_2 is therefore settled completely.

holds for any transcendental meromorphic function $f(z)$. In 1971, Mues
[1] improved this result to

$$\sum_{a\in\mathbb{C}} \delta(a, f^{(k)}) \leq \frac{k^2 + 5k + 4}{k^2 + 4k + 2}.$$

Theorem 4.9 is a further improvement on the results of Hayman and Mues.
When Mues obtained his result, he posed the following conjecture.

Problem M.　Let $f(z)$ be a transcendental meromorphic function
in the finite plane and k a positive integer. Then the relation

$$\sum_{a\in\mathbb{C}} \delta(a, f^{(k)}) \leq 1. \tag{4.5.18}$$

should be true.

The Mues conjecture appears to be true when $f(z)$ has finite order.
Estimate (4.5.18) is easily verified, if $f(z)$ satisfies one of the following
conditions:
  (1) the order of f is finite and $\Theta(\infty, f) \leq 1 - 1/k$;
  (2) f has only poles with multiplicity $\leq k$;
  (3) the order of f is finite and

$$\delta(\infty, f^{(k)}) + \sum_{a\in\mathbb{C}} \delta(a, f) \geq 2.$$

In connection with the Mues conjecture, the following problems were
recently raised by Frank and Goldberg, respectively.

Problem F.　Let $f(z)$ be a transcendental meromorphic function
in the finite plane and k be a positive integer. If ε is an arbitrary small
positive number, then the following estimate seems to be true.

$$k\overline{N}(r, f) < (1 + \varepsilon)N\left(r, \frac{1}{f^{(k+1)}}\right) + S(r, f),$$

where

$$S(r, f) = O(\log(rT(r, f))),$$

except for those r in a set of finite linear measure.

Problem G.   Let $f(z)$ be a transcendental meromorphic function in the finite plane. Then the following inequality should be correct.

$$\overline{N}(r, f) < N\left(r, \frac{1}{f''}\right) + S(r, f).$$

When $k \geq 2$, the Frank conjecture is much stronger than the Goldberg conjecture. The Mues conjecture is a direct consequence of either one of them when the order of f is finite.

Quite recently, Wang Yaofei (On the Mues conjecture and Picard exceptional values, preprint) has proved that (4.5.18) is true for every positive integer k, except for at most four integrers.

4.6 A New Method for Proving Normality

4.6.1 A lemma of Zalcman.   In this section, we introduce a new method which easily leads to a new criterion for normality. This method is based on the following lemma due to Zalcman [1].

Lemma 4.8   *A family $\mathcal{F}$ of meromorphic functions in the unit disk is not normal at zero, if and only if there exists a sequence $(f_j) \in \mathcal{F}$, a sequence of complex numbers z_j tending to zero, a sequence of positive numbers ρ_j tending to zero and a meromorphic function f on the plane with $f' \not\equiv 0$ such that $f_j(z_j + \rho_j z)$ tends to f uniformly on all compact subsets of $\mathbb{C}$ with respect to the spherical distance.*

Proof.   We first prove necessity. If $\mathcal{F}$ is not normal at the origin, we can find a sequence $(\tilde{z}_j)(j = 1, 2, \cdots)$ with $|\tilde{z}_j| \leq 1/(2j)$ and a sequence $(f_j) \in \mathcal{F}$ with

$$f_j^{\#}(\tilde{z}_j) = \frac{|f_j'(\tilde{z}_j)|}{1 + |f_j(\tilde{z}_j)|^2} \geq j^2.$$

If we define

$$M_j = \max_{|z| \leq \frac{1}{j}}(1 - |z|^2 j^2) f_j^{\#}(z) = (1 - |z_j|^2 j^2) f_j^{\#}(z_j), \qquad (4.6.1)$$

it follows from

$$M_j \geq (1 - |\tilde{z}_j|^2 j^2) f_j^{\#}(\tilde{z}_j) \geq \frac{3}{4} f_j^{\#}(\tilde{z}_j) \geq \frac{3}{4} j^2$$

that M_j tends to infinity.

Then, letting

$$\rho_j = \frac{1}{M_j}(1 - |z_j|^2 j^2) = \frac{1}{f_j^\#(z_j)}, \qquad (4.6.2)$$

one obtains

$$\frac{\rho_j}{\frac{1}{j} - |z_j|} = (1 + |z_j|j)\frac{j}{M_j} \le \frac{8}{3j} \to 0,$$

which implies that ρ_j tends to zero. The functions

$$g_j(z) = f_j(z_j + \rho_j z)$$

are defined on $|z| < R_j$, where

$$R_j = \frac{\frac{1}{j} - |z_j|}{\rho_j} \qquad (4.6.3)$$

tends to infinity. For every fixed R, $R < R_j$, when $|z| < R$ we get

$$|z_j + \rho_j z| < |z_j| + \rho_j R_j = \frac{1}{j},$$

and therefore from (4.6.1) and (4.6.2)

$$g_j^\#(z) = \rho_j f_j^\#(z_j + \rho_j z) \le \rho_j \frac{M_j}{1 - |z_j + \rho_j z|^2 j^2}$$

$$= \frac{1 + |z_j|j}{1 + |z_j + \rho_j z|j} \cdot \frac{1 - |z_j|j}{1 - |z_j + \rho_j z|j} \le 2\frac{1 - |z_j|j}{1 - |z_j|j - \rho_j|z|j}.$$

Noting that

$$\rho_j|z|j = \rho_j R_j \frac{|z|j}{R_j} = \frac{|z|j}{R_j}\left(\frac{1}{j} - |z_j|\right) \le \frac{R}{R_j}(1 - |z_j|j),$$

we have from (4.6.3)

$$g_j^\#(z) \le 2\frac{1 - |z_j|j}{1 - |z_j|j - \frac{R}{R_j}(1 - |z_j|j)} \to 2.$$

Hence by Theorem 2.2, $\{g_j\}$ is normal in $|z| < R$ $(R < R_j)$.

Applying the diagonal method, we find a subsequence of $(g_j)(j = 1, 2, \cdots)$ which tends to a function f uniformly on all compact subsets of $\mathbb{C}$ with respect to the spherical distance.

Since

$$\left(\frac{1}{g_j}\right)^{\#}(0) = g_j^{\#}(0) = \rho_j f_j^{\#}(z_j) = 1,$$

we have $f \neq \infty$, which implies that f is meromorphic and $f' \not\equiv 0$.

As for sufficiency, choose $z_0 \in \mathbb{C}$ with $f(z_0) \neq 0, \infty$ and $f'(z_0) \neq 0$. We have

$$\rho_j f_k^{\#}(z_j + \rho_j z_0) = \frac{|f_j(z_j + \rho_j z)'|}{1 + |f_j(z_j + \rho_j z)|^2}(z_0) \to \frac{|f'(z_0)|}{1 + |f(z_0)|^2} \neq 0,$$

which together with the fact that $z_j + \rho_j z_0 \to 0$ and Theorem 2.2, implies that $\mathcal{F}$ is not normal at the origin. $\square$

4.6.2 New method for proving normality. Before introducing a new and simple method for proving normality, which was successively used by Oshkin [1], Li and Schwick [1], we prove the following theorem of bounded type.

Theorem 4.12   *Let $f(z)$ be meromorphic in the unit disk and let n and k be two positive integers with $n \geq k+1$. If $f(0) \neq 0, \infty$, $(f^n)^{(k)}(0) - 1 \neq 0$ and $(f^n)^{(k+1)}(0) \neq 0$, then we have*

$$nT(r, f) < \overline{N}(r, f) + (k + 1)\overline{N}\left(r, \frac{1}{f}\right) + N\left(r, \frac{1}{(f^n)^{(k)} - 1}\right)$$

$$+ \log\left|\frac{(f^n)^{(k)}(0) - 1}{(f^n)^{(k+1)}(0)} f^n(0)\right|$$

$$+ C\Big\{ \log^+ T(R, f) + \log^+ \log^+ \frac{1}{|f(0)|} \qquad (4.6.4)$$

$$+ \log^+ \log^+ \frac{1}{|(f^n)^{(k)}(0) - 1|}$$

$$+ \log \frac{1}{R - r} + 1\Big\}$$

for $1/2 < r < R < 1$, where C is a constant depending only on n and k.

Proof. If we apply Theorem 4.3 to f^n, we have

$$nT(r,f) < \overline{N}(r,f^n) + N\left(r,\frac{1}{f^n}\right) + N\left(r,\frac{1}{(f^n)^{(k)}-1}\right)$$

$$-N\left(r,\frac{1}{(f^n)^{(k+1)}}\right) + \log\left|\frac{(f^n)^{(k)}(0)-1}{(f^n)^{(k+1)}(0)}\cdot f^n(0)\right|$$

$$+m\left(r,\frac{(f^n)^{(k)}}{f^n}\right) + m\left(r,\frac{(f^n)^{(k+1)}}{f^n}\right)$$

$$+m\left(r,\frac{(f^n)^{(k+1)}}{(f^n)^{(k)}-1}\right) + O(1).$$

Using the inequality

$$N\left(r,\frac{1}{f^n}\right) - N\left(r,\frac{1}{(f^n)^{(k+1)}}\right) \le (k+1)\overline{N}\left(r,\frac{1}{f}\right)$$

and Lemma 4.3 with $\rho = (r+R)/2$, we get

$$nT(r,f) < \overline{N}(r,f) + (k+1)\overline{N}\left(r,\frac{1}{f}\right) + N\left(r,\frac{1}{(f^n)^{(k)}-1}\right)$$

$$+\log\left|\frac{(f^n)^{(k)}(0)-1}{(f^n)^{(k+1)}(0)}f^n(0)\right| + m\left(r,\frac{(f^n)^{(k)}}{f^n}\right)$$

$$+m\left(r,\frac{(f^n)^{(k+1)}}{f^n}\right) + m\left(r,\frac{(f^n)^{(k+1)}}{(f^n)^{(k)}-1}\right) + O(1)$$

$$< \overline{N}(r,f) + (k+1)\overline{N}\left(r,\frac{1}{f}\right) + N\left(r,\frac{1}{(f^n)^{(k)}-1}\right)$$

$$+\log\left|\frac{(f^n)^{(k)}(0)-1}{(f^n)^{(k+1)}(0)}f^n(0)\right|$$

$$+C\left\{\log^+ T(R,f) + \log^+\log^+\frac{1}{|f(0)|} + \log^+ T(\rho,(f^n)^{(k)}-1)\right.$$

$$\left.+\log\frac{1}{R-r} + \log\frac{1}{\rho-r} + \log^+\log^+\frac{1}{|(f^n)^{(k)}(0)-1|} + 1\right\}$$

$$< \overline{N}(r,f) + (k+1)\overline{N}\left(r,\frac{1}{f}\right) + N\left(r,\frac{1}{(f^n)^{(k)}-1}\right)$$

$$+\log\left|\frac{(f^n)^{(k)}(0)-1}{(f^n)^{(k+1)}(0)}f^n(0)\right|$$

$$+C\left\{\log^+ T(R,f) + \log^+\log^+\frac{1}{|f(0)|} + \log^+\log^+\frac{1}{|(f^n)^{(k)}(0)-1|}\right.$$

$$\left.+\log\frac{1}{R-r} + 1\right\}. \quad \square$$

Now we prove the following criterion.

Theorem 4.13   *If $\mathcal{F}$ is a family of meromorphic functions in a region D and every $f \in \mathcal{F}$ satisfies*

$$(f^n)^{(k)}(z) \neq 1, \quad n \geq k + 3,$$

then $\mathcal{F}$ is normal in D.

Proof.   Without loss of generality, we may assume D is the unit disk. For $f \in \mathcal{F}$, $f(0) \neq 0, \infty$, $(f^n)^{(k)}(0) \neq 1$, $(f^n)^{(k+1)}(0) \neq 0$, and $1/2 < r < R < 1$, we have by Theorem 4.12 and the first fundamental theorem

$$nT(rf) < \overline{N}(r, f) + (k+1)T(r, f) + (k+1)\log \frac{1}{|f(0)|}$$

$$+ \log \left| \frac{(f^n)^{(k)}(0) - 1}{(f^n)^{(k+1)}(0)} \right| |f^n(0)|$$

$$+ C \Big\{ \log^+ T(R, f) + \log^+ \log^+ \frac{1}{|f(0)|}$$

$$+ \log^+ \log^+ \frac{1}{|(f^n)^{(k)}(0) - 1|} + \log \frac{1}{R - r} + 1 \Big\}.$$

Thus

$$\Big\{ n - (k - 2) \Big\} T(r, f) < \log \left| \frac{(f^n)^{(k)}(0) - 1}{(f^n)^{(k+1)}(0)} f^{n-(k+1)}(0) \right|$$

$$+ C \Big\{ \log^+ T(R, f) + \log^+ \log^+ \frac{1}{|f(0)|}$$

$$+ \log^+ \log^+ \frac{1}{|(f^n)^{(k)}(0) - 1|}$$

$$+ \log \frac{1}{R - r} + 1 \Big\}$$

$$< \log \frac{1}{|(f^n)^{(k+1)}(0)|} + \{n - (k+1)\} \log^+ |f(0)|$$

$$+ \log^+ |(f^n)^{(k)}(0) - 1| + C \Big\{ \log^+ T(R, f) + \log \frac{1}{R - r} + 1 \Big\}.$$

If $\mathcal{F}$ is not normal at the origin, there is a sequence (f_j) belonging to $\mathcal{F}$, $z_j \to 0$, $\rho_j \to 0$, and a nonconstant meromorphic function in the plane with $f_j(z_j + \rho_j z)$ tending to $f(z)$ uniformly on all compact subsets of $\mathbb{C}$

with respect to the spherical distance. Since every zero of $f(z)$ is a zero of $f(z)^n$ of order $\geq n$ and $n > k$, $f(z)^n$ is not a polynomial of degree k. Hence there exists $z_0 \in \mathbb{C}$ with $f(z_0) \neq 0$, $(f^n)^{(k)}(z_0) \neq 0$ and $(f^n)^{(k+1)}(z_0) \neq 0$. Therefore,

$$\log \frac{1}{|(f_j^n)^{(k+1)}(z_j + \rho_j z_0)|} + \{n - (k+1)\} \log^+ |f_j(z_j + \rho_j z_0)|$$

$$+ \log^+ |(f_j^n)^{(k)}(z_j + \rho_j z_0) - 1| + \log \frac{1}{\rho_j}$$

$$= \log \frac{1}{|(f_j^n(z_j + \rho_j z))^{(k+1)}(z_0)|} + \log \rho_j^{k+1}$$

$$+ \{n - (k+1)\} \log^+ |f_j(z_j + \rho_j z_0)| + \log \frac{1}{\rho_j}$$

$$+ \log^+ \left| \frac{1}{\rho_j^k}((f_j^n(z_j + \rho_j z))^{(k)}(z_0) - \rho_j^k \right|$$

$$\rightarrow \log \frac{1}{|(f^n)^{(k+1)}(z_0)|} + \{n - (k+1)\} \log^+ |f(z_0)| + \log |(f^n)^{(k)}(z_0)|.$$

This implies that

$$\log \frac{1}{|(f_j^n)^{(k+1)}(z_j + \rho_j z_0)|} + \{n - (k+1)\} \log^+ |f_j(z_j + \rho_j z_0)|$$

$$+ \log^+ |(f_j^n)^{(k)}(z_j + \rho_j z_0) - 1| \rightarrow -\infty.$$

Without loss of generality, we may again assume $z_j + \rho_j z_0 = 0$. Then

$$\{n - (k+2)\}T(r, f_j) < C\left\{ \log^+ T(R, f_j) + \log \frac{1}{R - r} + 1 \right\}.$$

From Lemma 2.4 and $n \geq k + 3$, we obtain

$$T(r, f_j) < C\left\{ \log \frac{2}{1 - r} + 1 \right\},$$

where C does not depend on f_j, and is different from the C above. Let b_j be a pole of $f_j(z)$ with $|b_j| < 1/2$. Thus

$$\log \frac{1}{\frac{2}{|b_j|}} \leq N\left(\frac{1}{2}, f_j\right) \leq T\left(\frac{1}{2}, f_j\right) < C_1,$$

so that

$$|b_j| > \frac{1}{2e^{C_1}}.$$

This means $f_j(z)$ is holomorphic in $|z| < 1/(2e^{C_1})$. Hence

$$\log M\left(\frac{1}{4e^{C_1}}, f_j\right) \le 3T\left(\frac{1}{2e^{C_1}}, f_j\right) < 3C\left\{\log \frac{2}{1 - \frac{1}{2e^{C_1}}} + 1\right\}.$$

Therefore, $\mathcal{F}$ is normal at the origin and the assertion of Theorem 4.13 follows immediately. $\square$

Recently, Pang Xuecheng [1] extended the method of this section and proved the following result.

Theorem 4.14　*Let $\mathcal{F}$ be a family of meromorphic functions in a region D. If every function $f(z) \in \mathcal{F}$ satisfies one of the following conditions, then $\mathcal{F}$ is normal in D:*

(i) $f^2 f' \ne 1$,

(ii) $f' - af^4 \ne b$, where a and b are two finite complex numbers with $a \ne 0$.

Chapter 5

Recent Studies on Borel Directions

The present chapter will be devoted to the following two problems:

(i) the distribution of the Borel directions of a meromorphic function;

(ii) the relationship between the Borel directions of a meromorphic function and those of its derivatives.

5.1 Distribution of Borel Directions

5.1.1 Preliminary lemmas. In Chapter 3 we have seen that every meromorphic function $f(z)$ of order λ $(0 < \lambda < \infty)$ in the finite plane, has at least one Borel direction $B : \arg z = \theta_0$ $(0 \le \theta_0 < 2\pi)$, such that

$$\varlimsup_{r \to \infty} \frac{\log n(r, \theta_0, \varepsilon, f = a)}{\log r} = \lambda$$

holds for any positive number ε and any complex value a, except for at most two values of a.

Clearly, if $\arg z = \theta_j (j = 1, 2, \cdots)$ is a sequence of Borel directions of $f(z)$ and $\lim_{j \to \infty} \theta_j = \theta_0$, then $\arg z = \theta_0$ must also be a Borel direction of $f(z)$. Thus the intersection points of all the Borel directions of $f(z)$ and the circumference of the unit disk, constitute a non-empty and closed set. Conversely, we may ask the following question.

Given a non-empty, closed set E on the circumference $|z| = 1$ and a finite positive number λ, is there a meromorphic function $f(z)$ of order λ such that its Borel directions coincide exactly with all the rays which connect the origin and every point of E?

Yang Lo and Zhang Guanghou [5] gave a positive answer in 1976. In order to prove their result, we need to introduce several lemmas.

Lemma 5.1   *If a and b are two complex numbers with moduli less than one, then we have*

$$\left|\frac{a+b}{1+\bar{a}b}\right| \leq \frac{|a|+|b|}{1+|a||b|}. \tag{5.1.1}$$

Proof.   The assertion follows directly from

$$\left(\frac{1+|a||b|}{|a|+|b|}\right)^2 - 1 = \frac{(1+|a||b|+|a|+|b|)(1+|a||b|-|a|-|b|)}{(|a|+|b|)^2}$$

$$\leq \frac{(1-|a|^2)(1-|b|^2)}{|a+b|^2} = \frac{(1+\bar{a}b)(1+a\bar{b})-(a+b)(\bar{a}+\bar{b})}{(a+b)(\bar{a}+\bar{b})}$$

$$= \left|\frac{1+\bar{a}b}{a+b}\right|^2 - 1. \quad \square$$

Lemma 5.2   *Let $g(z)$ be meromorphic on $|z| \leq R(0 < R < \infty)$. If $g(0) \neq 0, \infty$, then we have*

$$\log^+\left|\frac{g'(t)}{g(t)}\right| \leq 5\log 2 + \log^+ \rho + 3\log^+ \frac{1}{\rho-r} + \log^+ \mathcal{N}$$

$$+ \log^+ \frac{1}{\delta(t)} + \log^+ T(\rho,g) + \log^+\log^+ \frac{1}{|g(0)|} \tag{5.1.2}$$

for any point $t = re^{i\theta}$ and any $\rho(r < \rho \leq R)$, where $\mathcal{N} = n(\rho,g) + n(\rho,1/g)$, and $\delta(t)$ denotes the lower bound of the distances between t and all the zeros and poles of $g(z)$ on $|z| \leq \rho$.

Proof.   The Poisson-Jensen formula gives

$$\log g(t) = \frac{1}{2\pi}\int_0^{2\pi} \log|g(\rho e^{i\varphi})|\frac{\rho e^{i\varphi}+t}{\rho e^{i\varphi}-t}d\varphi$$

$$- \sum_j \log\frac{\rho^2-\bar{a}_j t}{\rho(t-a_j)} + \sum_k \log\frac{\rho^2-\bar{b}_k t}{\rho(t-b_k)} + iC,$$

where a_j and b_k are the zeros and poles of $g(z)$ on $|z| \leq \rho$ respectively.
    Differentiating both sides, we have

$$\frac{g'(t)}{g(t)} = \frac{1}{2\pi}\int_0^{2\pi} \log|g(\rho e^{i\varphi})|\frac{2\rho e^{i\varphi}}{(\rho e^{i\varphi}-t)^2}d\varphi$$

$$+ \sum_j\left(\frac{\bar{a}_j}{\rho^2-\bar{a}_j t} - \frac{1}{a_j-t}\right) - \sum_k\left(\frac{\bar{b}_k}{\rho^2-\bar{b}_k t} - \frac{1}{b_k-t}\right).$$

Since $|t - a_j| \geq \delta(t)$ and $|t - b_k| \geq \delta(t)$, we obtain

$$\left|\frac{g'(t)}{g(t)}\right| \leq \frac{2\rho}{(\rho - r)^2}\left\{m(\rho, g) + m\left(\rho, \frac{1}{g}\right)\right\} + \mathcal{N}\left(\frac{1}{\rho - r} + \frac{1}{\delta(t)}\right)$$

$$\leq \frac{4\rho}{(\rho - r)^2}\left\{T(\rho, g) + \log^+\frac{1}{|g(0)|}\right\} + \mathcal{N}\left(\frac{1}{\rho - r} + \frac{1}{\delta(t)}\right).$$

Thus (5.1.2) can be deduced by taking the positive logarithm of both sides of the above inequality. □

When $g(0) = \infty$ in Lemma 5.2, we choose $g(z) = C_\tau g_1(z)/z^\tau$, where τ is a positive integer and $g_1(0) = 1$, so that

$$\frac{g'(z)}{g(z)} = \frac{g_1'(z)}{g_1(z)} - \frac{\tau}{z},$$

and hence

$$\log^+\left|\frac{g'(t)}{g(t)}\right| \leq \log^+\left|\frac{g_1'(t)}{g_1(t)}\right| + \log^+\frac{\tau}{r} + \log 2.$$

Applying (5.1.2) to the first term of the right-hand side, we have

$$\log^+\left|\frac{g'(t)}{g(t)}\right| \leq 5\log 2 + \log^+\rho + 3\log^+\frac{1}{\rho - r} + \log^+\mathcal{N}$$

$$+ \log^+\frac{1}{\delta(t)} + \log^+ T(\rho, g_1) + \log^+\frac{\tau}{r} + \log 2.$$

Since

$$T(\rho, g_1) \leq T(\rho, g) + T\left(\rho, \frac{z^\tau}{C_\tau}\right) \leq T(\rho, g) + \tau\log\rho + \log^+\frac{1}{|C_\tau|},$$

we also get

$$\log^+\left|\frac{g'(t)}{g(t)}\right| \leq 6\log 2 + \log 3 + 2\log^+\tau + 2\log^+\rho + \log^+\frac{1}{r} + 3\log^+\frac{1}{\rho - r}$$

$$+ \log^+\mathcal{N} + \log^+\frac{1}{\delta(t)} + \log^+ T(\rho, g) + \log^+\log^+\frac{1}{|C_\tau|}.$$

Lemma 5.3   Let $f(z)$ be meromorphic in $|z| \leq R$ $(0 < R < \infty)$. Then we have

$$\log\left|\frac{f'(z)}{f(z)}\right| \leq \frac{R + r}{R - r}m\left(R, \frac{f'}{f}\right) + \{\bar{n}(R, \infty)$$

$$+ n(R, 0)\}\left(\log\frac{1}{d} + \log 2R\right) - \frac{(R - r)^2}{4R^2}n(r, f' = 0),$$

$$(5.1.3)$$

for any point z in $|z| \leq r(0 < r < R)$ having distances $\geq d$ from all the zeros and poles of $f(z)$.

Proof. A point is a pole of the function $f'(z)/f(z)$ if and only if it is a zero or pole of $f(z)$, and the poles of $f'(z)/f(z)$ are all simple poles. Moreover, it is clear that $f'(z)/f(z) = 0$ if and only if both $f(z) \neq 0$ and $f'(z) = 0$. Denote the zeros and poles of $f(z)$ by a_j and b_k respectively, and the zeros of $f'(z)$ on $|z| \leq R$ by c_l. Furthermore, let $\sum' \log |(R^2 - \bar{a}_j z)/R(z - a_j)|$ and $\sum'' \log |(R^2 - \bar{a}_j z)/R(z - a_j)|$ be the summations with every term involving a_j counted once and $m_j - 1$ times, respectively, where m_j is the multiplicity of a_j as a zero of $f(z)$. Similarly, $\sum' \log |(R^2 - \bar{b}_k z)/R(z - b_k)|$ denotes a similar reduced summation.

The Poisson-Jensen formula applied to $f'(z)/f(z)$ implies that

$$\log \left| \frac{f'(z)}{f(z)} \right| = \frac{1}{2\pi} \int_0^{2\pi} \log \left| \frac{f'(Re^{i\varphi})}{f(Re^{i\varphi})} \right| \frac{R^2 - |z|^2}{R^2 - 2R|z| \cos \varphi + |z|^2} d\varphi$$

$$+ \left(\sum' \log \left| \frac{R^2 - \bar{a}_j z}{R(z - a_j)} \right| + \sum' \log \left| \frac{R^2 - \bar{b}_k z}{R(z - b_k)} \right| \right) \qquad (5.1.4)$$

$$- \left(\sum \log \left| \frac{R^2 - \bar{c}_l z}{R(z - c_l)} \right| - \sum'' \log \left| \frac{R^2 - \bar{a}_j z}{R(z - a_j)} \right| \right).$$

Note that

$$\sum' \log \left| \frac{R^2 - \bar{a}_j z}{R(z - a_j)} \right| + \sum'' \log \left| \frac{R^2 - \bar{a}_j z}{R(z - a_j)} \right| = \sum_j \log \left| \frac{R^2 - \bar{a}_j z}{R(z - a_j)} \right|. \qquad (5.1.5)$$

Moreover, Lemma 5.1 implies that

$$\log \left| \frac{R^2 - \bar{c}_l z}{R(z - c_l)} \right| \geq \log \frac{R^2 + |c_l||z|}{R(|z| + |c_l|)} = \log \left\{ 1 + \frac{(R - |z|)(R - |c_l|)}{R(|z| + |c_l|)} \right\}$$

$$> \log \left\{ 1 + \frac{(R - r)^2}{2R^2} \right\} > \frac{(R - r)^2}{4R^2}$$

for $|z| \leq r$ and $|c_l| \leq r$ and hence

$$\sum_{|c_l| \leq R} \log \left| \frac{R^2 - \bar{c}_l z}{R(z - c_l)} \right| \geq \sum_{|c_l| \leq r} \log \left| \frac{R^2 - \bar{c}_l z}{R(z - c_l)} \right| > \frac{(R - r)^2}{4R^2} n(r, f' = 0).$$

$$(5.1.6)$$

Then, the conclusion of Lemma 5.3 follows from substituting (5.1.5) and (5.1.6) into (5.1.4). $\square$

5.1.2 Distribution of Borel directions

Theorem 5.1   *If λ is a positive number and E is a non-empty closed set of real numbers (mod 2π), then there exists a meromorphic function $f(z)$ of order λ such that all its Borel directions constitute exactly $\{\arg z = \theta : \theta \in E\}$.*

Proof.   For every positive integer j, we divide the plane into j equal angles by the rays $\arg z = 2k\pi/j (k = 1, 2, \cdots, j)$, but we exclude those rays of the form $\arg z = 2k_0\pi/j$ such that both of the closed angles $2(k_0 - 1)\pi/j \leq \arg z \leq 2k_0\pi/j$ and $2k_0\pi/j \leq \arg z \leq 2(k_0 + 1)\pi/j$ contain no point $e^{i\theta}$, where $\theta \in E$. The other rays are denoted by

$$L_{jl} : \arg z = \theta_{jl}, \quad j = 1, 2, \cdots, l = 1, 2, \cdots, L_j, 1 \leq L_j \leq j.$$

Set

$$a_{jl} = 2^j e^{i\theta_{jl}}, \quad j = 1, 2, \cdots, l = 1, 2, \cdots, L_j$$

and

$$m_j = \left[\frac{2^{j\lambda}}{j^3}\right], \quad j = 1, 2, \cdots$$

where $[x]$ denotes the greatest integer not exceeding x. Since

$$\sum_{j=1}^{\infty}\sum_{l=1}^{L_j}\frac{m_j}{|a_{jl}|^{\lambda}} = \sum_{j=1}^{\infty}\frac{L_j\left[\frac{2^{j\lambda}}{j^3}\right]}{2^{j\lambda}} \leq \sum_{j=1}^{\infty}\frac{1}{j^2} < \infty$$

and

$$\sum_{j=1}^{\infty}\sum_{l=1}^{L_j}\frac{m_j}{|a_{jl}|^{\lambda-\varepsilon}} \geq \sum_{j=1}^{\infty}\frac{\left[\frac{2^{j\lambda}}{j^3}\right]}{(2^j)^{\lambda-\varepsilon}} = \infty$$

for any positive number ε, the convergent exponent of the sequence $\{\underbrace{a_{jl}, a_{jl}, \cdots, a_{jl}}_{m_j} : j = 1, 2, \cdots; l = 1, 2, \cdots, L_j\}$ equals λ. Setting

$$q = \begin{cases} [\lambda] & \text{if } \lambda \text{ is not an integer,} \\ \lambda - 1 & \text{otherwise,} \end{cases}$$

we construct the canonical product

$$\Pi_1(z) = \prod_{j=1}^{\infty}\prod_{l=1}^{L_j}\left(1 - \frac{z}{a_{jl}}\right)^{m_j} e^{m_j\{\frac{z}{a_{jl}}+\frac{1}{2}(\frac{z}{a_{jl}})^2+\cdots+\frac{1}{q}(\frac{z}{a_{jl}})^q\}}. \qquad (5.1.7)$$

Moreover, set

$$b_{jl} = \left(2^j + \frac{1}{2^{j\lambda}}\right)e^{i\theta_{jl}}, \quad j = 1, 2, \cdots, l = 1, 2, \cdots, L_j.$$

We can see that

$$\sum_{j=1}^{\infty}\sum_{l=1}^{L_j}\frac{m_j}{|b_{jl}|^{\lambda}} \leq \sum_{j=1}^{\infty}\sum_{l=1}^{L_j}\frac{m_j}{|a_{jl}|^{\lambda}} < \infty$$

and

$$\sum_{j=1}^{\infty}\sum_{l=1}^{L_j}\frac{m_j}{|b_{jl}|^{\lambda-\varepsilon}} \geq \frac{1}{2^{\lambda-\varepsilon}}\sum_{j=1}^{\infty}\sum_{l=1}^{L_j}\frac{m_j}{|a_{jl}|^{\lambda-\varepsilon}} = \infty$$

for any positive number ε, so the convergent exponent of $\{\underbrace{b_{jl}, b_{jl}, \cdots, b_{jl}}_{m_j} :$

$j = 1, 2, \cdots, l = 1, 2, \cdots, L_j\}$ is also equal to λ. Construct the canonical product

$$\Pi_2(z) = \prod_{j=1}^{\infty}\prod_{l=1}^{L_j}\left(1 - \frac{z}{b_{jl}}\right)^{m_j}e^{m_j\{\frac{z}{b_{jl}}+\frac{1}{2}(\frac{z}{b_{jl}})^2+\cdots+\frac{1}{q}(\frac{z}{b_{jl}})^q\}}. \tag{5.1.8}$$

Let us prove that the function

$$f(z) = \frac{\Pi_1(z)}{\Pi_2(z)} \tag{5.1.9}$$

meets the requirement for Theorem 5.1.

First, the order of $f(z)$ is equal to λ since both $\Pi_1(z)$ and $\Pi_2(z)$ have order λ.

Next, $\arg z = \theta_0$, $\theta_0 \in E$, is a Borel direction of $f(z)$. For suppose otherwise, then there is a positive number ε_0 such that $f(z)$ has a finite non-zero exceptional value a_0 in $G : |\arg z - \theta_0| < \varepsilon_0$, i.e.,

$$n(r, \theta_0, \varepsilon_0, f = a_0) < r^{\tau}$$

for some $\tau(0 < \tau < \lambda)$, whenever r is sufficiently large.

Since $\theta_0 \in E$, there exists, for every sufficiently large integer j, a point a_{jl} such that the disk $K : |z - a_{jl}| < 4$ is contained in G. In the disk K, $f(z)$ has only one zero a_{jl} of order m_j and one pole b_{jl} of order m_j; in addition, in K, the number of zeros of $f(z) - a_0$ does not exceed r_j^{τ} with $r_j = 2^j + 4$.

Note that[1]

$$T(2r_j, f - a_0) < (2r_j)^{\lambda+1}, \log^+ \log^+ \frac{1}{|f(0) - a_0|} < (2r_j)^{\lambda+1},$$

$$n(2r_j, f = a_0) < (2r_j)^{\lambda+1}, n(2r_j, f = \infty) < (2r_j)^{\lambda+1}$$

for sufficiently large j. Let (γ) be the union of all the disks having their centers at each a_0-point and pole of $f(z)$ in $|z| < 2r_j$ and each having radius

$$d = \frac{1}{8(2r_j)^{\lambda+1}}.$$

According to Lemma 5.2,

$$\log \left| \frac{f'(z)}{f(z) - a_0} \right| \leq 5 \log 2 + \log^+ (2r_j) + 3 \log^+ \frac{1}{r_j} + \log^+ n(2r_j, f = a_0)$$

$$+ \log^+ n(2r_j, f = \infty) + \log 2 + \log(8(2r_j)^{\lambda+1})$$

$$+ \log^+ T(2r_j, f - a_0) + \log^+ \log^+ \frac{1}{|f(0) - a_0|}$$

$$\leq (5\lambda + 6) \log(2r_j) + 9 \log 2$$

$$(5.1.10)$$

for any point z in the region $(|z| \leq r_j) \setminus (\gamma)$.

By our choice of d, there are two circumferences $|z - a_{jl}| = r_1(1 < r_1 < 2)$ and $|z - a_{jl}| = r_2(3 < r_2 < 4)$, not intersecting (γ), so that Lemma 5.3 gives us

$$\log \left| \frac{f'(z)}{f(z) - a_0} \right| \leq \frac{r_2 + r_1}{r_2 - r_1} m \left(r_2, a_{jl}, \frac{f'}{f - a_0} \right)$$

$$+ \left\{ \bar{n}(K, f = \infty) + n(K, f = a_0) \right\} \left(\log \frac{1}{d} + \log 2r_2 \right)$$

$$- \frac{(r_2 - r_1)^2}{4r_2^2} n \left(|z - a_{jl}| \leq r_1, \frac{1}{f'} \right),$$

whenever $|z - a_{jl}| = r_1$. Since $|z - a_{jl}| = r_2$ is contained in $|z| \leq r_j$ and

---

1) When $f(0) = a_0$, $f(0) - a_0$ must be replaced by the first non-zero coefficient of the Taylor expansion of $f(z) - a_0$ in a neighborhood of the origin.

has no common point with (γ), we deduce from (5.1.10) that

$$\log\left|\frac{f'(z)}{f(z) - a_0}\right| \le 6\Big\{(5\lambda + 6)\log(2r_j) + 9\log 2\Big\}$$

$$+\Big(1 + r_j^\tau\Big)\Big\{\log 8(2r_j)^{\lambda+1} + \log 8\Big\}$$

$$-\frac{1}{4} \cdot \frac{1}{16}\Big(\frac{2^{j\lambda}}{j^3} - 2\Big).$$

Noting that $\tau < \lambda$ and $r_j = 2^j + 4$ yields

$$\log\left|\frac{f'(z)}{f(z) - a_0}\right| < 0 \tag{5.1.11}$$

for any point z on $|z - a_{jl}| = r_1$, provided j is sufficiently large.
  On the other hand, according to the formula

$$n(|z - a_{jl}| \le r_1, f = \infty) - n(|z - a_{jl}| \le r_1, f = a_0)$$

$$= -\frac{1}{2\pi i}\int_{|z-a_{jl}|=r_1}\frac{f'}{f - a_0}dz,$$

we have

$$\Big[\frac{2^{j\lambda}}{j^3}\Big] - r_j^\tau \le 2\max_{|z-a_{jl}|=r_1}\left|\frac{f'(z)}{f(z) - a_0}\right|. \tag{5.1.12}$$

Comparing (5.1.11) and (5.1.12), we obtain

$$\Big[\frac{2^{j\lambda}}{j^3}\Big] \le r_j^\tau + 2.$$

Since $\tau < \lambda$, this inequality does not hold as j tends to infinity. This contradiction shows that the ray $\arg z = \theta_0$ is a Borel direction of $f(z)$.
  Finally, let us prove that $\arg z = \theta_1$, $\theta_1 \bar{\in} E$, is not a Borel direction of $f(z)$. For this purpose, we need to estimate $|f'(z)/f(z)|$. From

$$\frac{f'(z)}{f(z)} = \sum_{j=1}^{\infty}\sum_{l=1}^{L_j}m_j\Big\{\frac{a_{jl} - b_{jl}}{(a_{jl} - z)(b_{jl} - z)} + \frac{b_{jl} - a_{jl}}{a_{jl}b_{jl}}$$

$$+z\Big(\frac{1}{a_{jl}} + \frac{1}{b_{jl}}\Big)\Big(\frac{b_{jl} - a_{jl}}{a_{jl}b_{jl}}\Big) + \cdots$$

$$+z^{q-1}\Big(\frac{1}{a_{jl}^{q-1}} + \cdots + \frac{1}{b_{jl}^{q-1}}\Big)\Big(\frac{b_{jl} - a_{jl}}{a_{jl}b_{jl}}\Big)\Big\},$$

we have

$$\left|\frac{f'(z)}{f(z)}\right| \leq \sum_{j=1}^{\infty}\sum_{l=1}^{L_j}\left\{\frac{1}{d^2|z|^2} + \frac{1}{2^{2j}} + \frac{2|z|}{2^{3j}} + \cdots + \frac{q|z|^{q-1}}{2^{(q+1)j}}\right\}\frac{m_j}{2^{j\lambda}}$$

$$\leq \left\{\frac{1}{d^2|z|^2} + 1 + 2|z| + \cdots + q|z|^{q-1}\right\}\sum_{j=1}^{\infty}\frac{1}{j^2},$$

whenever $|z|$ is sufficiently large and $\max_{j,l}\{|z-a_{jl}|, |z-b_{jl}|\} \geq d|z|$. Hence

$$\left|\frac{f'(z)}{f(z)}\right| \leq \begin{cases} \dfrac{2}{d^2|z|^2} & \text{when } q = 0, \\[4mm] 2q|z|^{q-1} & \text{when } q \geq 1. \end{cases} \tag{5.1.13}$$

Because $\theta_1 \bar{\in} E$ and E is a closed set, there is an angle $G_1 : |\arg z - \theta_1| < \delta$ such that the $L_{j1}(j \geq j_0, 1 \leq l \leq L_j)$ are not contained in G_1 whenever j is sufficiently large. Thus $f(z)$ has neither zero nor pole in the region $(|z| > r) \cap G_1$ for every large number r. Suppose our assertion is false, i.e., suppose $\arg z = \theta_1$ is a Borel direction of $f(z)$. Then, according to Theorem 3.11, there exists a sequence of filling disks

$$\Gamma_k : |z - z_k| < \varepsilon_k|z_k|, z_k = |z_k|e^{i\theta_0},$$

$$\lim_{k\to\infty}|z_k| = \infty, \quad \lim_{k\to\infty}\varepsilon_k = 0, \quad k = 1, 2, \cdots,$$

such that $f(z)$ takes every complex value $|z_k|^{\lambda-\delta_k}$ times in Γ_k, except for those values in the union of two spherical circles each with radius 2^{-k}, where $\lim_{k\to\infty}\delta_k = 0$. Let Γ'_k be the disk $|z - z_k| < 2\varepsilon_k|z_k|$, $k = 1, 2, \cdots$.

Since Γ'_k is contained in $|\arg z - \theta_1| < \delta/2$ for k sufficiently large, we have from (5.1.13),

$$\left|\frac{f'(z)}{f(z)}\right| \leq \begin{cases} \dfrac{2}{\left(\sin\dfrac{\delta}{2}\right)^2|z|^2} \leq \dfrac{32}{\delta^2|z_k|^2} & \text{when } q = 0, \\[6mm] 2q|z|^{q-1} \leq 3q|z_k|^{q-1} & \text{when } q \geq 1, \end{cases}$$

whenever $z \in \Gamma'_k$.

For every sufficiently large integer k, there is a point z'_k in Γ'_k such that $|f(z'_k)| \leq 1$. Otherwise, $|f(z)| > 1$ in Γ'_k, and the unit disk would be covered by two spherical circles each with radius 2^{-k}, since Γ_k is a sequence of filling disks of $f(z)$. Furthermore, for every small positive number

η, there is a point z_k'' in Γ_k' with the property that $|f(z_k'')| \geq e^{|z_k|^{\lambda-\eta}}$. Otherwise, we would have $|f(z)| < e^{|z_k|^{\lambda-\eta}}$ in Γ_k', whence

$$T(2\varepsilon_k|z_k|, z_k, f) \leq \log^+ M(2\varepsilon_k|z_k|, z_k, f) < |z_k|^{\lambda-\eta}.$$

Consequently,

$$n(\Gamma_k, f = a)$$

$$\leq \frac{1}{\log 2} N\left(2\varepsilon_k|z_k|, z_k, \frac{1}{f-a}\right)$$

$$\leq \frac{1}{\log 2}\left\{T(2\varepsilon_k|z_k|, z_k, f) + \log^+|a| + \log 2 + \log\frac{1}{|f(z_k) - a|}\right\}$$

$$\leq \frac{1}{\log 2}\left\{|z_k|^{\lambda-\eta} + \log\frac{1}{|f(z_k), a|} + \log 2\right\}$$

for any complex value a. This contradicts the definition of Γ_k, provided that k is sufficiently large.

Therefore

$$|z_k|^{\lambda-\eta} \leq \left|\log|f(z_k'')| - \log|f(z_k')|\right|$$

$$\leq \left|\int_{z_k'z_k''}\frac{f'}{f}dz\right| \leq 2\varepsilon_k|z_k| \cdot \max_{z \in z_k'z_k''}\left|\frac{f'(z)}{f(z)}\right|$$

$$\leq \begin{cases} \dfrac{64\varepsilon_k}{\delta^2|z_k|} & \text{when } q = 0, \\[3mm] 6q\varepsilon_k|z_k|^q & \text{when } q \geq 1. \end{cases}$$

Since $q < \lambda$ and η can be arbitrarily small, this inequality is impossible for sufficiently large k. This contradiction shows that $\arg z = \theta_1$ cannot be a Borel direction of $f(z)$. $\quad\square$

For an entire function of a finite and positive order, D. Drasin and A. Weitsman [2] obtained a complete result on the distribution of its Borel directions. Let us introduce the following definition.

Definition 5.1 *A finite sequence*

$$\theta_1 < \theta_2 < \cdots < \theta_n \leq \theta_1 + 2\pi$$

of real numbers is called a chain of order λ for $\lambda > 1/2$ if

(i) $\theta_{j+1} - \theta_j \le \dfrac{\pi}{\lambda}, \quad j = 1, 2, \cdots, n-1,$

(ii) $\theta_n - \theta_1 \ge \dfrac{\pi}{\lambda},$ *and*

(iii) *strict inequality holds in* (i) *and* (ii) *except possibly for* $n = 2$.

Theorem 5.1′ (Drasin and Weitsman [2])   *Let $0 < \lambda < \infty$ be given and let E be a non-empty closed set in $[0, 2\pi]$. If $\lambda \le 1/2$, then E is without any further restrictions, the set of all the Borel directions of an entire function of order λ. For $\lambda > 1/2$, a necessary and sufficient condition for E to be the set of all Borel directions of an entire function of order λ is that each element in E be a member of a chain of order $\lambda : \theta_1, \theta_2, \cdots, \theta_n$ with $\theta_j \pmod{2\pi} \in E(j = 1, 2, \cdots, n)$.*

5.2 Common Borel Directions of a Meromorphic Function and Its Derivatives

5.2.1 On the Milloux theorem.   For every meromorphic function $f(z)$ of positive and finite order in the plane, Theorem 3.8 states that $f(z)$ has at least one Borel direction, and that $f'(z)$ also has a Borel direction since the order of $f'(z)$, by Theorem 4.2, is also equal to λ. In 1928, G. Valiron posed an interesting but difficult problem: Do a meromorphic function and its derivatives necessarily have a common Borel direction? After the partial results of A. Rauch [2] and Chuang Chi-tai [1], H. Milloux [3] established the following striking theorem.

Theorem 5.2   *If $f(z)$ is an entire function of order λ $(0 < \lambda < \infty)$, then every Borel direction of its derivative $f'(z)$ is also a Borel direction of $f(z)$.*

The original proof by Milloux is very long and complicated. Later on, Zhang Guang-hou gave a simple proof and extended the Milloux theorem to the case of meromorphic functions having a Borel exceptional value ∞. However, Zhang's handling of initial values remains complicated.

Now let us establish a general theorem, from which Milloux's theorem and Zhang's theorem follow immediately. Its proof is direct and simple (See Yang Lo [2]).

5.2.2 Fundamental theorem.

Theorem 5.3 *Suppose that $f(z)$ is meromorphic and of order $\lambda(0 < \lambda < \infty)$ in the finite plane, and takes infinity as a Borel exceptional value in $|\arg z| < \gamma_0$. Let*

$$\left.\begin{array}{c} \Gamma_k : |z - R_k| < \varepsilon_k R_k, \quad R_{k+1} > 2R_k, \\[2mm] \lim_{k \to \infty} \varepsilon_k = 0, \quad k = 1, 2, \cdots \end{array}\right\} \tag{5.2.1}$$

be a sequence of filling disks of order λ of $f'(z)$ such that $f'(z)$ takes every complex value at least $R_k^{\lambda - \varepsilon_k'}$ times in Γ_k, except for those values in the union of two spherical disks each with radius δ_k on the Riemann sphere, where $\lim_{k \to \infty} \varepsilon_k' = \lim_{k \to \infty} \delta_k = 0$.

Let

$$\beta_k = \left(\sup_{r \geq R_k^{1/2}} \frac{\log T(r, f)}{\log r} \right) - \lambda. \tag{5.2.2}$$

If

$$\varepsilon_k \geq \max \left(\frac{2\varepsilon_k'}{\lambda}, \frac{2\beta_k}{\lambda}, \frac{1}{(\log R_k)^{\frac{1}{2}}} \right), \tag{5.2.3}$$

then the regions

$$G_k : \left(\frac{R_k^{1-\eta_k}}{2} < |z| < 2R_k^{1+\eta_k} \right) \bigcap (|\arg z| < 20\pi\eta_k), \tag{5.2.4}$$

$$\eta_k = 4\pi\varepsilon_k^{\frac{1}{2}}, \quad k = 1, 2, \cdots \tag{5.2.5}$$

must contain a subsequence (G_{k_j}) as filling regions of order λ for $f(z)$, i.e. $f(z)$ takes every complex number at least $R_{k_j}^{\lambda - \varepsilon_{k_j}''}$ times in G_{k_j}, except for those numbers in the union of two spherical disks each with the radius δ_{k_j}' on the Riemann sphere, where $\lim_{j \to \infty} \varepsilon_{k_j}'' = \lim_{j \to \infty} \delta_{kj}' = 0$.

Proof.   Let us suppose otherwise, i.e., no subsequence of G_k is a sequence of filling regions for $f(z)$. We shall prove that this is absurd.

Since most of the inequalities in the following discussion are valid only for sufficiently large values of the index k, we shall assume throughout that k is sufficiently large.

Since (Γ_k) is a sequence of filling disks for $f'(z)$, there exists a sequence of complex numbers $\{a_k\}$ such that

$$0 < |a_k| < 1, \quad n(\Gamma_k, f' = a_k) > R_k^{\lambda - \varepsilon'_k}. \tag{5.2.6}$$

In the interval $[R_k^{1-\eta_k}, R_k^{1+\eta_k}]$, we take the points

$$r'_{k,p} = R_k^{1-\eta_k}(1 + \eta_k)^p, \quad p = 0, 1, 2, \cdots, P,$$

where $P = [2\eta_k \log R_k / \log(1+\eta_k)]+1$, and $[2\eta_k \log R_k / \log(1+\eta_k)]$ denotes the integral part of $2\eta_k \log R_k / \log(1 + \eta_k)$.

Setting

$$S_{k,p} : |z - r'_{k,p}| < 2\eta_k r'_{k,p}, \quad S'_{k,p} : |z - r'_{k,p}| < 40\eta_k r'_{k,p}$$

and

$$G'_k : (R_k^{1-\eta_k} < |z| < R_k^{1+\eta_k})\bigcap(|\arg z| < \eta_k), \tag{5.2.7}$$

it is easy to see that

$$G'_k \subset \left(\bigcup_{p=0}^{P} S_{k,p} \right) \subset \left(\bigcup_{p=0}^{P} S'_{k,p} \right) \subset G_k. \tag{5.2.8}$$

Since (G_k) does not contain any subsequence as filling regions of order λ for $f(z)$, there is a subsequence (G_{k_j}) having the following properties:

For every positive integer j, there are three distinct complex numbers $\alpha_{l,k_j}(l = 1,2,3)$ such that $|\alpha_{l_1,k_j}, \alpha_{l_2,k_j}| > \delta(1 \leq l_1 \neq l_2 \leq 3)$ and $\sum_{l=1}^{3} n(G_{k_j}, f = \alpha_{l,k_j}) < R_{k_j}^{\tau_1}$, where δ and $\tau_1(\tau_1 < \lambda)$ are two positive numbers independent of j.

In fact, we take two sequences of positive numbers ε''_j and δ'_j which tend to zero. If the preceding assertion is not true, then a subsequence $(G_{k,1})$ of (G_k) can be found such that all the complex numbers satisfying the inequality $n(G_{k,1}, f = \alpha) < R_{k,1}^{\lambda-\varepsilon''_1}$ belong to the union of two spherical circles with radii δ'_1 on the Riemann sphere. Similarly, there is a subsequence $(G_{k,2})$ of $(G_{k,1})$ such that all the complex numbers satisfying the inequality $n(G_{k,2}, f = \alpha) < R_{k,2}^{\lambda-\varepsilon''_2}$ belong to the union of two spherical circles with radii δ'_2. By continuing this procedure and taking the diagonal sequence $(G_{k,k})$, the complex numbers satisfying the inequality $n(G_{k,k}, f = \alpha) < R_{k,k}^{\lambda-\varepsilon''_k}$ belong to the union of two spherical circles with radii δ'_k. This means the subsequence $(G_{k,k})$ is a sequence of filling regions of order λ for $f(z)$, and we derive a contradiction.

In the following we shall use (G_k) instead of (G_{kj}) for the sake of brevity. It is obvious that we can take $\alpha_{3,k} = \infty (k = 1, 2, \cdots)$. Hence, for every k, there are three distinct complex numbers $\alpha_{l,k} (l = 1, 2, 3)$ such that

$$\alpha_{3,k} = \infty, \quad \max\left\{|\alpha_{1,k}|, |\alpha_{2,k}|, \frac{1}{|\alpha_{1,k} - \alpha_{2,k}|}\right\} \le \frac{2}{\delta},$$

and

$$\sum_{l=1}^{3} n(G_k, f = \alpha_{l,k}) < R_k^{\tau_1},$$

where δ and $\tau_1 (\tau_1 < \lambda)$ are two positive numbers independent of k.

By putting

$$h_k(z) = f(z) - a_k z$$

and

$$G_{k,p}(t) = h_k(r'_{k,p} + 40\eta_k r'_{k,p} t),$$

we see that $G_{k,p}(t)$ is meromorphic in $|t| < 1$, and

$$\sum_{l=1}^{3} n(|t| < 1, G_{k,p}(t) = P_{l,k,p}(t)) < R_k^{\tau_1},$$

where the functions $P_{l,k,p}(t) = \alpha_{l,k} - a_k r'_{k,p} - 40 a_k \eta_k r'_{k,p} t$ $(l = 1, 2, 3)$ have no zeros or poles in $|t| < 1$ with

$$\iint_{|t|<1} \log^+\left(\sum_{l=1}^{2} |P_{l,k,p}(t)| + \sum_{1 \le l_1 \neq l_2 \le 3} \frac{1}{|P_{l_1,k,p}(t) - P_{l_2,k,p}(t)|}\right) d\sigma_t$$

$$= O(\log R_k). \tag{5.2.9}$$

According to Theorem 3.5, the inequality

$$n\left(|t| < \frac{1}{20}, G_{k,p} = \alpha\right) < AR_k^{\tau_1},$$

i.e.,

$$n(S_{k,p}, h_k = \alpha) < AR_k^{\tau_1},$$

holds for all the complex numbers α, except for those values of α in a spherical circle with radius $e^{-R_k^{\tau_1}}$.

Since

$$P = \left[\frac{2\eta_k \log R_k}{\log(1 + \eta_k)}\right] + 1 \le 4 \log R_k + 1$$

and (5.2.8) holds, there is a finite complex number b_k, which lies outside the P exceptional circles with spherical radii $e^{-R_k^{\tau_1}}$, such that

$$|b_k| < 1, \quad |f(0) - b_k| > \frac{1}{2}^{1)}$$

---

1) When $f(0) = \infty$, we choose b_k such that $|b_k| < 1$ and $n(G'_k, h_k = b_k) < R_k^{\tau}$, $(\tau < \lambda)$.

and

$$n(G'_k, h_k = b_k) < R^\tau_k, \quad \tau < \lambda. \tag{5.2.10}$$

Let

$$\xi_k = \frac{2\eta_k}{\pi} \tag{5.2.11}$$

and suppose the function

$$\zeta = \zeta_k(z) = \frac{z^{\frac{1}{\xi_k}} - R_k^{\frac{1}{\xi_k}}}{z^{\frac{1}{\xi_k}} + R_k^{\frac{1}{\xi_k}}}, \tag{5.2.12}$$

maps $|\arg z| < \eta_k$ onto $|\zeta| < 1$. Then the inverse of (5.2.12) is

$$z = z_k(\zeta) = R_k \left(\frac{1+\zeta}{1-\zeta} \right)^{\xi_k}. \tag{5.2.13}$$

We denote $h_k(z_k(\zeta))$ by $H_k(\zeta)$.

If $|\zeta| \leq 1 - 2/R_k^2$, then its original image z satisfies

$$R_k^{1-\eta_k} \leq |z| \leq R_k^{1+\eta_k} \tag{5.2.14}$$

according to (5.2.13) and (5.2.11). Since $f(z)$ adopts ∞ as a Borel exceptional value in $|\arg z| < \gamma_0$, we have by (5.2.7), (5.2.10) and (5.2.14),

$$n\left(|\zeta| \leq 1 - \frac{2}{R_k^2}, H_k = \infty\right) + n\left(|\zeta| \leq 1 - \frac{2}{R_k^2}, H_k = b_k\right) \tag{5.2.15}$$

$$\leq n(G'_k, h_k = \infty) + n(G'_k, h_k = b_k) < R^{\tau'}_k, \quad \tau' < \lambda.$$

Furthermore, if ζ is the image of an arbitrary point $z = re^{i\theta} \in \Gamma_k$, then

$$|\zeta| = \left\{ 1 - \frac{4r^{\frac{1}{\xi_k}} R_k^{\frac{1}{\xi_k}} \cos \frac{\theta}{\xi_k}}{r^{\frac{2}{\xi_k}} + 2r^{\frac{1}{\xi_k}} R_k^{\frac{1}{\xi_k}} \cos \frac{\theta}{\xi_k} + R_k^{\frac{2}{\xi_k}}} \right\}^{\frac{1}{2}} \tag{5.2.16}$$

$$\leq \left\{ 1 - \frac{4(1 - \varepsilon_k)^{\frac{1}{\xi_k}} \cos \frac{\theta}{\xi_k}}{((1 + \varepsilon_k)^{\frac{1}{\xi_k}} + 1)^2} \right\}^{\frac{1}{2}}.$$

Noting that (5.2.11) and (5.2.5) give

$$\frac{\varepsilon_k}{\xi_k} = \frac{\varepsilon_k^{\frac{1}{2}}}{8} \to 0,$$

we have

$$\frac{\theta}{\xi_k} \le \frac{\frac{\pi}{2}\varepsilon_k}{\xi_k} \to 0.$$

Since $(1 - \varepsilon_k)^{\frac{1}{\varepsilon_k}} \to e^{-1}$,

$$(1 - \varepsilon_k)^{\frac{1}{\xi_k}} = \left\{(1 - \varepsilon_k)^{\frac{1}{\varepsilon_k}}\right\}^{\frac{\varepsilon_k}{\xi_k}} \to 1$$

and

$$1 \le (1 + \varepsilon_k)^{\frac{1}{\xi_k}} \le \frac{1}{(1 - \varepsilon_k)^{\frac{1}{\xi_k}}} \to 1.$$

Therefore (5.2.16) means that the image of Γ_k under the mapping $\zeta = \zeta_k(z)$ is contained in $|\zeta| < 1/2$.

Setting

$$\nu_k = 8\varepsilon_k, \tag{5.2.17}$$

from (5.2.17), (5.2.11), (5.2.5) and (5.2.3) we deduce that

$$R_k^{\frac{\nu_k}{\xi_k}} \ge R_k^{\frac{e^{\frac{1}{2}}}{k}} \ge e^{(\log R_k)^{\frac{3}{4}}} \to \infty. \tag{5.2.18}$$

Consequently, the image of Γ_k is contained in $|\zeta| < 1 - 6/R_k^{\nu_k/\xi_k}$ and we have by (5.2.6),

$$n\left(|\zeta| \le 1 - \frac{6}{R_k^{\frac{\xi_k}{k}}}, H_k' = 0\right) \ge n(\Gamma_k, f' = a_k) > R_k^{\lambda - \varepsilon_k'}. \tag{5.2.19}$$

In $|\zeta| \le 1 - 2/R_k^{\frac{\pi}{2}}$, we construct disks with centers at a pole or a b_k-point of $H_k(\zeta)$ and their radii $d_k = 1/R_k^{\lambda+3}$. The union of these disks is denoted by $(\gamma)_{\zeta,k}$. Then we select $r_{1,k}$ and $r_{2,k}$ such that

$$r_{1,k} = 1 - \frac{6}{R_k^{\frac{\nu_k}{\xi_k}}}, \tag{5.2.20}$$

$$1 - \frac{4}{R_k^{\frac{\pi}{2}}} < r_{2,k} < 1 - \frac{3}{R_k^{\frac{\pi}{2}}}, \tag{5.2.21}$$

$$(|\zeta| = r_{2,k}) \bigcap (\gamma)_{\zeta,k} = \emptyset.$$

According to Lemma 5.3, for any point ζ in the region $(|\zeta| \leq r_{1,k}) \setminus (\gamma)_{\zeta,k}$, we have

$$\log\left|\frac{H_k'(\zeta)}{H_k(\zeta) - b_k}\right| \leq \frac{r_{2,k} + r_{1,k}}{r_{2,k} - r_{1,k}} m\left(r_{2,k}, \frac{H_k'}{H_k - b_k}\right) + \left\{\bar{n}(r_{2,k}, H_k = \infty)\right.$$

$$+ n(r_{2,k}, H_k = b_k)\} \left(\log 2 + \log\frac{1}{d_k}\right)$$

$$-\frac{(r_{2,k} - r_{1,k})^2}{4r_{2,k}^2} n(r_{1,k}, H_k' = 0).$$

$$(5.2.22)$$

Now let us estimate $m(r_{2,k}, H_k'/(H_k - b_k))$ starting from

$$m\left(r_{2,k}, \frac{H_k'}{H_k - b_k}\right) = \frac{1}{2\pi} \int_0^{2\pi} \log^+\left\{\frac{|h_k'(z_k(r_{2,k}e^{i\varphi}))|}{|h_k(z_k(r_{2,k}e^{i\varphi})) - b_k|}|z_k'(r_{2,k}e^{i\varphi})|d\varphi\right.$$

$$\leq \frac{1}{2\pi} \int_0^{2\pi} \log^+\left|\frac{h_k'(z_k(r_{2,k}e^{i\varphi}))}{h_k(z_k(r_{2,k}e^{i\varphi})) - b_k}\right| d\varphi$$

$$+ \frac{1}{2\pi} \int_0^{2\pi} \log^+ |z_k'(r_{2,k}e^{i\varphi})|d\varphi.$$

$$(5.2.23)$$

It is clear that (5.2.13) implies that

$$\frac{\xi_k R_k(1 - r_{2,k})^{\xi_k - 1}}{2^{\xi_k}} \leq |z_k'(r_{2,k}e^{i\varphi})| \leq \frac{\xi_k R_k 2^{\xi_k}}{(1 - r_{2,k})^{\xi_k + 1}}, \qquad (5.2.24)$$

so that

$$\frac{1}{2\pi} \int_0^{2\pi} \log^+ |z_k'(r_{2,k}e^{i\varphi})|d\varphi \leq \log^+ \frac{\xi_k R_k 2^{\xi_k}}{(1 - r_{2,k})^{\xi_k + 1}} \leq 3\log R_k. \quad (5.2.25)$$

Next, let us estimate the integral

$$\frac{1}{2\pi} \int_0^{2\pi} \log^+ |h_k'(z_k(r_{2,k}e^{i\varphi}))/(h_k(z_k(r_{2,k}e^{i\varphi})) - b_k)|d\varphi$$

by using Lemma 5.2. Set

$$g(z) = h_k(z) - b_k, \quad t = z_k(r_{2,k}e^{i\varphi}), \quad R = \frac{2^{\xi_k + 1} R_k}{(1 - r_{2,k})^{\xi_k}}. \qquad (5.2.26)$$

Since

$$\frac{R_k(1 - r_{2,k})^{\xi_k}}{2^{\xi_k}} \leq |z_k(r_{2,k}e^{i\varphi})| \leq \frac{2^{\xi_k} R_k}{(1 - r_{2,k})^{\xi_k}},$$

(5.2.21) and (5.2.10) hold, we have

$$R_k^{1-\eta_k} < |t| = r < R < 2R_k^{1+\eta_k},$$

$$R - r \geq \frac{2^{\xi_k} R_k}{(1 - r_{2,k})^{\xi_k}} \geq \frac{R_k^{1+\eta_k}}{2}, \tag{5.2.27}$$

$$\mathcal{N} = n\left(\frac{2^{\xi_k+1} R_k}{(1 - r_{2,k})^{\xi_k}}, h_k = \infty\right) + n\left(\frac{2^{\xi_k+1} R_k}{(1 - r_{2,k})^{\xi_k}}, h_k = b_k\right)$$

$$< R_k^{\lambda+1}, \tag{5.2.28}$$

$$T(R, g) = T\left(\frac{2^{\xi_k+1} R_k}{(1 - r_{2,k})^{\xi_k}}, f - a_k z - b_k\right) < R_k^{\lambda+1} \tag{5.2.29}$$

and

$$|g(0)| = |f(0) - b_k| > \frac{1}{2}. \tag{5.2.30}$$

(When $f(0) = \infty$, we note that $\lim_{z\to 0} f(z)z^\gamma = \lim_{z\to 0} g(z)z^\gamma = C_\gamma$ is a finite non-zero number.)

Then it only remains to estimate the quantity $\delta(t)$ in Lemma 5.2. If $\zeta = re^{i\varphi}(1/2 < r < 1)$ is a point in the ζ plane, then for its original image z we have

$$\arg z = \xi_k \arg\frac{1+\zeta}{1-\zeta} = \xi_k \arcsin \frac{2r\sin\varphi}{\{(1 - r^2)^2 + 4r^2\sin^2\varphi\}^{\frac{1}{2}}}$$

$$\leq \xi_k \cdot \frac{\pi}{2} \cdot \frac{2r}{1+r^2} = \eta_k \frac{1}{1 + \frac{(1-r)^2}{2r}} \leq \eta_k\left\{1 - \frac{(1-r)^2}{4r}\right\}.$$

In particular, for a point ζ on $|\zeta| = r_{2,k}$, its original image z must satisfy

$$\arg z \leq \eta_k\left\{1 - \frac{\left(\frac{3}{R_k^{\pi/2}}\right)^2}{4\left(1 - \frac{3}{R_k^{\pi/2}}\right)}\right\} \leq \eta_k\left(1 - \frac{2}{R_k^\pi}\right). \tag{5.2.31}$$

If x_l is a pole or b_k-point of $h_k(z)$ in the region $\{(|z| \leq R) \setminus (|\arg z| < \eta_k)\}$, then

$$|t - x_l| \geq R_k^{1-\eta_k} \sin\frac{2\eta_k}{R_k^\pi} \geq \frac{1}{R_k^\pi} \tag{5.2.32}$$

by (5.2.27) and (5.2.31).

For an arbitrary point ζ in $|\zeta| \leq 1$, by analogy with the inequality (5.2.24), we obtain from (5.2.11), (5.2.5) and (5.2.3):

$$|z_k'(\zeta)| \geq \frac{\xi_k R_k}{2^{\xi_k}(1-|\zeta|)^{\xi_k-1}} \geq \frac{\xi_k R_k}{2} = 4\varepsilon_k^{\frac{1}{2}} R_k \geq \frac{4R_k}{(\log R_k)^{\frac{1}{4}}} \geq 1. \quad (5.2.33)$$

Suppose x_l' is a pole or b_k-point of $h_k(z)$ in $|\arg z| < \eta_k$ and its image $\zeta_k(x_l')$ is in $|\zeta| \leq 1 - 2/R_k^{\frac{\pi}{2}}$. Since $r_{2,k}e^{i\varphi}$ is the image of t by (5.2.27), we have

$$|r_{2,k}e^{i\varphi} - \zeta_k(x_l')| = \left| \int_{tx_l'} \zeta_k'(z)dz \right| \leq \left(\max_{z \in tx_l'} |\zeta_k'(z)| \right) |t - x_l'|$$

$$\leq \left(\max_{|\zeta| \leq 1} \frac{1}{|z_k'(\zeta)|} \right) |t - x_l'|$$

$$= \left(\frac{1}{\min_{|\zeta| \leq 1} |z_k'(\zeta)|} \right) |t - x_l'| \leq |t - x_l'|,$$

and hence

$$|t - x_l'| \geq d_k = \frac{1}{R_k^{\lambda+3}}, \quad (5.2.34)$$

since $(|\zeta| = r_{2,k}) \cap (\gamma)_{\zeta,k} = \emptyset$ by (5.2.21).

Similarly, if x_l'' is a pole or b_k-point of $h_k(z)$ in $|\arg z| < \eta_k$ and its image $\xi_k(x_l'')$ is outside of $|\zeta| \leq 1 - 2/R_k^{\frac{\pi}{2}}$, we also have

$$|r_{2,k}e^{i\varphi} - \zeta_k(x_l'')| \leq \left(\max_{z \in tx_l''} |\zeta_k'(z)| \right) |t - x_l''|$$

$$\leq \left(\frac{1}{\min_{|\zeta| \leq 1} |z_k'(\zeta)|} \right) |t - x_l''| \leq |t - x_l''|,$$

so

$$|t - x_l''| \geq \frac{1}{R_k^{\frac{\pi}{2}}}. \quad (5.2.35)$$

Then combining the inequalities (5.2.32), (5.2.34) and (5.2.35) gives us

$$\log \frac{1}{\delta(t)} = \max \left\{ \log \frac{1}{|t - x_l|}, \log \frac{1}{|t - x_l'|}, \log \frac{1}{|t - x_l''|} \right\}$$

$$\quad (5.2.36)$$

$$= O(\log R_k).$$

Therefore, substituting the estimates (5.2.27), (5.2.28), (5.2.29), (5.2.30) and (5.2.36) into (5.1.2) (or (5.1.2)', if $f(0) = \infty$) we obtain by Lemma 5.2,

$$\log^+ \left| \frac{h'_k(z_k(r_{2,k}e^{i\varphi}))}{h_k(z_k(r_{2,k}e^{i\varphi})) - b_k} \right| = O(\log R_k). \tag{5.2.37}$$

Hence

$$m\left(r_{2,k}, \frac{H'_k}{H_k - b_k}\right) = O(\log R_k), \tag{5.2.38}$$

by (5.2.23), (5.2.25) and (5.2.37).

Finally, we transform (5.2.22) by using (5.2.15), (5.2.20), (5.2.21), (5.2.38) and

$$\frac{(r_{2,k} - r_{1,k})^2}{4r_{2,k}^2} n(r_{1,k}, H'_k = 0) \geq \left(\frac{1}{R_k^{\nu_k/\xi_k}}\right)^2 n(r_{1,k}, H'_k = 0)$$

$$> R_k^{\lambda - \epsilon'_k - 2\nu_k/\xi_k} \tag{5.2.39}$$

to obtain

$$\log \left| \frac{H'_k(\zeta)}{H_k(\xi) - b_k} \right| < -\frac{1}{2} R_k^{\lambda - \epsilon'_k - \frac{2\nu_k}{\xi_k}} \tag{5.2.40}$$

for $\zeta \in \{(|\zeta| \leq r_{1,k}) \setminus (\gamma)_{\zeta,k}\}$.

Now we return to the z plane and consider

$$D_k : (R_k^{1-\frac{\tau_k}{4}} < |z| < R_k^{1+\frac{\tau_k}{4}}) \cap (|\arg z| < \nu_k). \tag{5.2.41}$$

When $z = re^{i\theta} \in D_k$, by (5.2.18) its image ζ must satisfy

$$|\zeta| \leq \left\{1 - \frac{4r^{\frac{1}{\xi_k}} R_k^{\frac{1}{\xi_k}} \cos\frac{\theta}{\xi_k}}{(r^{\frac{1}{\xi_k}} + R_k^{\frac{1}{\xi_k}})^2}\right\}^{\frac{1}{2}} \leq 1 - \frac{(R_k^{1-\frac{\nu_k}{4}})^{\frac{1}{\xi_k}} R_k^{\frac{1}{\xi_k}}}{\{2(R_k^{1+\frac{\nu_k}{4}})^{\frac{1}{\xi_k}}\}^2} < r_{1,k}.$$

Denoting the original image of $(\gamma)_{\zeta,k}$ with respect to $\zeta = \zeta_k(z)$ by $(\gamma)_{z,k}$, we obtain for $z \in (D_k \setminus (\gamma)_{z,k})$

$$\log \left| \frac{h'_k(z)}{h_k(z) - b_k} \right| = \log \left| \frac{H'_k(\zeta)}{H_k(\zeta) - b_k} \right| + \log \frac{1}{|z'_k(\zeta)|} < -\frac{1}{2} R_k^{\lambda - \epsilon'_k - \frac{2\nu_k}{\xi_k}}, \tag{5.2.42}$$

where $\log 1/|z'_k(\zeta)| \leq 0$ by (5.2.33).

On the other hand, for an arbitrary point

$$z_{0,k} \in \left\{(D_k \cap (|z| \leq 2R_k^{1-\frac{\nu_k}{4}})) \setminus (\gamma)_{z,k}\right\},$$

the Poisson-Jensen formula gives us

$$\log|h_k(z_{0,k}) - b_k| \leq \frac{3R_k^{1-\frac{\nu_k}{4}} + 2R_k^{1-\frac{\nu_k}{4}}}{3R_k^{1-\frac{\nu_k}{4}} - 2R_k^{1-\frac{\nu_k}{4}}} m(3R_k^{1-\frac{\nu_k}{4}}, h_k - b_k)$$

$$\text{(5.2.43)}$$

$$+ \sum_\mu \log\left|\frac{(3R_k^{1-\frac{\nu_k}{4}})^2 - \bar{c}_\mu z_{0,k}}{3R_k^{1-\frac{\nu_k}{4}}(z_{0,k} - c_\mu)}\right|,$$

where the c_μ's denote all the poles of $h_k(z)$ in $|z| \leq 3R_k^{1-\frac{\nu_k}{4}}$ with their multiplicities counted.

If $c_\mu \bar{\in}(|\arg z| < \eta_k)$, then we have

$$|z_{0,k} - c_\mu| \geq R_k^{1-\frac{\nu_k}{4}} \sin(\eta_k - \nu_k)$$

$$\geq R_k^{1-\frac{\nu_k}{4}}\frac{\eta_k}{\pi} \geq 4R_k^{1-\frac{\nu_k}{4}} \varepsilon_k^{\frac{1}{2}} \qquad \text{(5.2.44)}$$

$$\geq \frac{4R_k^{1-\frac{\nu_k}{4}}}{(\log R_k)^{\frac{1}{4}}} \geq 1$$

by (5.2.41), (5.2.17), (5.2.5) and (5.2.3).

If $c_\mu \in (|\arg z| < \eta_k)$, its image ζ_μ must be in $|\zeta| \leq r_{1,k}$ since $|c_\mu| \leq 3R_k^{1-\frac{\tau_k}{4}}$. Denote the image of $z_{0,k}$ on the plane ζ by $\zeta_{0,k}$. It is clear that $\zeta_{0,k}$ is outside of $(\gamma)_{\zeta,k}$. Thus

$$d_k \leq |\zeta_{0,k} - \zeta_\mu| = \left|\int_{z_{0,k}c_\mu} \zeta_k'(z)dz\right|$$

$$\leq \left(\max_{z \in z_{0,k}C\mu} |\zeta_k'(z)|\right)|z_{0,k} - c_\mu|$$

$$\leq \left(\max_{|\zeta| \leq 1} \frac{1}{|z_k'(\zeta)|}\right)|z_{0,k} - c_\mu| \qquad \text{(5.2.45)}$$

$$= \left(\frac{1}{\min_{|\zeta| \leq 1} |z_k'(\zeta)|}\right)|z_{0,k} - c_\mu| \leq |z_{0,k} - c_\mu|.$$

By substituting (5.2.44) and (5.2.45) into (5.2.43), we have

$$\log |h_k(z_{0,k}) - b_k| < 5m(3R_k^{1-\frac{\nu_k}{4}}, h_k - b_k)$$

$$+ n(3R_k^{1-\frac{\nu_k}{4}}, h_k = \infty) \log \frac{6R_k^{1-\frac{\nu_k}{4}}}{d_k}$$

$$< \left(5 + \frac{\log \dfrac{6R_k^{1-\frac{\nu_k}{4}}}{d_k}}{\log \dfrac{4}{3}}\right) T(4R_k^{1-\frac{\nu_k}{4}}, h_k - b_k).$$

Since $h_k(z) = f(z) - a_k z$, we have by (5.2.6), (5.2.10), (5.2.17), (5.2.2) and (5.2.3),

$$\log |h_k(z_{0,k}) - b_k| < (\lambda + 5)(\log R_k)T(4R_k^{1-2\varepsilon_k}, f)$$

$$< 4^{\lambda+1}(\lambda + 5)(\log R_k)R_k^{\lambda - 2\lambda\xi_k + \beta_k - 2\xi_k\beta_k} \qquad (5.2.46)$$

$$< R_k^{\lambda - \lambda\varepsilon_k}.$$

Every contour in $(\gamma)_{z,k}$ can be covered by a corresponding disk with radius d_k', where

$$d_k' \le \left(\max_{|\zeta| \le 1 - \frac{2}{R_k^{\pi/2}}} |z_k'(\zeta)|\right) d_k \le \frac{2^{\xi_k}\xi_k R_k}{\left(\dfrac{2}{R_k^{\pi/2}}\right)^{\xi_k+1}} \cdot \frac{1}{R_k^{\lambda+3}} \le \frac{1}{R_k^{\lambda+\frac{1}{4}}},$$

and hence the total sum of their radii does not exceed

$$\left\{n\left(|\zeta| \le 1 - \frac{2}{R_k^{\frac{\pi}{2}}}, H_k = \infty\right) + n\left(|\zeta| \le 1 - \frac{2}{R_k^{\frac{\pi}{2}}}, H_k = b_k\right)\right\} d_k' < \frac{1}{R_k^{\frac{1}{4}}},$$

$$(5.2.47)$$

according to (5.2.15).

An arbitrary point z in $D_k \setminus (\gamma)_{z,k}'$, may be joined to the point $z_{0,k}$ by a segment. If this segment is intercepted by $(\gamma)_{z,k}'$, then we replace that segment by the corresponding arcs and obtain a curve L_k. By (5.2.41) and (5.2.47), the length of L_k does not exceed $2R_k^{1+\nu_k/4}$, so

$$\left|\log \frac{h_k(z) - b_k}{h_k(z_{0,k}) - b_k}\right| = \left|\int_{L_k} \frac{h_k'(u)}{h_k(u) - b_k} du\right| < e^{-\frac{1}{2}R_k^{\lambda - \varepsilon_k' - 2\frac{\nu_k}{\xi_k}}} \cdot 2R_k^{1+\frac{\nu_k}{4}} < 1,$$

$$(5.2.48)$$

and consequently

$$\log|h_k(z) - b_k| \leq \log|h_k(z_{0,k}) - b_k| + 1 < R_k^{\lambda - \lambda\varepsilon_k} + 1. \qquad (5.2.49)$$

Combining this inequality with (5.2.42), we obtain

$$\log|h_k'(z)| < R_k^{\lambda - \lambda\varepsilon_k} + 1 - \frac{1}{2}R_k^{\lambda - \varepsilon_k' - \frac{2\nu_k}{\xi_k}} \qquad (5.2.50)$$

for $z \in (D_k \setminus (\gamma)'_{z,k})$.

By (5.2.47), we can choose a point z_k in D_k such that $|z_k - R_k| < 1$ and $z_k \bar{\in} (\gamma)'_{z,k}$. Obviously, D_k contains the disk $|z - z_k| < 4\varepsilon_k R_k$. In the annulus $3\varepsilon_k R_k < |z - z_k| < 4\varepsilon_k R_k$, we can also choose a circle $|z - z_k| = r_k$ which does not intersect $(\gamma)'_{z,k}$. According to (5.2.50) and (5.2.6), we have

$$\log^+|f'(z)| \leq \log^+|h_k'(z)| + \log^+|a_k| + \log 2 < R_k^{\lambda - \lambda\varepsilon_k} \qquad (5.2.51)$$

for every point z on $|z - z_k| = r_k$; hence

$$m(r_k, z_k, f') < R_k^{\lambda - \lambda\varepsilon_k}. \qquad (5.2.52)$$

In the angle $|\arg z| < \gamma_0$, $f'(z)$ has ∞ as a Borel exceptional value, i.e.,

$$n(r_k, z_k, f') < R_k^{\tau_1}, \quad \tau_1 < \lambda.$$

Since $z_k \bar{\in} (\gamma)'_{z,k}$, the distances between z_k and every pole of $f(z)$ are not less than d_k, as shown in (5.2.45). Thus

$$N(r_k, z_k, f') \leq \int_{d_k}^{r_k} \frac{n(t, z_k, f')}{t} dt < R_k^{\tau_1} \log \frac{r_k}{d_k} < R_k^{\tau}, \quad \tau < \lambda. \qquad (5.2.53)$$

Therefore, combining (5.2.52) with (5.2.53), we have

$$T(r_k, z_k, f') < 2R_k^{\lambda - \lambda\varepsilon_k}. \qquad (5.2.54)$$

On the other hand, for any complex number α with $\alpha \neq f'(z_k)$ we have

$$n(\Gamma_k, f' = \alpha) \leq n\left(\frac{3}{2}\varepsilon_k R_k, z_k, f' = \alpha\right) \leq \frac{1}{\log 2} N\left(r_k, z_k, \frac{1}{f' - \alpha}\right)$$

$$\leq \frac{1}{\log 2}\left\{T(r_k, z_k, f') + \log^+|\alpha| + \log\frac{1}{|f'(z_k) - \alpha|} + \log 2\right\}$$

$$\leq \frac{1}{\log 2}\left\{T(r_k, z_k, f') + \log\frac{1}{|f'(z_k), \alpha|} + \log 2\right\},$$

where $|f'(z_k), \alpha|$ denotes the spherical distance of $f'(z_k)$ and α. Substituting (5.2.54) into this inequality, we obtain

$$n(\Gamma_k, f' = \alpha) < \frac{3}{\log 2} R_k^{\lambda - \lambda \varepsilon_k}, \tag{5.2.55}$$

except for those α in a spherical circle of radius $e^{-R_k^{\lambda/2}}$. But according to the hypothesis of the Theorem, (Γ_k) is a sequence of filling disks of order λ for $f'(z)$, so

$$n(\Gamma_k, f' = \alpha) > R_k^{\lambda - \varepsilon_k'}. \tag{5.2.56}$$

for all the complex numbers α, except for those α in the union of two spherical circles each with radius δ_k.

Comparing (5.2.55) and (5.2.56), we derive the fact that $R_k^{\lambda \varepsilon_k - \varepsilon_k'} < 3/\log 2$. But (5.2.3) implies that

$$R_k^{\lambda \varepsilon_k - \varepsilon_k'} \geq R_k^{\frac{\lambda \varepsilon_k}{2}} \geq e^{\frac{\lambda}{2}(\log R_k)^{\frac{1}{2}}} \to \infty.$$

This yields the contradiction and we complete the proof of Theorem 5.3.
□

5.2.3 Corollaries.   From Theorem 5.3, we can obtain a series of corollaries.

Corollary 1.   Let $f(z)$ be a meromorphic function of order $\lambda(0 < \lambda < \infty)$ in the finite plane. Suppose $B : \arg z = \theta_0$ $(0 \leq \theta_0 < 2\pi)$ is a Borel direction of order λ of $f'(z)$ and $f(z)$ has ∞ as a Borel exceptional value in $|\arg z - \theta_0| < \gamma_0$. Then there exists a sequence of positive numbers R_{k_j} tending to infinity and a sequence of positive numbers η_{k_j} tending to zero such that

$$\left(\frac{R_{k_j}^{1-\eta_{k_j}}}{2} < |z| < 2R_{k_j}^{1+\eta_{k_j}}\right) \cap \left(|\arg z - \theta_0| < \eta_{k_j}\right) \quad (j = 1, 2, \cdots) \tag{5.2.57}$$

is a sequence of filling regions for both $f(z)$ and $f'(z)$.

Without loss of generality, we may assume that $\theta_0 = 0$. Since $B : \arg z = 0$ is a Borel direction of order λ of $f'(z)$, according to Theorem 3.11 there exists a sequence of filling disks of order λ,

$$\Gamma_k^* : |z - z_k| < \varepsilon_k^* z_k, \quad \arg z_k = 0, z_{k+1} > 2z_k,$$

$$\lim_{k \to \infty} \varepsilon_k^* = 0 \quad (k = 1, 2, \cdots),$$

such that $f'(z)$ takes every complex number α at least $z_k^{\lambda-\varepsilon_k'}$ times in Γ_k^*, except for those numbers in the union of two spherical circles with radii δ_k on the Riemann sphere, where $\lim_{k\to\infty} \varepsilon_k' = \lim_{k\to\infty} \delta_k = 0$.

Choose

$$R_k = z_k$$

and

$$\varepsilon_k = \max\left\{\varepsilon_k^*, \frac{2\varepsilon_k'}{\lambda}, \frac{2\beta_k}{\lambda}, \frac{1}{(\log R_k)^{1/2}}\right\},$$

where β_k is given by (5.2.2). It is obvious that every $\Gamma_k : |z - R_k| < \varepsilon_k R_k$ contains the corresponding disk Γ_k^*. Thus (Γ_k) is a sequence of filling disks of order λ of $f'(z)$ and satisfies the conditions of Theorem 5.3. Setting $\eta_k = 4\pi\varepsilon_k^{1/2}$, then

$$\left(\frac{R_k^{1-\eta_k}}{2} < |z| < 2R_k^{1+\eta_k}\right) \cap (|\arg z| < \eta_k), \quad k = 1, 2, \cdots$$

must contain a subsequence of filling regions for both $f(z)$ and $f'(z)$.

Corollary 2.   *Hypotheses as in Corollary 1, B is a Borel direction of order λ of $f(z)$.*

In fact, it follows from Corollary 1 that

$$G_{k_j} : \left(\frac{R_{k_j}^{1-\eta_{k_j}}}{2} < |z| < 2R_{k_j}^{1+\eta_{k_j}}\right) \cap (|\arg z - \theta_0| < \eta_{k_j}), \quad j = 1, 2, \cdots$$

is a sequence of filling regions of order λ of $f(z)$, i.e. $f(z)$ takes every complex number α at least $R_{k_j}^{\lambda-\varepsilon_{k_j}''}$ times, except for those numbers in the union of two spherical disks each with radius δ_{k_j}' on the Riemann sphere, where $\lim_{j\to\infty} \varepsilon_{k_j}'' = \lim_{j\to\infty} \delta_{k_j}' = 0$. Without loss of generality, we may assume that $\sum_{j=1}^{\infty} \delta_{k_j}'$ is less than a preassigned positive number ρ_0.

Consequently, the inequality $n(G_{k_j}, f = \alpha) > R_{k_j}^{\lambda-\varepsilon_{k_j}''}$ holds for all positive integers k and all complex numbers α, except for those α in the union of a sequence of disks the total sum of whose radii is less than ρ_0. For such "normal" (i.e., nonexceptional) numbers α and any positive number ε, we have

$$\lambda \geq \varlimsup_{r \to \infty} \frac{\log n(r, \theta_0, \varepsilon, f = \alpha)}{\log r} \geq \varlimsup_{j \to \infty} \frac{\log n(2R_{k_j}^{1+\eta_{k_j}}, \theta_0, \varepsilon, f = \alpha)}{\log(2R_{k_j})^{1+\eta_{k_j}}}$$

$$\geq \varlimsup_{j \to \infty} \frac{\log(R_{k_j}^{\lambda-\varepsilon_{k_j}''})}{(1+\eta_{k_j})\log(2R_{k_j})} = \lambda,$$

and hence

$$\lim_{\varepsilon \to 0} \left\{ \varlimsup_{r \to \infty} \frac{\log n(r, \theta_0, \varepsilon, f = \alpha)}{\log r} \right\} = \lambda \qquad (5.2.58)$$

holds for all the "normal" numbers α. Therefore, it follows from Theorem 3.9 that (5.2.58) holds for any complex number α, except for at most two numbers, i.e., B is a Borel direction of $f(z)$.

Corollary 3.   *Suppose $f(z)$ is meromorphic and of order $\lambda(0 < \lambda < \infty)$ in the finite plane. Suppose furthermore that $f(z)$ has ∞ as a Borel exceptional value. Then there exists at least a common Borel direction for $f(z)$ and all its derivatives.*

In fact, $f^{(k)}(z)(k = 0, 1, 2, \cdots; f^{(0)}(z) \equiv f(z))$ has at least one Borel direction $B_k : \arg z = \theta_k(k = 0, 1, 2, \cdots; 0 \leq \theta_k < 2\pi)$ by Theorems 3.8 and 4.2. The set $\{\theta_k\}$ has at least one accumulation point θ^*. Then it is clear from Corollary 2 that $\arg z = \theta^*$ is a common Borel direction of $f(z)$ and all its derivatives.

5.3 Angular Distribution of Meromorphic Functions Together with Their Derivatives

5.3.1 Preliminary lemma.   Comparing the theorems due to Picard (Theorem 1.7), Montel (Theorem 2.5) and Valiron (Theorem 3.8), one cannot escape the impression that a Picard type theorem would always lead to a criterion of normality as well as the existence of a singular disection.

It is then natural to ask: Is there a singular direction corresponding to the Hayman inequality (Theorem 4.5) and Gu's criterion (Theorem 4.7)?

In order to discuss this problem, we need the following lemma:

Lemma 5.4   *Let $f(z)$ be meromorphic in $|z| \le R(0 < R < \infty)$ and k be a positive integer. If $f(0) \ne 0, \infty$, $f^{(k)}(0) \ne 1$, $f^{(k+1)}(0) \ne 0$ and*

$$(k+2)\frac{f^{(k+1)}(0)}{f^{(k)}(0) - 1} - (k+1)\frac{f^{(k+2)}(0)}{f^{(k+1)}(0)} \ne 0,$$

then we have

$$T(r, f) < C\left\{ N\left(r, \frac{1}{f}\right) + \overline{N}\left(r, \frac{1}{f^{(k)} - 1}\right) + 1 + \log \frac{R}{R - r} + \log^+ R \right.$$

$$+ \log^+ \frac{1}{R} + \log^+ |f(0)| + \log^+ |f^{(k)}(0) - 1| + \log^+ \frac{1}{|f^{(k+1)}(0)|}$$

$$\left. + \log^+ \frac{1}{\left|(k+2)\dfrac{f^{(k+1)}(0)}{f^{(k)}(0) - 1} - (k+1)\dfrac{f^{(k+2)}(0)}{f^{(k+1)}(0)}\right|} \right\}$$

for $0 < r < R$.

We omit the proof of Lemma 5.4, since it is similar to that of Lemma 4.5.

5.3.2 Fundamental theorem.   Now let us prove the following fundamental theorem.

Theorem 5.4   *Suppose $f(z)$ is meromorphic in $|z| \le 1$ and*

$$N = n(1, f = 0) + n(1, f^{(k)} = 1) + 2,$$

$$N' = n(1, f = \infty) + 1.$$

Then we have

$$n\left(\frac{1}{64}, f = a\right) < C_k\left\{ N\log(N + N') + \log^+ \log^+ |f(z_0)| + \log \frac{1}{|f(\zeta), a|} \right\}$$

$$\tag{5.3.1}$$

for all complex values a, where z_0 and ζ are two points in $|z| < 3/64$ and the distances between z_0 and each pole of $f(z)$ are greater than $1/(1024N')$.

Proof.   Let $(\gamma)_1$ be the union of the disks centered at each zero of $f(z)$ or $f^{(k)}(z) - 1$ and having radius $1/(1024N)$, and let $(\gamma)_2$ be the union of the disks centered at each pole of $f(z)$ and having radius $1/(1024N')$. Let

$$(\gamma) = (\gamma)_1 \cup (\gamma)_2,$$

so that the sum of the radii of the disks in (γ) is less than $1/256$.

We take a point z_0 in $|z| < 1/64$ with $z_0 \bar{\in} (\gamma)$. Clearly, there exist r_1 and r_2 such that

$$\frac{5}{128} \le r_1 < r_2 \le \frac{6}{128}, r_2 - r_1 = \frac{1}{256(N + N')},$$

and for the region $D : r_1 \le |z - z_0| \le r_2$,

$$D \cap (\gamma) = \emptyset.$$

This gives rise to two mutually exclusive cases.

Case A. The inequality

$$\sum_{j=0}^{k+1} |f^{(j)}(z)| > \frac{1}{12}$$

holds uniformly in D. In this case we have

$$m\left(r, z_0, \frac{1}{f}\right) < C\left\{ \sum_{j=1}^{k+1} m\left(r, z_0, \frac{f^{(j)}}{f}\right) + 1 \right\}$$

for $r_1 \le r < \rho \le r_2$. The choice of the point z_0 implies that

$$N\left(r, z_0, \frac{1}{f}\right) \le N \log(1024N).$$

Hence

$$T\left(r, z_0, \frac{1}{f}\right) < C_k\left\{ N \log N + \log \frac{1}{\rho - r} \right.$$

$$\left. + \log^+ \log^+ \frac{1}{|f(z_0)|} + \log^+ T(\rho, z_0, f) \right\}$$

for $r_1 \le r < \rho \le r_2$. Applying the Jensen formula and Lemmas 2.3 and 2.4, we obtain

$$T\left(\frac{5}{128}, z_0, f\right) < \log^+ |f(z_0)| + C_k\{N \log N + \log^+ \log^+ |f(z_0)|\}.$$

Thus

$$n\left(\frac{1}{64}, f = a\right) \le n\left(\frac{1}{32}, z_0, f = a\right) \le \frac{1}{\log \frac{5}{4}} T\left(\frac{5}{128}, z_0, \frac{1}{f - a}\right)$$

$$< C_k\left\{ N \log N + \log^+ \log^+ |f(z_0)| + \log \frac{1}{|f(z_0), a|} \right\}.$$

<u>Case B.</u>   There is a point z_1 in D such that

$$\sum_{j=0}^{k+1} |f^{(j)}(z_1)| \leq \frac{1}{12}.$$ (5.3.2)

We now have two subcases.

B.1   There exists a point z_2 in D such that

$$|f^{(k+1)}(z_2)| \geq \frac{1}{4}.$$

Choose a curve $L_{z_1 z_2}$ from z_1 to z_2 in D of length less than $1/6$. We see that there are two points z_1^* and z_2^* on $L_{z_1 z_2}$ satisfying

$$\left.\begin{array}{l} |f^{(k+1)}(z_1^*)| = \dfrac{1}{12}, \quad |f^{(k+1)}(z_2^*)| = \dfrac{1}{4}, \\[2mm] \dfrac{1}{12} \leq |f^{(k+1)}(z)| \leq \dfrac{1}{4}, \quad z \in L_{z_1^* z_2^*}, \\[2mm] |f^{(k+1)}(z)| \leq \dfrac{1}{4}, \quad z \in L_{z_1 z_2^*}, \end{array}\right\}$$ (5.3.3)

where $L_{z_1^* z_2^*}$ denotes the part of $L_{z_1 z_2}$ between z_1^* and z_2^*, and $L_{z_1 z_2^*}$ is defined similarly. Since the length of $L_{z_1^* z_2^*}$ is less than $1/6$, we can take a point $\zeta \in L_{z_1^* z_2^*}$ such that

$$\left| \frac{f^{(k+2)}(\zeta)}{f^{(k+1)}(\zeta)} \right| > \left| \int_{L_{z_1^* z_2^*}} \frac{f^{(k+2)}(t)}{f^{(k+1)}(t)} dt \right|$$ (5.3.4)

$$\geq \log|f^{(k+1)}(z_2^*)| - \log|f^{(k+1)}(z_1^*)| = \log 3.$$

On the other hand,

$$|f^{(k)}(z)| \leq |f^{(k)}(z_1)| + \left| \int_{L_{z_1 z}} f^{(k+1)}(t) dt \right| \leq \frac{1}{12} + \frac{1}{4} \cdot \frac{1}{6} < \frac{1}{6}, \quad z \in L_{z_1 z_2^*}.$$ (5.3.5)

Consequently, integration yields

$$|f(\zeta)| < \frac{1}{6}.$$ (5.3.6)

Now it follows from (5.3.3)–(5.3.6) that

$$\left| (k+2) \frac{f^{(k+1)}(\zeta)}{f^{(k)}(\zeta) - 1} - (k+1) \frac{f^{(k+2)}(\zeta)}{f^{(k+1)}(\zeta)} \right|$$ (5.3.7)

$$> (k+1)\log 3 - (k+2)\frac{\frac{1}{4}}{1 - \frac{1}{6}} > 1$$

for any positive integer k. Choose r^* such that $6/16 < r^* < 7/16$ and

$$(|z - \zeta| = r^*) \cap (\gamma)_1 = \emptyset.$$

Thus, from the definition of $(\gamma)_1$, we have

$$N(r^*, \zeta, f = 0) + N(r^*, \zeta, f^{(k)} = 1) < N \log(1024N).$$

Applying Lemma 5.4 to $f(z)$ in $|z - \zeta| \leq r^*$, we deduce that

$$T\left(\frac{1}{4}, \zeta, f\right) < C_k N \log N.$$

Hence

$$n\left(\frac{1}{64}, f = a\right) \leq n\left(\frac{1}{8}, \zeta, f = a\right) \leq \frac{1}{\log 2} T\left(\frac{1}{4}, \zeta, \frac{1}{f - a}\right)$$

$$< C_k\left\{N \log N + \log \frac{1}{|f(\zeta), a|}\right\}.$$

B.2 $|f^{(k+1)}(z)| < 1/4$ holds uniformly in D. Then

$$|f^{(k)}(z)| \leq |f^{(k)}(z_1)| + \left|\int_{L_{z_1 z}} f^{(k+1)}(t)dt\right| < \frac{1}{12} + \frac{1}{6} \cdot \frac{1}{4} < \frac{1}{6}$$

and

$$\left|\frac{f^{(k+1)}(z)}{f^{(k)}(z) - 1}\right| < \frac{\frac{1}{4}}{1 - \frac{1}{6}} < \frac{1}{3} \tag{5.3.8}$$

for any point z in D. Since $|f^{(k)}(z)| < 1/6$ for all the points z in D, an integration gives

$$|f(z)| < \frac{1}{6}, \quad z \in D. \tag{5.3.9}$$

From (5.3.8), we have

$$n(r_1, z_0, f^{(k)}) = n\left(r_1, z_0, \frac{1}{f^{(k)} - 1}\right),$$

so that

$$n(r_1, z_0, f) \leq n(r_1, z_0, f^{(k)}) = n\left(r_1, z_0, \frac{1}{f^{(k)} - 1}\right) \leq N.$$

Hence

$$N(r_1, z_0, f) \leq N \log(1024N').$$

Now (5.3.9) gives

$$T(r_1, z_0, f) \leq N \log(1024N').$$

Thus

$$n\left(\frac{1}{64}, f = a\right) \leq n\left(\frac{1}{32}, z_0, f = a\right) \leq C\left\{T(r_1, z_0, f) + \log \frac{1}{|f(z_0), a|}\right\}$$

$$\leq C\left\{N \log(N' + N) + \log \frac{1}{|f(z_0), a|}\right\}$$

and we finish the proof.   □

5.3.3   New singular directions.   Now we can prove the following alternative result.

Theorem 5.5   *If $f(z)$ is meromorphic and of order λ $(0 < \lambda < \infty)$ in the finite plane, then there exists a sequence of disks*

$$\Gamma_j : |z - z_j| < \varepsilon_j |z_j|, \quad \lim_{j \to \infty} \varepsilon_j = 0,$$

$$|z_{j+1}| > 2|z_j|, \, \arg z_j = \theta_0 \text{ (a constant)}, \quad j = 1, 2, \cdots,$$

satisfying either of the following conditions:

(i) *For each $j = 1, 2, \cdots$, there is an η_j with $\lim_{j \to \infty} \eta_j = 0$ such that $f(z)$ takes every complex value at least $|z_j|^{\lambda - \eta_j}$ times in Γ_j, except for those values in a spherical disk having its center at infinity and radius $e^{-|z_j|^{\lambda - \eta_j}}$.*

(ii) *For each $j = 1, 2, \cdots$ and each positive integer k, $f^{(k)}(z)$ takes every complex value at least $|z_j|^{\lambda - \eta_j}$ times in Γ_j, except for those values in the union of two spherical disks having centers at zero and infinity, respectively, and each having radius $e^{-|z_j|^{\lambda - \eta_j}}$.*

Proof.   Since $f(z)$ is meromorphic and of order λ $(0 < \lambda < \infty)$, there exists a sequence of filling disks (Theorem 3.7)

$$\Gamma'_j : |z - z_j| < \varepsilon'_j |z_j|, \lim_{j \to \infty} \varepsilon'_j = 0,$$

$$|z_{j+1}| > 2|z_j|, \, \arg z_j = \theta_0, \quad j = 1, 2, \cdots,$$

such that

$$n(\Gamma'_j, f = a) > |z_j|^{\lambda - \eta'_j} \tag{5.3.10}$$

holds for every complex value a, except for some values in the union of two spherical disks S_j' and S_j'' each with radius $e^{-|z_j|^{\lambda-\eta_j'}}$, where $\lim_{j\to\infty}\eta_j' = 0$.

Set $\varepsilon_j = 64\varepsilon_j'$. Then $\Gamma_j : |z - z_j| < \varepsilon_j|z_j|(j = 1, 2, \cdots)$ meets the requirements of Theorem 5.5, for otherwise there exist a positive number τ, a positive integer k and two sequences of complex values a_j and b_j $(j = 1, 2, \cdots)$ such that

$$\tau < \lambda,$$

$$|a_j, \infty| \geq e^{-|z_j|^{\tau}}, |b_j, 0| \geq e^{-|z_j|^{\tau}}, |b_j, \infty| \geq e^{-|z_j|^{\tau}}, \tag{5.3.11}$$

$$n(\Gamma_j, f = a_j) \leq |z_j|^{\tau}, n(\Gamma_j, f^{(k)} = b_j) \leq |z_j|^{\tau}. \tag{5.3.12}$$

By definition of spherical distance (§3.1.2) and (5.3.11), we have

$$e^{-|z_j|^{\tau}} < |b_j| < e^{|z_j|^{\tau}}, |a_j| < e^{|z_j|^{\tau}}. \tag{5.3.13}$$

Next, for every fixed j, the function

$$g(\zeta) = \frac{f(z_j + 64\varepsilon_j|z_j|\zeta) - a_j}{b_j(64\varepsilon_j|z_j|)^k}$$

is clearly meromorphic in $|\zeta| < 1$ and

$$n(1, g = 0) + n(1, g^{(k)} = 1) \leq 2|z_j|^{\tau}$$

by (5.3.12). Therefore, Theorem 5.4 implies that

$$n\left(\frac{1}{64}, g = \frac{a - a_j}{b_j(64\varepsilon_j|z_j|)^k}\right) < C_k\Big\{ |z_j|^{\tau}\log^+ |z_j| + \log^+\log^+ |g(t_j)|$$

$$+ \log\frac{1}{\left|g(\zeta_j), \dfrac{a - a_j}{b_j(64\varepsilon_j|z_j|)^k}\right|}\Big\},$$

where t_j and ζ_j are two points in $|\zeta| < 3/64$ and the distance between t_j and every pole of $g(\zeta)$ is greater than $1/(1024(n(1, g) + 1))$. Thus

$$n(\Gamma_j', f = a) < C_k\Big\{ |z_j|^{\tau}\log^+ |z_j| + \log^+\log^+ |f(z_j^*)|$$

$$+ \log\frac{1}{\left|\dfrac{f(z_j') - a_j}{b_j(64\varepsilon_j|z_j|)^k}, \dfrac{a - a_j}{b_j(64\varepsilon_j|z_j|)^k}\right|}\Big\}, \tag{5.3.14}$$

where z_j^* and z_j' are two points in $|z - z_j| < 3\varepsilon_j|z_j|$ and the distances between z_j^* and every pole of $f(z)$ are greater than

$$\frac{64\varepsilon_j|z_j|}{1024(n(|z_j| + 64\varepsilon_j|z_j|, f = \infty) + 1)}.$$

Let us estimate the last two terms on the right-hand side of (5.3.14). When j is sufficiently large, we have

$$|z_j^*| \leq |z_j| + 3\varepsilon_j|z_j| < 2|z_j|.$$

The Poisson-Jensen formula (Theorem 1.1) gives us

$$\log^+ |f(z_j^*)| \leq \frac{3|z_j| + 2|z_j|}{3|z_j| - 2|z_j|} m(3|z_j|, f) + \sum_{\substack{f(\beta_k)=\infty \\ |\beta_k| \leq 3|z_j|}} \left| \frac{(3|z_j|)^2 - \bar{\beta}_k z_j^*}{3|z_j|(z_j^* - \beta_k)} \right|$$

$$\leq 5m(3|z_j|, f) + n(3|z_j|, f = \infty) \log \frac{2(3|z_j|)}{\dfrac{64\varepsilon_j|z_j|}{1024(n(2|z_j|, f = \infty) + 1)}}$$

$$< CT(4|z_j|, f) \log T(4|z_j|, f),$$

provided j is sufficiently large. Thus

$$\log^+ \log^+ |f(z_j^*)| = O(\log |z_j|). \tag{5.3.15}$$

Since

(i) $$|x\alpha, x\beta| \geq \min\left(|x|, \frac{1}{|x|}\right)|\alpha, \beta|,$$

(ii) $$|\alpha - x, \beta - x| \geq \frac{1}{2}|x, \infty|^2|\alpha, \beta|,$$

where x, α and β are three complex numbers, we have for the last term of spherical distance in (5.3.14),

$$\left| \frac{f(z_j') - a_j}{b_j(64\varepsilon_j|z_j|)^k}, \frac{a - a_j}{b_j(64\varepsilon_j|z_j|)^k} \right|$$

$$\geq \min\left(|b_j|(64\varepsilon_j|z_j|)^k, \frac{1}{|b_j|(64\varepsilon_j|z_j|)^k}\right)|f(z_j') - a_j, a - a_j| \tag{5.3.16}$$

$$\geq \min\left(|b_j|(64\varepsilon_j|z_j|)^k, \frac{1}{|b_j|(64\varepsilon_j|z_j|)^k}\right) \cdot \frac{1}{2}|a_j, \infty|^2|f(z_j'), a|.$$

Without loss of generality, we may assume that

$$\varepsilon_j |z_j| \geq 1, \qquad (5.3.17)$$

since the radius of Γ_j can be replaced by 1 if necessary. Moreover, for sufficiently large j we have

$$(\varepsilon_j |z_j|)^k < e^{|z_j|^\tau}. \qquad (5.3.18)$$

If we let a be a complex number satisfying

$$|f(z_j'), a| > e^{-|z_j|^\tau}, \qquad (5.3.19)$$

then (5.3.11), (5.3.13), (5.3.16), (5.3.17) (5.3.18) and (5.3.19), taken together, give us

$$\left| \frac{f(z_j') - a_j}{b_j (64 \varepsilon_j |z_j|)^k}, \frac{a - a_j}{b_j (64 \varepsilon_j |z_j|)^k} \right| > \frac{1}{2} e^{-5|z_j|^\tau} \qquad (5.3.20)$$

Clearly, there is a complex number $a \bar\in (S_j' \cup S_j'')$ which satisfies not only (5.3.19) but also (5.3.20).

Comparing (5.3.10), (5.3.14), (5.3.15) and (5.3.20), we obtain

$$|z_j|^{\lambda - \eta_j} < C_k \left\{ |z_j|^\tau \log |z_j| + \log |z_j| + |z_j|^\tau \right\}.$$

This inequality contradicts the fact that $\tau < \lambda$, since $\lim_{j \to \infty} \eta_j = 0$, and the proof is complete. □

As a consequence of Theorem 5.5, we have the follwing theorem and corollary:

Theorem 5.6 *If $f(z)$ is meromorphic and of order λ $(0 < \lambda < \infty)$ in the finite plane, then there exists a direction $\arg z = \theta_0 (0 \leq \theta_0 < 2\pi)$ such that for any positive number ε and any two finite complex numbers a and $b(b \neq 0)$, we have*

$$\varlimsup_{r \to \infty} \frac{\log\{n(r, \theta_0, \varepsilon, f = a) + n(r, \theta_0, \varepsilon, f^{(k)} = b)\}}{\log r} = \lambda.$$

Corollary. *Let $f(z)$ be meromorphic and of order λ $(0 < \lambda < \infty)$ in the finite plane. If $\arg z = \theta_0 (0 \leq \theta_0 < 2\pi)$ is a Borel direction of $f(z)$, and $f(z)$ has a finite Borel exceptional value in the angle $|\arg z - \theta_0| < \varepsilon_0 (\varepsilon_0 > 0)$, then $\arg z = \theta_0$ is a common Borel direction of $f(z)$ and all its derivatives.*

Theorems 5.5 and 5.6 are due to Yang Lo and Zhang Qingde [1].

Chapter 6

Deficient Values and Borel Directions
of Meromorphic Functions

In Chapter 1, we have seen that while Borel direction are basic to angular distribution theory, the concept of a deficient value is important in modular distribution theory, where the definition of a deficient value depends only on the modulii of the points at which the function takes on this value. It would therefore not appear initially to be the case that there is any relationship between Borel directions and deficient values. However, Yang Lo and Zhang Guanghou have shown that if a meromorphic function of finite positive order has a deficient value, then the distribution of its Borel directions has to follow a certain rule. In the general case, the deficient values and Borel directions are closely related to each other in number.

6.1 Precise Order and Three Lemmas

6.1.1 Precise order. In the following section, we shall need a very important tool, viz., precise order. This was introduced originally by G. Valiron.

Definition 6.1 *Let $f(z)$ be meromorphic in the finite plane and λ a finite positive number. A non-negative continuous function $\lambda(r)$ in $[0, \infty)$ is called a precise order of $T(r, f)$ or $f(z)$, if*

(1) $\lim_{r \to \infty} \lambda(r) = \lambda$,

(2) $\lambda'(r)$ *exists everywhere in $[0, \infty)$ except on a countable set of points, and $\lim_{r \to \infty} r\lambda'(r) \log r = 0$,*

(3) $r^{\lambda(r)} \geq T(r, f)$ *always holds for sufficiently large* r, *and the equality is true for a sequence* (r_j) *of positive numbers tending to infinity.* $r^{\lambda(r)}$ *is called a type function of* $T(r, f)$ *or* $f(z)$.

Theorem 6.1 *Let* $f(z)$ *be meromorphic and of finite positive order* λ *in the finite plane. Then* $f(z)$ *has a precise order.*

Proof. Let us consider two different cases as follows.

(i) $T(r, f) > r^\lambda$ holds for a sequence of values r tending to infinity. It is clear that the function

$$\varphi(r) = \max_{x \geq r} \frac{\log^+ T(x, f)}{\log x} \tag{6.1.1}$$

is continuous, non-negative and non-increasing in $[0, \infty)$, and that $\lim_{r \to \infty} \varphi(r) = \lambda$; moreover, the set M of values r statisfying $\varphi(r) = (\log^+ T(r, f))/\log r$ is unbounded, and $T(r, f) > r^\lambda$ for every $r \in M$.

Choose r_1 and t_1 such that $r_1 > e^{e^e}$, $r_1 \in M$, and t_1 is the minimum integer with $t_1 > 1 + r_1$ and $\varphi(t_1) < \varphi(r_1)$.

Consider[1]

$$y_1(x) = \varphi(r_1) - \log_3 x + \log_3 t_1, \quad x \geq t_1$$

and

$$y_2(x) = \varphi(x), \quad x \geq t_1.$$

Since $y_1(t) = \varphi(r_1) > \varphi(t_1) = y_2(t_1)$ and $\lim_{x \to \infty} y_1(x) = -\infty$, the curves $y = y_1(x)$ and $y = y_2(x)$ must intersect for $x > t_1$. Let u_1 be the abscissa of the first point of intersection.

Let $r_2 = \min\{M \cap [u_1, \infty)\}$ and repeat the above procedure. Since

$$r_j - r_{j-1} \geq u_{j-1} - r_{j-1} \geq t_{j-1} - r_{j-1} > 1,$$

---

[1] We denote $\log\log \alpha$ and $\log\log\log \alpha$ by $\log_2 \alpha$ and $\log_3 \alpha$ respectively.

we have $\lim_{j \to \infty} r_j = \infty$. Define

$$
\lambda(r) = \begin{cases}
\varphi(r_1), & 0 \leq r \leq t_1, \\
\varphi(r_1) - \log_3 r + \log_3 t_1, & t_1 < r \leq u_1, \\
\varphi(r_2), & u_1 < r \leq t_2, \\
\varphi(r_2) - \log_3 r + \log_3 t_2, & t_2 < r \leq u_2, \\
\cdots, & \cdots
\end{cases}
$$

It is clear that the function $\lambda(r)$ is continuous and non-negative in $[0, \infty)$, and that $\lambda'(r)$ exists everywhere except at t_j and u_j. Moreover, in every small interval, either $\lambda'(r) \equiv 0$ or $\lambda'(r) = -1/(r \log r \log_2 r)$. Thus $\lim_{r \to \infty} r \lambda'(r) \log r = 0$.

When $r \geq r_1$, we have $\lambda(r) \geq \varphi(r) \geq (\log^+ T(r, f))/\log r$. On the other hand,

$$
\lambda(r_j) = \varphi(r_j) = \frac{\log^+ T(r_j, f)}{\log r_j}, \quad j = 1, 2, \cdots.
$$

Since $\lambda(r)$ is non-increasing, we have $\lim_{r \to \infty} \lambda(r) = \lim_{r \to \infty} \varphi(r) = \lambda$.

(ii) $T(r, f) \leq r^\lambda$ holds for sufficiently large r.

If there is a sequence (r_j) of positive numbers tending to infinity such that $T(r_j, f) = r_j^\lambda$, then $\lambda(r) \equiv \lambda$ is a precise order of $f(z)$. Therefore we only need to consider the case in which $T(r, f) < r^\lambda$ holds for $r \geq r_0 > e^{e^e}$.

The function

$$
\psi(r) = \max_{r_0 \leq x \leq r} \frac{\log^+ T(x, f)}{\log x} \tag{6.1.2}
$$

is continuous and non-decreasing. The set of values r with $\psi(r) = (\log r)^{-1} \times (\log^+ T(r, f))$ is unbounded.

There exists a positive and sufficiently large number $r_1 > r_0$ such that if s_1 denotes the maximum abscissa of the points of intersection of the curves

$$
y_1(x) = \lambda + \log_3 x - \log_3 r_1
$$

and

$$
y_2(x) = \psi(x),
$$

then we have $r_0 < s_1 < r_1$ and $L \cap [r_0, s_1] \neq \emptyset$. Setting

$$t_1 = \max\{L \cap [r_0, s_1]\},$$

we see that $r_0 \leq t_1 \leq s_1 \leq r_1$.

Furthermore, there exists $r_2 > r_1 + 1$ such that if s_2 is the maximum abscissa of the points of intersection of the curves

$$y_1(x) = \lambda + \log_3 x - \log_3 r_2$$

and

$$y_2(x) = \psi(x),$$

then we have $r_1 < s_2$ and $L \cap [r_1, s_2] \neq \emptyset$. Let

$$t_2 = \max\{L \cap [r_1, s_2]\}$$

and u_1 the point on $[s_1, r_1]$ with $\lambda + \log_3 u_1 - \log_3 r_1 = \psi(t_2)$. Then we have

$$r_0 \leq t_1 \leq s_1 \leq u_1 \leq r_1 < t_2 \leq s_2 \leq \cdots.$$

Define

$$\lambda(r) = \begin{cases} \psi(t_1), & 0 \leq r \leq s_1, \\ \lambda + \log_3 r - \log_3 r_1, & s_1 \leq r \leq u_1, \\ \psi(t_2), & u_1 \leq r \leq s_2, \\ \lambda + \log_3 r - \log_3 r_2, & s_2 \leq r \leq u_2, \\ \cdots & \cdots \end{cases}$$

According to the definition of t_j, u_j and $\psi(r)$, the function $\lambda(r)$ is continuous and non-negative in $[0, \infty)$. Moreover we have

$$\lambda(r) \geq \psi(r) \geq \frac{\log^+ T(r, f)}{\log r}$$

and

$$\lambda(t_j) = \psi(t_j) = \frac{\log^+ T(t_j, f)}{\log t_j}.$$

Finally $\lambda'(r)$ exists everywhere except at the points s_j and u_j, and $\lim\limits_{r \to \infty} \{r \lambda'(r) \log r\} = 0$.

6.1.2. Three lemmas.   Now we establish three lemmas which play a very important role in the present Chapter.

Lemma 6.1[1]   *Let $f(z)$ be meromorphic and of order λ $(0 < \lambda < +\infty)$ in the finite plane and let $a_\nu (\nu = 1, 2, \cdots, p)$ be $p(1 \leq p < \infty)$ distinct complex numbers. Suppose $\delta(a_\nu, f) = \delta_\nu > 0$ $(\nu = 1, 2, \cdots, p)$ and $\delta = \min_{1 \leq \nu \leq p} \delta_\nu$. If $\lambda(r)$ is a precise order of $f(z)$ and $U(r) = r^{\lambda(r)}$, then there exists a sequence of positive numbers $R_j (j = 1, 2, \cdots)$ tending to infinity such that, for every sufficiently large j and $\nu = 1, 2, \cdots, p$, the set $E_{j\nu}$ of values $\varphi (0 \leq \varphi < 2\pi)$ with*

$$\begin{cases} \log \dfrac{1}{|f(R_j e^{i\varphi}) - a_\nu|} > \dfrac{\delta}{2^{\lambda+4}} U(R_j), & \text{when } a_\nu \neq \infty, \\[3mm] \log |f(R_j e^{i\varphi})| > \dfrac{\delta}{2^{\lambda+4}} U(R_j), & \text{when } a_\nu = \infty \end{cases} \tag{6.1.3}$$

satisfies

$$\text{mes } E_{j\nu} > K(\delta, p, \lambda) > 0, \tag{6.1.4}$$

where $K(\delta, p, \lambda)$ is a positive constant depending only on δ, p and λ, e.g.,

$$K(\delta, p, \lambda) = \frac{\delta \pi}{\left(5 + \dfrac{\log 25pe}{\log \dfrac{4}{3}}\right) 4^{\lambda+2}}. \tag{6.1.5}$$

Proof.   According to the definition of $\lambda(r)$, we can choose a sequence of positive numbers r_j tending to infinity such that

$$\lim_{j \to \infty} \frac{T(r_j, f)}{U(r_j)} = 1.$$

Let us consider the function $1/(f(z) - a_\nu)$, $\nu = 1, 2, \cdots, p$, (when $a_\nu = \infty$, $1/(f(z) - a_\nu)$ should be replaced by $f(z)$). Applying the Boutroux-Cartan theorem for every fixed j, to the poles $b_{\nu l}$ $(l = 1, 2, \cdots, n(3r_j, f = a_\nu))$ of $1/(f(z) - a_\nu)$ in the disk $|z| \leq 3r_j$, we have

$$\prod_{l=1}^{n(3r_j, f=a_\nu)} |z - b_{\nu l}| > \left(\frac{h}{p} r_j\right)^{n(3r_j, f=a_\nu)}, \quad \nu = 1, 2, \cdots, p,$$

---

1) Although Lemma 6.1 can also be derived from the spread relation (Theorem 7.4 of Chapter 7), the proof here is much simpler, and it does not require the precise lower bound established by the spread relation.

except on at most $n(3r_j, f = a_\nu)$ disks $(\gamma)_\nu$ the total sum of whose radii does not exceed $2ehr_j/p$. Now let us set $h = 1/5e$ and $(\gamma) = \cup_{\nu=1}^p (\gamma)_\nu$.

In the annulus $r_j \le |z| \le 2r_j$, there is a cricle $|z| = R_j$, not intersecting (γ), on which an arbitrary point z satisfies

$$\log \frac{1}{|f(z) - a_\nu|} \le \frac{3r_j + R_j}{3r_j - R_j} m\left(3r_j, \frac{1}{f - a_\nu}\right)$$

$$+ \sum_{l=1}^{n(3r_j, f=a_\nu)} \log\left|\frac{(3r_j)^2 - \bar{b}_{\nu l} z}{3r_j(z - b_{\nu l})}\right|$$

$$\le 5m\left(3r_j, \frac{1}{f - a_\nu}\right) + n(3r_j, f = a_\nu)\log 5r_j$$

$$+ \log \frac{1}{\left(\frac{h}{p}r_j\right)^{n(3r_j, f=a_\nu)}} \qquad (6.1.6)$$

$$\le 5m\left(3r_j, \frac{1}{f - a_\nu}\right) + \frac{\log \dfrac{5p}{h}}{\log \dfrac{4}{3}} N\left(4r_j, \frac{1}{f - a_\nu}\right)$$

$$\le KT\left(4r_j, \frac{1}{f - a_\nu}\right),$$

where $K = 5 + (\log 25pe)/(\log 4/3)$.

Considering the set

$$E_{j\nu}^* = E\left\{\varphi : 0 \le \varphi < 2\pi, \log^+ \frac{1}{|f(R_j e^{i\varphi}) - a_\nu|} > \frac{1}{2}m\left(R_j, \frac{1}{f - a_\nu}\right)\right\}$$

$$(6.1.7)$$

and its complement $CE_{j\nu}^*$ with respect to $[0, 2\pi)$, we deduce from (6.1.6) and (6.1.7) that

$$m\left(R_j, \frac{1}{f - a_\nu}\right) = \frac{1}{2\pi}\int_{E_{j\nu}^*} \log^+ \frac{1}{|f(R_j e^{i\varphi}) - a_\nu|} d\varphi$$

$$+ \frac{1}{2\pi}\int_{CE_{j\nu}^*} \log^+ \frac{1}{|f(R_j e^{i\varphi}) - a_\nu|} d\varphi$$

$$\le \frac{1}{2\pi} KT\left(4r_j, \frac{1}{f - a_\nu}\right) \text{mes } E_{j\nu}^* + \frac{1}{2}m\left(R_j, \frac{1}{f - a_\nu}\right),$$

so

$$m\left(R_j, \frac{1}{f - a_\nu}\right) \le \frac{K}{\pi}T\left(4r_j, \frac{1}{f - a_\nu}\right) \text{mes } E_{j\nu}^*, \quad \nu = 1, 2, \cdots, p. \quad (6.1.8)$$

When j is sufficiently large, we have

$$m\left(R_j, \frac{1}{f - a_\nu}\right) > \frac{\delta_\nu}{2} T(R_j, f) \geq \frac{\delta_\nu}{2} T(r_j, f) > \frac{\delta}{4} U(r_j), \qquad (6.1.9)$$

since $\delta(a_\nu, f) = \delta_\nu \geq \delta$. Moreover[1],

$$T\left(4r_j, \frac{1}{f - a_\nu}\right) = T(4r_j, f - a_\nu) + \log \frac{1}{|f(0) - a_\nu|}$$

$$\leq T(4r_j, f) + \log^+ |a_\nu| + \log \frac{1}{|f(0) - a_\nu|} + \log 2,$$

$$\nu = 1, 2, \cdots, p,$$

and hence

$$T\left(4r_j, \frac{1}{f - a_\nu}\right) < 2U(4r_j) < 4^{\lambda+1} U(r_j), \quad \nu = 1, 2, \cdots, p, \qquad (6.1.10)$$

by the properties of precise order. Then (6.1.8), (6.1.9) and (6.1.10) taken together give

$$\text{mes } E_{j\nu}^* > K(\delta, p, \lambda) = \frac{\delta\pi}{\left(5 + \dfrac{\log 25pe}{\log \frac{4}{3}}\right) 4^{\lambda+2}} > 0, \quad \nu = 1, 2, \cdots, p.$$

For the value φ in $E_{j\nu}^*$, we have

$$\log \frac{1}{|f(R_j e^{i\varphi}) - a_\nu|} > \frac{1}{2} m\left(R_j, \frac{1}{f - a_\nu}\right) > \frac{\delta_\nu}{8} U(r_j) \geq \frac{\delta}{2^{\lambda+4}} U(R_j);$$

hence $E_{j\nu}^* \subset E_{j\nu}$, and the proof is complete.   □

Lemma 6.2   *Let $f(z)$ be meromorphic on $|z| \leq R(< \infty)$. If*

$$\mathcal{N} = n(R, f = 0) + n(R, f = 1) + n(R, f = \infty)$$

and the shortest distance between the origin O and these $\mathcal{N}$ points equals $d(0 < d < 1)$, then we have

$$T(r, f) < \frac{CR(\mathcal{N} + 1)}{R - r} \log \frac{2R^2(\mathcal{N} + 1)}{d(R - r)} + \log^+ |f(0)| \qquad (6.1.11)$$

---

1) When $f(0) = a_\nu$, we should substitute $\log \dfrac{1}{|C_k|}$ for $\log \dfrac{1}{|f(0) - a_\nu|}$, where C_k is the first non-zero coefficient of the Taylor expansion of $f(z)$ at the origin.

for $0 < r < R$, where C is a constant not depending on $f(z)$.

Proof.   It is sufficient to prove that

$$T(r, f) < \frac{CR(\mathcal{N}+1)}{R-r}\left\{\log\frac{2R^2(\mathcal{N}+1)}{d(R-r)} + \log^+|f(0)|\right\}. \qquad (6.1.12)$$

In fact, if $|f(0)| \leq 1$, then (6.1.11) and (6.1.12) are the same. Otherwise, from the fact that

$$T(r, f) = T\left(r, \frac{1}{f}\right) + \log|f(0)|$$

and applying (6.1.12) to the function $1/f(z)$, the inequality (6.1.11) is also proved.

We consider the case of $R = 1$ first. Let (γ) be the union of the disks centered at each zero of $f(z)$, $f(z) - 1$ and $1/f(z)$ and having radius $(d(1-r))/(16(\mathcal{N}+1))$. It is obvious that $O \bar{\in}(\gamma)$. There exists a cricle $|z| = \rho$ satisfying $r < \rho < (1+r)/2$ and $(|z| = \rho) \cap (\gamma) = \emptyset$.

Let $|f(\zeta)| = \max_{|z|=\rho}|f(z)|$. If we join the origin O and ζ by a segment, then the intersection of this segment and (γ) is replaced by the corresponding arcs, and thus we obtain a curve L with its length not exceeding $\rho + \pi d(1-r)/16 < 1$.

Now let us examine two mutually exclusive cases.

(i) $|f'(z)| < 1$ holds uniformly on L.

In this case, we have

$$|f(\zeta)| \leq |f(0)| + \left|\int_L f'(z)dz\right| < |f(0)| + 1,$$

and hence

$$m(\rho, f) < \log^+|f(0)| + \log 2.$$

Combining this inequality with

$$N(\rho, f) \leq \mathcal{N}\log\frac{1}{d},$$

we obtain

$$T(\rho, f) < \mathcal{N}\log\frac{1}{d} + \log^+|f(0)| + \log 2.$$

Since $T(r, f) \leq T(\rho, f)$, the inequality (6.1.12) follows immediately.

(ii) There exists a point z_0 on L with $|f'(z_0)| \geq 1$ such that $|f'(z)| < 1$ holds uniformly on L from the origin to z_0. When $|f'(0)| \geq 1$, we take z_0 to be the origin.

Let us examine the function $f(z)$ in the disk $|z - z_0| < (3+r)/4 - |z_0|$. Since $|\zeta - z_0| \leq \rho - |z_0| + (1-r)/8$, the Poisson-Jensen formula gives us

$$m(\rho, f) \leq \log^+ |f(\zeta)|$$

$$< \frac{2}{\left(\frac{3+r}{4} - |z_0|\right) - \left(\rho - |z_0| + \frac{1-r}{8}\right)} m\left(\frac{3+r}{4} - |z_0|, z_0, f\right)$$

$$+ \sum_{\nu} \log \left| \frac{\left(\frac{3+r}{4} - |z_0|\right)^2 - (\overline{b_\nu} - \overline{z_0})(\zeta - z_0)}{\left(\frac{3+r}{4} - |z_0|\right)(\zeta - b_\nu)} \right|$$

$$< \frac{16}{1-r} m\left(\frac{3+r}{4} - |z_0|, z_0, f\right) + \sum_{\nu} \log \frac{2}{|\zeta - b_\nu|},$$

(6.1.13)

where b_ν denotes the poles of $f(z)$ lying in $|z - z_0| < (3+r)/4 - |z_0|$.

In order to estimate $m((3+r)/4 - |z_0|, z_0, f)$, we apply Theorem 3.2 to $f(z)$ and note that $|f(z_0)| \leq |f(0)| + 1$, $|f'(z_0)| \geq 1$ and $z_0 \bar{\in} (\gamma)$. Thus

$$m\left(\frac{3+r}{4} - |z_0|, z_0, f\right) < C\left\{ N \log \frac{2(N+1)}{d(1-r)} + \log^+ |f(0)| + \log \frac{2}{1-r} \right\}.$$

Incorporating this estimate into (6.1.13), we obtain

$$m(\rho, f) < \frac{C(N+1)}{1-r} \left\{ \log \frac{2(N+1)}{d(1-r)} + \log^+ |f(0)| \right\}.$$

Since

$$N(\rho, f) \leq N \log \frac{1}{d},$$

we also have (6.1.12) in the case of $R = 1$.

When $R \neq 1$, by setting $f(z) = f(Rt) = g(t)$, we see that $g(t)$ is meromorphic on $|t| \leq 1$ and

$$N = n(1, g = 0) + n(1, g = 1) + n(1, g = \infty).$$

The shortest distance between the origin and these N points equals d/R. According to the above proof, we obtain

$$T\left(\frac{r}{R}, g\right) < \frac{C(N+1)}{1 - \frac{r}{R}} \left\{ \log \frac{2(N+1)}{\frac{d}{R}\left(1 - \frac{r}{R}\right)} + \log^+ |g(0)| \right\}.$$

Hence

$$T(r, f) < \frac{CR(\mathcal{N} + 1)}{R - r} \left\{ \log \frac{2R^2(\mathcal{N} + 1)}{d(R - r)} + \log^+ |f(0)| \right\}. \quad \square$$

Lemma 6.3 *Let $f(z)$ be meromorphic and of order λ $(0 < \lambda < \infty)$ in the finite plane and let $B_1 : \arg z = \varphi_1$ and $B_2 : \arg z = \varphi_2$ $(0 \leq \varphi_1 < \varphi_2 \leq 2\pi + \varphi_1)$ be two rays such that no Borel direction of $f(z)$ is located in the angle $\varphi_1 < \arg z < \varphi_2$. Suppose there is a sequence of positive numbers R_j tending to infinity and a complex number a_0 (finite or infinite) with the following property: For any positive number ε, the set E_j of values $\varphi(\varphi_1 < \varphi < \varphi_2)$ for which*

$$\begin{cases} \log \dfrac{1}{|f(R_j e^{i\varphi}) - a_0|} > R_j^{\lambda - \varepsilon} & \text{if } a_0 \neq \infty, \\[2mm] \log |f(R_j e^{i\varphi})| > R_j^{\lambda - \varepsilon} & \text{if } a_0 = \infty, \end{cases} \tag{6.1.14}$$

satisfies mes $E_j > K_1 > 0$ *(K_1 not depending on ε), provided j is sufficiently large. Then, given a positive constant K_2 and a sufficiently small positive number α, there is a sequence of curves L_j satisfying the following conditions:*

(i) *L_j is located in the region $\varphi_1 + 8\alpha \leq \arg z \leq \varphi_2 - 8\alpha$ with $R_j - 1 \leq |z| \leq R_j$, and has the extremal points $R_j e^{i(\varphi_1 + \alpha_j')}$ and $R_j e^{i(\varphi_2 - \alpha_j')}$ $(8\alpha \leq \alpha_j' \leq 9\alpha)$. The set of values φ for which $R_j e^{i\varphi} \in A_j - L_j$, has measure less than K_2, i.e.,*

$$\text{mes}\{\varphi : R_j e^{i\varphi} \in A_j - L_j\} < K_2.$$

(ii) *For every positive number η and $z \in L_j$, the following inequalities hold:*

$$\begin{cases} \log \dfrac{1}{|f(z) - a_0|} > R_j^{\lambda - \eta} & \text{when } a_0 \neq \infty, \\[2mm] \log |f(z)| > R_j^{\lambda - \eta} & \text{when } a_0 = \infty, \end{cases}$$

provided j is sufficiently large.

Proof. We may assume that $a_0 = \infty$. Otherwise, it is sufficient to consider the function $1/(f(z) - a_0)$.

Choose α such that $0 < \alpha < 1/32 \min(K_1, K_2)$. The set E_j of values φ in the angle $G: \varphi_1 + 8\alpha \le \arg z \le \varphi_2 - 8\alpha$ for which

$$\log|f(R_j e^{i\varphi})| > R_j^{\lambda - \varepsilon}, \tag{6.1.15}$$

satisfies mes $E_j > K_1/2$.

Now we divide the angle G into $N = [\pi/\alpha] + 1$ equal subangles $G_\nu (\nu = 1, 2, \cdots, N)$ with their magnitude not exceeding 2α. Among these, there exists at least one subangle, say $G_{j\nu_0}$, such that

$$\text{mes } E_{j\nu_0} \ge \frac{K_1}{2N} \ge \frac{\alpha}{2(\pi + \alpha)} K_1,$$

where $E_{j\nu_0}$ is the set of values φ satisfying (6.1.15) in $G_{j\nu_0}$.

Since $f(z)$ has no Borel direction in $\varphi_1 < \arg z < \varphi_2$, by Theorem 3.10 there are three distinct complex numbers $\beta_l (l = 1, 2, 3)$ and a real number $\tau, 0 < \tau < \lambda$, such that

$$\sum_{l=1}^{3} n\left\{ (|z| \le r) \cap (\varphi_1 + \alpha \le \arg z \le \varphi_2 - \alpha), f = \beta_l \right\} < r^\tau, \quad r > r_0.$$

Setting

$$\mathcal{N}_j = \sum_{l=1}^{3} n\left\{ (|z| \le (1 + 6\alpha)R_j) \cap (\varphi_1 + \alpha \le \arg z \le \varphi_2 - \alpha), f = \beta_l \right\}$$

$$+ n\left\{ (z| \le (1 + 6\alpha)R_j) \cap (\varphi_1 + \alpha \le \arg z \le \varphi_2 - \alpha), f = \infty \right\}$$

and

$$h = \min\left(\frac{K_2}{8}, \frac{\alpha}{8(\pi + \alpha)} K_1 \right) \le \frac{1}{4} \text{mes } E_{j\nu_0},$$

we have $\mathcal{N}_j < R_j^{\lambda + \varepsilon}$ for sufficiently large j.

Denote by $(\gamma)_j$ the union of all the circles in the region

$$(|z| \le (1 + 6\alpha)R_j) \cap (\varphi_1 + \alpha \le \arg z \le \varphi_2 - \alpha),$$

where the circles have centers at the points where $f(z) = \beta_l (l = 1, 2, 3)$ or ∞ and the same radius $h/(\mathcal{N}_j + 1)$; denote by L_j the curve obtained from the arc $(|z| = R_j) \cap (\varphi_1 + 8\alpha \le \arg z \le \varphi_2 - 8\alpha)$ by replacing its intersection with whichever circle in $(\gamma)_j$ by the corresponding arcs of this circle in $|z| \le R_j$. The curve L_j is clearly situated in $(\varphi_1 + 8\alpha \le \arg z \le$

$\varphi_2 - 8\alpha) \cap (R_j - 1 \leq |z| \leq R_j)$. Since the total sum of the radii of the circles in $(\gamma)_j$ is less than $h \leq \alpha/4$, there exists α'_j, $8\alpha \leq \alpha'_j \leq 9\alpha$, such that $R_j e^{i(\varphi_1 + \alpha'_j)}$ and $R_j e^{i(\varphi_2 - \alpha'_j)}$, extreme points of L_j, lie on the outside of all the circles in $(\gamma)_j$. Thus

$$\text{mes } \{\varphi : R_j e^{i\varphi} \in A_j - L_j\} < 2h + 18\alpha < K_2.$$

Now let us prove the second assertion of Lemma 6.3. Let z_0 be the point of intersection of $|z| = R_j$ with the bisector of $G_{j\nu_0}$. In the disk $\Gamma_j : |z - z_0| \leq \alpha R_j$, there exists a point $z_1 \in L_j$ satisfying the inequality

$$\log |f(z_1)| > R_j^{\lambda - \varepsilon}.$$

If z_2 is an arbitrary point of L_j in Γ_j, then we have

$$\log |f(z_1)| \leq \frac{3\alpha R_j + 2\alpha R_j}{3\alpha R_j - 2\alpha R_j} m(3\alpha R_j, z_2, f)$$

$$+ \sum_l \log \left| \frac{(3\alpha R_j)^2 - \overline{(\beta_l - z_2)}(z_1 - z_2)}{3\alpha R_j (\beta_l - z_1)} \right|$$

$$\leq 5m(3\alpha R_j, z_2, f) + n(3\alpha R_j, z_2, f)\left\{ \log \frac{N_j + 1}{h} + \log 6\alpha R_j \right\}$$

$$< C(\log R_j) T(4\alpha R_j, z_2, f),$$

where C is a constant depending only on λ, α, K_1 and K_2.

For sufficiently large j, Lemma 6.2 gives us

$$T(4\alpha R_j, z_2, f) < T\left(4\alpha R_j, z_2, \frac{f - \beta_1}{f - \beta_3} \cdot \frac{\beta_2 - \beta_3}{\beta_2 - \beta_1}\right) + \log |f(z_2) - \beta_3| + C$$

$$< C\left(\sum_{l=1}^{3} n(5\alpha R_j, z_2, f = \beta_l) + 1\right)$$

$$\times \left\{ \log \frac{\left(N_j + 1\right)\left(\sum\limits_{l=1}^{3} n(5\alpha R_j, z_2, f = \beta_l) + 1\right)}{h} \right.$$

$$+ \log(5\alpha R_j) + 10 \bigg\}$$

$$+ \log^+ \left| \frac{f(z_2) - \beta_1}{f(z_2) - \beta_2} \right| + \log |f(z_2) - \beta_3| + C.$$

From the above analysis and noting that

$$n(5\alpha R_j, z_2, f = \beta_l)$$
$$\leq n\left\{(|z| \leq (1+6\alpha)R_j) \cap (\varphi_1 + \alpha \leq \arg z \leq \varphi_2 - \alpha), f = \beta_l\right\}$$
$$< (1+6\alpha)^\tau R_j^\tau,$$

we get finally

$$R_j^{\lambda-\varepsilon} < \log|f(z_1)| \leq C \log R_j(R_j^\tau \log R_j + 2\log^+|f(z_2)|),$$

and hence

$$\log|f(z_2)| > R_j^{\lambda-2\varepsilon}$$

holds for sufficiently large j.

Starting from the disk Γ_j in the region G, we can construct a chain of at most $2N$ disks to cover L_j by moving its centre on $|z| = R_j$ in succession with the argument increasing (or decreasing) by one α every time. Therefore, in view of the above discussion, we can deduce that the inequality

$$\log|f(z)| > R_j^{\lambda-(2N+1)\varepsilon} > R_j^{\lambda-\eta} \qquad (6.1.16)$$

holds everywhere on L_j, provided that j is sufficiently large.

6.2   Distribution of Borel Directions of Meromorphic Functions with Deficient Values

6.2.1. Main result.   For an arbitrary meromorphic function $f(z)$ of finite positive order, the only distribution rule for its Borel directions is that they intersect the unit circumference in a non-empty closed set, as indicated in §5.1. However, in the case when $f(z)$ has a deficient value, the following gives more information (Yang Lo and Zhang Guanghou [2]).

Theorem 6.2   *Let $f(z)$ be meromorphic and of order $\lambda(0 < \lambda < \infty)$ in the finite plane, and suppose $f(z)$ has a deficient value a_0 (finite or infinite). When $f(z)$ has more than one Borel direction, there exist two Borel directions with an acute angle between them, which is less than or equal to π/λ. When $f(z)$ has only one Borel direction, then $\lambda \leq 1/2$.*

Proof.   We may assume that $a_0 = \infty$. Otherwise it would be sufficient to consider the function $1/(f(z) - a_0)$ instead of $f(z)$.

If $f(z)$ has an infinite number of Borel directions (countable or un-
countable), then they accumulate at (at least) one direction. Therefore, it
is clear that there exist two Borel directions with the angle between them
less than any preassigned positive number, and the theorem is true in this
case.

If $f(z)$ has $q(1 < q < \infty)$ Borel directions, then Lemma 6.1 shows that
there exist two successive Borel directions B_{m_0}: $\arg z = \varphi_{m_0}$, B_{m_0+1} :
$\arg z = \varphi_{m_0+1}$ and a sequence of positive numbers R_j such that

$$\text{mes} \left\{ \varphi : \varphi_{m_0} < \varphi < \varphi_{m_0+1}, \log |f(R_j e^{i\varphi})| \geq \frac{\delta(a_0, f)}{2^{\lambda+4}} U(R_j) \right\}$$
$$> \frac{1}{q} K(\delta, 1, \lambda),$$

(6.2.1)

provided j is sufficiently large. Therefore, we can find a sequence of curves
L_j satisfying the conclusion of Lemma 6.3.

Assuming the magnitude of the angle between B_{m_0} and B_{m_0+1} is
greater than π/λ, we shall derive a contradiction.

Without loss of generality, we may assume that

$$\varphi_{m_0+1} = -\varphi_{m_0}, \quad \varphi_{m_0+1} > 0.$$

Set $\varphi_{m_0+1} - \alpha = k\pi/2$, where α is a sufficiently small positive number such
that $k\pi > \pi/\lambda$. Choose the number b such that $1 < b < 2$ and $f(b) \neq \infty$.
Now we construct the transformation

$$\zeta = \frac{z^{\frac{1}{k}} - b^{\frac{1}{k}}}{z^{\frac{1}{k}} + b^{\frac{1}{k}}},$$

(6.2.2)

which maps the angle $D : \varphi_{m_0} + \alpha < \arg z < \varphi_{m_0+1} - \alpha$ onto the unit disk
$|\zeta| < 1$. The function $f(z)$ becomes $g(\zeta) = f(z(\zeta))$.

Let z be an arbitrary point on L_j. Its image ζ must satisfy

$$|\zeta| = \left| \frac{|z|^{\frac{1}{k}} e^{i\frac{\varphi}{k}} - b^{\frac{1}{k}}}{|z|^{\frac{1}{k}} e^{i\frac{\varphi}{k}} + b^{\frac{1}{k}}} \right| = \left\{ 1 - \frac{4|z|^{\frac{1}{k}} b^{\frac{1}{k}} \cos \frac{\varphi}{k}}{|z|^{\frac{2}{k}} + 2|z|^{\frac{1}{k}} b^{\frac{1}{k}} \cos \frac{\varphi}{k} + b^{\frac{2}{k}}} \right\}^{\frac{1}{2}}.$$

By noting that

$$|\varphi| \leq \varphi_{m_0+1} - \alpha'_j \leq \varphi_{m_0+1} - 8\alpha = k\left(\frac{\pi}{2} - \frac{7}{k}\alpha \right),$$

we have

$$|\zeta| \le \left\{ 1 - \frac{4|z|^{\frac{1}{k}} b^{\frac{1}{k}} \sin \frac{7}{k}\alpha}{(|z|^{\frac{1}{k}} + b^{\frac{1}{k}})^2} \right\}^{\frac{1}{2}} \le \left\{ 1 - |z|^{-\frac{1}{k}} \sin \frac{7}{k}\alpha \right\}^{\frac{1}{2}}$$

$$\le 1 - \frac{1}{2}|z|^{-\frac{1}{k}} \sin \frac{7}{k}\alpha \le 1 - \frac{\sin \frac{7}{k}\alpha}{2} R_j^{-\frac{1}{k}}.$$

Thus the image l_j of L_j under the transformation (6.2.2) is located in $|\zeta| \le r_j$, where

$$r_j = 1 - \frac{\sin \frac{7}{k}\alpha}{2} R_j^{-\frac{1}{k}}. \tag{6.2.3}$$

The inverse of (6.2.2) is

$$z = z(\zeta) = b\left(\frac{1+\zeta}{1-\zeta}\right)^k.$$

Because there is no Borel direction between B_{m_0} and B_{m_0+1}, we can find three distinct complex numbers $\beta_\nu (\nu = 1, 2, 3)$ and a number $\tau (0 < \tau < \lambda)$ such that

$$\sum_{\nu=1}^{3} n\left\{D(R), f = \beta_\nu\right\} < R^\tau, \quad R > R_0,$$

where $D(R)$ denotes the region $(|z| \le R) \cap (\varphi_{m_0} + \alpha \le \arg z \le \varphi_{m_0+1} - \alpha)$. Moreover, τ can be chosen so that $1 < k\tau < k\lambda$.

When $|\zeta| \le r$, we have $|z| \le b2^k/(1-r)^k$, hence

$$\sum_{\nu=1}^{3} n(|\zeta| \le r, g = \beta_\nu) \le \sum_{\nu=1}^{3} n\left\{D\left(b\frac{2^k}{(1-r)^k}\right), f = \beta_\nu\right\}$$

$$< (2^k b)^\tau \left(\frac{1}{1-r}\right)^{k\tau}.$$

Thus

$$\int_{r_0}^{1} \sum_{\nu=1}^{3} n(r, g = \beta_\nu)(1-r)^{k\tau - 1 + \varepsilon} dr < +\infty$$

for any $\varepsilon > 0$.

Since $k\tau > 1$ and

$$(k\tau - 1 + \varepsilon) \int_{r_0}^{r} \sum_{\nu=1}^{3} N(t, g = \beta_\nu)(1 - t)^{k\tau - 2 + \varepsilon} dt$$

$$= (1 - r_0)^{k\tau - 1 + \varepsilon} \sum_{\nu=1}^{3} N(r_0, g = \beta_\nu) - (1 - r)^{k\tau - 1 + \varepsilon}$$

$$\times \sum_{\nu=1}^{3} N(r, g = \beta_\nu) + \int_{r_0}^{r} \sum_{\nu=1}^{3} n(t, g = \beta_\nu)(1 - t)^{k\tau - 1 + \varepsilon} \frac{dt}{t},$$

we obtain

$$\int_{r_0}^{1} \sum_{\nu=1}^{3} N(r, g = \beta_\nu)(1 - r)^{k\tau - 2 + \varepsilon} dr < +\infty.$$

Furthermore, the second fundamental theorem of Nevanlinna yields

$$\int_{r_0}^{1} T(r, g)(1 - r)^{k\tau - 2 + \varepsilon} dr < +\infty. \tag{6.2.4}$$

This means the order of $g(\zeta)$ in $|\zeta| < 1$ does not exceed $k\tau - 1$.

On the other hand, put

$$N_j = n\left(\frac{1 + r_j}{2}, g = \infty\right), \quad h_j = \frac{1 - r_j}{4}.$$

Let $(\gamma)_j$ denote the family of all the exceptional disks with centers at a pole of $g(\zeta)$ in $|\zeta| \leq (1 + r_j)/2$ and the same radius $h_j/(n_j + 1)$. Then the total sum of the radii of the disks in $(\gamma)_j$ is less than h_j.

There exists a point ζ_2 on l_j (hereinafter we omit the lower index j of ζ_2 and ζ_1 for brevity), outside the disks in $(\gamma)_j$, since the images ζ_1 and $\bar{\zeta}_1$ of the extreme points $R_j e^{i(-\varphi_{m_0} + 1 + \alpha'_j)}$ and $R_j e^{i(\varphi_{m_0} + 1 - \alpha'_j)}$ of L_j under (6.2.2) satisfy

$$|\zeta_1 - \bar{\zeta}_1| = \left| \frac{2R_j^{\frac{1}{k}} b^{\frac{1}{k}} \left(e^{i\frac{\varphi_{m_0} + 1 - \alpha'_j}{k}} - e^{i\frac{-\varphi_{m_0} + 1 + \alpha'_j}{k}}\right)}{R_j^{\frac{2}{k}} + R_j^{\frac{1}{k}} b^{\frac{1}{k}} \left(e^{i\frac{\varphi_{m_0} + 1 - \alpha'_j}{k}} + e^{i\frac{-\varphi_{m_0} + 1 + \alpha'_j}{k}}\right) + b^{\frac{2}{k}}} \right|$$

$$\tag{6.2.5}$$

$$\geq \frac{4R_j^{\frac{1}{k}} b^{\frac{1}{k}} \cos \frac{8\alpha}{k}}{(R_j^{\frac{1}{k}} + b^{\frac{1}{k}})^2} \geq \frac{R_j^{-\frac{1}{k}}}{2} > 4h_j.$$

Then the Poisson-Jensen formula gives

$$\log |g(\zeta_2)| \leq \frac{\frac{1+r_j}{2}+r_j}{\frac{1+r_j}{2}-r_j} m\left(\frac{1+r_j}{2},g\right) + \sum_\nu \log \left| \frac{\left(\frac{1+r_j}{2}\right)^2 - \bar{b}_\nu \zeta_2}{\frac{1+r_j}{2}(\zeta_2 - b_\nu)} \right|$$

$$\leq \frac{4}{1-r_j} m\left(\frac{1+r_j}{2},g\right) + n\left(\frac{1+r_j}{2},g=\infty\right)\log\frac{2(\mathcal{N}+1)}{h_j},$$

where the $b_\nu (\nu = 1,2,\cdots,n((1+r_j)/2, g = \infty))$ are the poles of $g(\zeta)$ in $|\zeta| \leq (1+r_j)/2$. Since

$$n\left(\frac{1+r_j}{2}, g=\infty\right) < \frac{4}{1-r_j}T\left(\frac{3+r_j}{4},g\right)$$

and

$$\mathcal{N}_j \leq n\left\{ \left(|z| \leq b\frac{4^k}{(1-r_j)^k}\right) \cap (\varphi_{m_0}+\alpha \leq \arg z \leq \varphi_{m_0+1}-\alpha), f=\infty\right\}$$

$$< \frac{1}{(1-r_j)^{k\lambda+\varepsilon}},$$

we deduce that

$$\log |g(\zeta_2)| \leq \frac{C}{1-r_j}\left(\log\frac{1}{1-r_j}\right)T\left(\frac{3+r_j}{4},g\right), \qquad (6.2.6)$$

provided j is sufficiently large. Moreover,

$$\log |g(\zeta_2)| > R_j^{\lambda-\eta} = \left(\frac{\sin\frac{7}{k}\alpha}{2(1-r_j)}\right)^{k(\lambda-\eta)}$$

from Lemma 6.3, so we get finally

$$T\left(\frac{3+r_j}{4},g\right) > C\left(\frac{1}{1-r_j}\right)^{k\lambda-1-k\eta}\frac{1}{\log\frac{1}{1-r_j}}. \qquad (6.2.7)$$

This means that the order of $g(\zeta)$ in $|\zeta| < 1$ is not less than $k\lambda - 1$.

Thus $k\lambda - 1 \leq k\tau - 1$, i.e. $\lambda \leq \tau$ which contradicts the hypothesis $\tau < \lambda$. Therefore, the magnitude of the acute angle between B_{m_0} and B_{m_0+1} does not exceed π/λ.

Finally, when $f(z)$ has only one Borel direction B_1: $\arg z = \varphi_1$ $(0 \leq \varphi_1 < 2\pi)$, we may assume that $\varphi_1 = \pi$. If $\lambda > 1/2$, the above discussion can be repeated in the angle $|\arg z| < \pi - \alpha$, where $\pi - \alpha = k/2\pi$ and $k\pi > \pi/\lambda$. A contradiction is again derived.   □

6.2.2 Discussion.

Corollary.   *Let $f(z)$ be meromorphic and of order $\lambda(0 < \lambda < \infty)$ in the finite plane, and suppose $f(z)$ has a deficient value. If λ is larger than $1/2$, then there are at least two Borel directions with the acute angle between them not exceeding π/λ.*

It is easy to see that this Corollary is equivalent to Theorem 6.2 and implies the following result of Valiron and Cartwright:

An entire function of finite order $\lambda > 1/2$ has at least two Borel directions.

In the statement of Theorem 6.2, we can replace 'a deficient value' of the function $f(z)$ by a 'Borel exceptional value' to obtain a similar theorem.

Theorem 6.3   *Let $f(z)$ be meromorphic and of order $\lambda(0 < \lambda < \infty)$ in the finite plane, and suppose $f(z)$ has a Borel exceptional value a_0 (finite or infinite). When $f(z)$ has more than one Borel direction, there exist two Borel directions with an acute angle between them which is less than or equal to π/λ. When $f(z)$ has only one Borel direction, then $\lambda \le 1/2$.*

In fact, the condition that $f(z)$ has a deficient value is used only in (6.1.9) of Lemma 6.1. When $f(z)$ takes a_0 as a Borel exceptional value, we may assume that $a_0 = \infty$. Thus

$$N(R_j, f) < R_j^\tau < (2r_j)^\tau < \frac{1}{4}U(r_j),$$

for some $\tau, 0 < \tau < \lambda$, and for sufficiently large integers j. From this inequality and

$$T(R_j, f) \ge T(r_j, f) \ge \frac{1}{2}U(r_j),$$

we have

$$m(R_j, f) > \frac{1}{4}U(r_j).$$

The proofs of Lemma 6.1 and Theorem 6.2 can be completed as before.

Furthermore, the condition that $f(z)$ have a deficient value or a Borel exceptional value in Theorems 6.2 and 6.3, can be replaced by a corresponding condition for its derivative. More precisely, we have the following theorem:

Theorem 6.4   *Let $f(z)$ be meromorphic and of order λ ($0 < \lambda < \infty$) in the finite plane and k a positive integer. If $f^{(k)}(z)$ takes a_0 as a deficient value or Borel exceptional value, then the conclusion of Theorem 6.2 remains true.*

For the proof of Theorem 6.4, the reader may consult the original paper (Yang Lo and Zhang Guanghou [2]).

6.3 Deficient Values and Borel Directions
of Meromorphic Functions

6.3.1   Deficient values and Borel directions.   We have seen in Theorem 1.9 that for a transcendental meromorphic function $f(z)$ in the finite plane, its deficient values are countable and the total sum of deficiencies does not exceed 2. It is therefore natural that R. Nevanlinna and many others were interested in the following problem. Under what conditions does the number of deficient values of $f(z)$ have a finite upper bound? For instance, R. Nevanlinna [2] conjectured that a meromorphic function of finite order has a finite number of deficient values. In particular, he conjectured that an entire function of finite order λ has at most 2λ finite deficient values.

There have been many papers on this subject. First, A. Pfluger [1] proved that if the total sum of deficiencies of an entire function $f(z)$ of finite order λ is exactly equal to 2, then it has at most $\lambda + 1$ deficient values. Later, G. Valiron [4] proved that a meromorphic function of order zero has at most one deficient value . On the other hand, A. A. Goldberg [1] constructed a meromorphic function of finite order having an infinite number of deficient values.

In the early sixties, A. Edrei and W. H. J. Fuchs [4, 5] studied this problem for a special class of meromorphic functions which satisfy some restrictions on the location of the zeros and poles. In 1966, N. U. Arakelyan

[1] constructed, for any $\lambda > 1/2$, an entire function of order λ having an infinite number of deficient values. Later, A. Weitsman [1] extended Pfluger's result to meromorphic functions.

Here, we shall exphasize the relation between the number of deficient values and the number of Borel directions (Yang Lo and Zhang Guanghou [3]).

Theorem 6.5 *Let $f(z)$ be meromorphic and of order λ in the finite plane, where $0 < \lambda < \infty$. If p denotes the number of deficient values of $f(z)$ and q the number of Borel directions of $f(z)$, then we have $p \leq q$.*

Proof.   According to Theorem 3.8, $f(z)$ has at least one Borel direction. Thus $q \geq 1$.

When $q = \infty$, the conclusion of Theorem 6.5 is obvious.

When $1 \leq q < \infty$, let us suppose Theorem 6.5 is not true, i.e. $p \geq q+1$, and derive a contradiction.

Take $q+1$ deficient values $a_\nu(\nu = 1, 2, \cdots, q+1)$ of $f(z)$ and let $\delta(a_\nu, f) = \delta_\nu$.

We may assume that the $a_\nu(\nu = 1, 2, \cdots, q+1)$ are all finite. Otherwise we can consider the function $1/(f(z) - a_0)$ (a_0 is not a deficient value of $f(z)$) instead of $f(z)$. Clearly $f(z)$ and $1/(f(z)-a_0)$ have the same number of deficient values and the same Borel directions.

Denote by B_m: $\arg z = \varphi_m(m = 1, 2, \cdots, q; 0 \leq \varphi_1 < \varphi_2 < \cdots < \varphi_q < 2\pi; \varphi_{q+1} = 2\pi + \varphi_1)$ the q Borel directions of $f(z)$, and by $\lambda(r)$ a precise order of $f(z)$ with $U(r) = r^{\lambda(r)}$. Applying Lemma 6.1 to $f(z)$ and the $q + 1$ deficient values $a_\nu(\nu = 1, 2, \cdots, q + 1)$, we obtain a sequence of positive numbers R_j tending to infinity such that

$$\text{mes } E_{j\nu} = \text{mes } \left\{ \varphi : 0 \leq \varphi < 2\pi, \log \frac{1}{|f(R_j e^{i\varphi}) - a_\nu|} > \frac{\delta}{2^{\lambda+4}} U(R_j) \right\}$$

$$> K(\delta, q+1, \lambda), \quad \nu = 1, 2, \cdots, q+1.$$

$$(6.3.1)$$

for sufficiently large j, where $\delta = \min_{1 \leq \nu \leq q+1} \delta_\nu$ and

$$K(\delta, q+1, \lambda) = \frac{8\pi}{\left(5 + \dfrac{\log 25(q+1)e}{\log \dfrac{4}{3}} \right) 4^{\lambda+2}}. \qquad (6.3.2)$$

Among the q angles $\varphi_m < \arg z < \varphi_{m+1}(m = 1, 2, \cdots, q)$, there exists at least one angle, say $G_1 : \varphi_{m_1} < \arg z < \varphi_{m_1+1}$, and a subsequence (R_{j1}) of (R_j), such that the set of values φ for which $\varphi_{m_1} < \varphi < \varphi_{m_1+1}$ and

$$\log \frac{1}{|f(R_{j1}e^{i\varphi}) - a_1|} > \frac{\delta}{2^{\lambda+4}} U(R_{j1}) > R_{j1}^{\lambda-\varepsilon}, \quad \varepsilon > 0, \tag{6.3.3}$$

has measure greater than $(1/q)K(\delta, q + 1, \lambda)$.

Then applying Lemma 6.3 to $f(z)$, its deficient value a_1, the two Borel directions

$$B_{m_1} : \arg z = \varphi_{m_1},$$

and

$$B_{m_1+1} : \arg z = \varphi_{m_1+1},$$

and the sequence (R_{j1}) and positive constants

$$K_1 = K_2 = \frac{1}{q}K(\delta, q + 1, \lambda),$$

we obtain a sequence of curves $L_{j1}^{(1)}$ such that for every positive number η, the inequality

$$\log \frac{1}{|f(z) - a_1|} > R_{j1}^{\lambda-\eta} \tag{6.3.4}$$

holds everywhere on $L_{j1}^{(1)}$, and moreover

$$\text{mes } \{\varphi : R_{j1}e^{i\varphi} \in A_{j1}^{(1)} - L_{j1}^{(1)}\} < \frac{1}{q}K(\delta, q + 1, \lambda).$$

It is clear that for sufficiently large j,

$$\max \left\{ e^{-R_{j1}^{\lambda-\varepsilon}}, e^{-R_{j1}^{\lambda-\eta}} \right\} < \frac{d}{2},$$

where $d = \min\limits_{1 \leq \gamma \leq q+1} |a_\mu - a_\nu|$. Thus no point z on $L_{j1}^{(1)}$ satisfies the inequality

$$\log \frac{1}{|f(z) - a_\nu|} > R_{j1}^{\lambda-\varepsilon}, \quad \nu = 2, 3, \cdots, q + 1. \tag{6.3.5}$$

In fact, it follows from comparing (6.3.4) and (6.3.5) that if there exists a point $z \in L_{j1}^{(1)}$ satisfying (6.3.5) for some a_{ν_0} ($2 \leq \nu_0 \leq q + 1$), then it leads to a contradiction:

$$d \leq |a_1 - a_{\nu_0}| \leq |a_1 - f(z)| + |f(z) - a_{\nu_0}| < e^{-R_{j1}^{\lambda-\eta}} + e^{-R_{j1}^{\lambda-\varepsilon}} < d.$$

Thus the points $z = R_{j1}e^{i\varphi} \in G_1$ satisfying the inequality

$$\log \frac{1}{|f(R_{j1}e^{i\varphi}) - a_2|} > R_{j1}^{\lambda-\varepsilon} \tag{6.3.6}$$

must be contained in $A_{j1}^{(1)} - L_{j1}^{(1)}$. Therefore the set of values $\varphi(\varphi_1 < \varphi < \varphi_2)$ for which (6.3.6) holds has measure less than $(1/q)K(\delta, q+1, \lambda)$. Since (R_{j1}) is a subsequence of (R_j), the measure of the set of values $\varphi(0 \le \varphi < 2\pi)$ satisfying (6.3.6) is greater than $K(\delta, q+1, \lambda)$.

Among the $q-1$ angles $\varphi_m < \arg z < \varphi_{m+1}(m = 1, 2, \cdots, m_1-1, m_1+1, \cdots, q)$, there is at least one angle, say $G_2 : \varphi_{m_2} < \arg z < \varphi_{m_2+1}$, and a subsequence (R_{j2}) of (R_{j1}) such that the set of values φ for which

$$\log \frac{1}{|f(R_j e^{i\varphi}) - a_2|} > \frac{\delta}{2^{\lambda+4}}U(R_{j2}) > R_{j2}^{\lambda-\varepsilon}, \tag{6.3.7}$$

has measure larger than $(1/q)K(\delta, q+1, \lambda)$.

Again, applying Lemma 6.3 to $f(z)$, its deficient value a_2, two Borel directions

$$B_{m_2} : \arg z = \varphi_{m_2}$$

and

$$B_{m_2+1} : \arg z = \varphi_{m_2+1},$$

the sequence (R_{j2}) and positive constants

$$K_1 = K_2 = \frac{1}{q}K(\delta, q+1, \lambda),$$

we obtain a sequence of curves $L_{j2}^{(2)}$, so that each $z \in L_{j2}^{(2)}$ satisfies

$$\log \frac{1}{|f(z) - a_2|} > R_{j2}^{\lambda-\eta} \tag{6.3.8}$$

for every positive number η, and

$$\text{mes}\{\varphi : R_{j2}e^{i\varphi} \in A_{j2}^{(2)} - L_{j2}^{(2)}\} < \frac{1}{q}K(\delta, q+1, \lambda).$$

Similarly, no point z on $L_{j2}^{(2)}$ satisfies the inequalities

$$\log \frac{1}{|f(z) - a_\nu|} > R_{j2}^{\lambda-\varepsilon}, \quad \nu = 1, 3, 4, \cdots, q+1. \tag{6.3.9}$$

Thus, the point $z = R_{j2}e^{i\varphi} \in G_\nu(\nu = 1, 2)$ satisfying the inequality

$$\log \frac{1}{|f(R_{j2}e^{i\varphi}) - a_3|} > R_{j2}^{\lambda-\varepsilon} \tag{6.3.10}$$

must be contained in $A_{j2}^{(\nu)} - L_{j2}^{(\nu)}$. Consequently the measure of the set of values $\varphi(\varphi_{m_\nu} < \varphi < \varphi_{m_\nu+1})$ $(\nu = 1, 2)$ satisfying (6.3.10) is less than $(1/q)K(\delta, q+1, \lambda)$. Since (R_{j2}) is a subsequence of (R_j), the set of values $\varphi(0 \le \varphi < 2\pi)$ satisfying (6.3.10) has measure greater than $K(\delta, q+1, \lambda)$.

Among the $q - 2$ angles $\varphi_m < \arg z < \varphi_{m+1}(m \ne m_1, m_2)$, there is at least one, say G_3, having similar properties.

Continuing this procedure, we obtain finally q angles $G_\nu : \varphi_{m_\nu} < \arg z < \varphi_{m_\nu+1}(\nu = 1, 2, \cdots, q)$, not intersecting one another, such that for every G_ν, there is a sequence of positive numbers $R_{j\nu}$ and a sequence of curves $L_{j\nu}^{(\nu)}$, with

$$\log \frac{1}{|f(z) - a_\nu|} > R_{j\nu}^{\lambda-\eta} \tag{6.3.11}$$

and

$$\log \frac{1}{|f(z) - a_\mu|} \le R_{j\nu}^{\lambda-\varepsilon}, \quad \mu = 1, 2, \cdots, \nu - 1, \nu + 1, \cdots, q + 1, \tag{6.3.11}$$

for every $z \in L_{j\nu}^{(\nu)}$ and two arbitrary small positive numbers η and ε. Moreover, (R_{j1}) is a subsequence of (R_j), $(R_{j\nu+1})$ is a subsequence of $(R_{j\nu})$ $(\nu = 1, 2, \cdots, q - 1)$, and

$$\text{mes } \{\varphi : R_{j\nu}e^{i\varphi} \in A_{j\nu}^{(\nu)} - L_{j\nu}^{(\nu)}\} < \frac{1}{q}K(\delta, q + 1, \lambda),$$

where $A_{j\nu}^{(\nu)}$ denotes the arc $\{R_{j\nu}e^{i\varphi} : \varphi_{m_\nu} < \varphi < \varphi_{m_\nu+1}\}$.

Thus, in every angle G_ν $(\nu = 1, 2, \cdots, q)$, the set of values φ for which

$$\log \frac{1}{|f(R_{jq}e^{i\varphi}) - a_{q+1}|} > R_{jq}^{\lambda-\varepsilon}, \tag{6.3.13}$$

has measure less than or equal to

$$\text{mes}\{\varphi : R_{jq}e^{i\varphi} \in A_{jq}^{(\nu)} - L_{jq}^{(\nu)}\} < \frac{1}{q}K(\delta, q + 1, \lambda).$$

The q Borel directions $B_m : \arg z = \varphi_m(m = 1, 2, \cdots, q)$ divide the plane into q contiguous angular sectors, and therefore each of the q angles coincides with one of the G'_νs ($\nu = 1, 2, \cdots, q$). Thus

$$\left(\bigcup_{\nu=1}^{q} G_\nu \right) \cup \left(\bigcup_{m=1}^{q} B_m \right)$$

covers the plane. Consequently the set of values $\varphi (0 \le \varphi < 2\pi)$ satisfying (6.3.13) has measure

$$\le \sum_{\nu=1}^{q} \text{mes } \{\varphi : R_{jq}e^{i\varphi} \in A_{jq}^{(\nu)} - L_{jq}^{(\nu)}\} < K(\delta, q+1, \lambda).$$

However, since (R_{jq}) is a subsequence of (R_j), it follows from (6.3.1) that the set of values $\varphi (0 \le \varphi < 2\pi)$ satisfying (6.3.13) must have measure greater than $K(\delta, q+1, \lambda)$. With this contradiction our proof is complete.
□

6.3.2 Complement.   The inequality $p \le q$ in Theorem 6.5 is sharp in the following sense: Given a positive integer n, there is a meromorphic function $f(z)$ of finite positive order, such that the number of its deficient values and the number of its Borel directions are both equal to n.

In fact, when $n = 1$, the function

$$f(z) = \prod_{k=1}^{n} \left(1 - \frac{z}{k^2} \right) \tag{6.3.14}$$

provides such an example. Following the calculation in Lemma 7.9 in the next chapter, we can derive that the order of $f(z)$ equals $1/2$ and

$$\log |f(re^{i\theta})| = \left(\pi \cos \frac{\theta - \pi}{2} \right) r^{\frac{1}{2}} + o(r^{\frac{1}{2}}) \tag{6.3.15}$$

in the angle $\varepsilon < \arg z < 2\pi - \varepsilon$ for every positive number ε. Thus $f(z)$ has no finite deficient value by Lemma 6.1 and (6.3.15). On the other hand, $f(z)$ has at least one Borel direction, according to Theorem 3.8. However, (6.3.15) indicates that $f(z)$ cannot take any direction other than the positive real axis as a Borel direction. Therefore, $p = q = 1$.

When $n = 2$, the function e^z affords us a good example. It is clear that e^z has order 1, two deficient values, zero and infinity, and two Borel directions, the positive and negative imaginary axes. Thus $p = q = 2$.

When $n > 2$, we consider the function

$$f(z) = \frac{J_{\frac{1}{n}}\left(\frac{2z^{\frac{n}{2}}}{n}\right)}{J_{-\frac{1}{n}}\left(\frac{2z^{\frac{n}{2}}}{n}\right)}, \qquad (6.3.16)$$

where

$$J_{\frac{1}{n}}\left(\frac{2z^{\frac{n}{2}}}{n}\right) = \frac{1}{n^{\frac{1}{n}}z^{\frac{1}{2}}}\sum_{k=0}^{\infty}\frac{(-1)^k z^{nk+1}}{n^{2k}k!\,\Gamma\left(\frac{1}{n}+k+1\right)}$$

and

$$J_{-\frac{1}{n}}\left(\frac{2z^{\frac{n}{2}}}{n}\right) = \frac{n^{\frac{1}{n}}}{2^{\frac{1}{2}}}\sum_{k=0}^{\infty}\frac{(-1)^k z^{nk}}{n^{2k}k!\,\Gamma\left(-\frac{1}{n}+k+1\right)}.$$

By a calculation, we can see that $f(z)$ has order $n/2$, exactly n deficient values $e^{(2k+1)\pi i/n}$ $(k = 0, 1, \cdots, n-1)$ and exactly n Borel directions $\arg z = 2k\pi/n$ $(k = 0, 1, \cdots, n-1)$. Consequently $p = q = n$.

For meromorphic functions of higher order, we can obtain a more precise result.

Theorem 6.6   *Let $f(z)$ be meromorphic and of order λ $(0 < \lambda < \infty)$ in the finite plane, p the number of its deficient values and q the number of its Borel directions. If $\lambda > p/2$, then $p \le q - 1$.*

Proof.   Theorem 6.6 clearly holds when $q = \infty$. When $q < \infty$, let us assume that Theorem 6.6 is not true, i.e., assume $p > q - 1$. Then we can choose q deficient values $a_\nu(\nu = 1, 2, \cdots, q)$ with $\delta(a_\nu, f) > 0$. Set $\delta = \min_{1 \le \nu \le q}\delta(a_\nu, f)$ and denote by $B_m : \arg z = \varphi_m(m = 1, 2, \cdots, q, 0 \le \varphi_1 < \varphi_2 < \cdots < \varphi_q < 2\pi, \varphi_{q+1} = 2\pi + \varphi_1)$, the q Borel directions of $f(z)$.

By a discussion similar to the proof of Theorem 6.5, there are q distinct angles $G_\nu : \varphi_{m_\nu} < \arg z < \varphi_{m_\nu+1}(\nu = 1, 2, \cdots, q)$ corresponding to q deficient values a_ν. For every G_ν, Lemma 6.3 gives us a sequence of positive numbers $R_{j\nu}$ tending to infinity and a sequence of curves $L_{j\nu}^{(\nu)}$. According to the proof of Lemma 6.3, the magnitude of G_ν does not exceed π/λ, i.e.,

$$\varphi_{m_\nu+1} - \varphi_{m_\nu} \le \frac{\pi}{\lambda}, \qquad \nu = 1, 2, \cdots, q.$$

Thus

$$\sum_{\nu=1}^{q} (\varphi_{m_\nu+1} - \varphi_{m_\nu}) \leq q \cdot \frac{\pi}{\lambda} \leq p \cdot \frac{\pi}{\lambda} < 2\pi.$$

Since the finite plane is divided into q angles $G_\nu (\nu = 1, 2, \cdots, q)$ by the q Borel directions $B_m(m = 1, 2, \cdots, q)$, we have

$$\sum_{\nu=1}^{q} (\varphi_{m_\nu+1} - \varphi_{m_\nu}) = 2\pi.$$

This contradiction proves that p does not exceed $q - 1$.   □

6.4 Deficient Values, Borel Directions and the Order of Entire Functions

6.4.1 Several lemmas.   In order to obtain further results for entire functions, we need to establish several lemmas.

Lemma 6.4   Let $f(z)$ be an entire function of order λ $(0 < \lambda < \infty)$ with no Borel direction in the angle $\varphi_1 < \arg z < \varphi_2$ $(0 \leq \varphi_1 < \varphi_2 \leq 2\pi + \varphi_1)$, and $\lambda(r)$ its precise order with $U(r) = r^{\lambda(r)}$. Suppose there are positive numbers η and K', a finite complex number a_0, and a sequence of positive numbers R_j tending to infinity such that

$$\text{mes } E\{\varphi : \varphi_1 < \varphi < \varphi_2, \log |f(R_j e^{i\varphi}) - a_0| < -\eta U(R_j)\} > K'. \quad (6.4.1)$$

Then, for any two positive numbers α and Q with

$$\alpha < \min \left(\frac{K'}{32}, \frac{\varphi_2 - \varphi_1}{20} \right) \quad \text{and} \quad 1 < Q < \frac{1}{4\alpha},$$

we have

$$\log |f(z) - a_0| < -\frac{\eta}{\left\{ 4 \left(5 + 4 \log \frac{2}{h} \right) \right\}^{2N_1 + N_2}} U(R_j) \quad (6.4.2)$$

and

$$\log |f^{(k)}(z)| < -\frac{\eta}{\left\{ 4 \left(5 + 4 \log \frac{2}{h} \right) \right\}^{2N_1 + N_2}} U(R_j) + \log k! \left\{ \left(\frac{2Q}{\alpha R_j} \right)^k \right\},$$

$$k = 1, 2, \cdots,$$

$$(6.4.3)$$

in the region

$$D_j : \left(\frac{R_j}{Q} \leq |z| \leq QR_j\right) \cap (\varphi_1 + 10\alpha \leq \arg z \leq \varphi_2 - 10\alpha),$$

where

$$N_1 = \left[\frac{\pi}{\alpha}\right] + 1, \quad N_2 = \left[\frac{2(Q-1)}{\alpha}\right] + 4Q \qquad (6.4.4)$$

and

$$h = \frac{K'}{80(2e+1)(\pi + \alpha)}. \qquad (6.4.5)$$

Proof.   We divide $G : \varphi_1 + 8\alpha \leq \arg z \leq \varphi_2 - 8\alpha$ into $N_1 = [\pi/\alpha] + 1$ equal angles $G_\nu(\nu = 1, 2, \cdots, N_1)$, such that the magnitude of every G_ν does not exceed 2α. Since we have clearly

$$\text{mes } \{\varphi : \varphi_1 + 8\alpha \leq \varphi \leq \varphi_2 - 8\alpha, \log |f(R_j e^{i\varphi}) - a_0| < -\eta U(R_j)\} > \frac{K'}{2},$$

there is an angle G_{ν_0} $(\nu_0 = \nu_0(j))$ with

$$\text{mes } \{\varphi : R_j e^{i\varphi} \in G_{\nu_0}, \log |f(R_j e^{i\varphi}) - a_0| < -\eta U(R_j)\} > \frac{K'}{2N_1}$$

$$\geq \frac{\alpha}{2(\pi + \alpha)} K'.$$

Since $f(z)$ has no Borel direction in $\varphi_1 < \arg z < \varphi_2$, we can find two finite distinct complex values $b_l(l = 1, 2)$ and a positive number $\tau, 0 < \tau < \lambda$, by Theorem 3.10, such that

$$\sum_{l=1}^{2} n\{(|z| \leq r) \cap (\varphi_1 + \alpha \leq \arg z \leq \varphi_2 - \alpha), f = b_l\} < r^\tau$$

for sufficiently large r.

Let

$$n_1 = \sum_{l=1}^{2} n\left\{(|z| \leq (1 + 6\alpha)R_j) \cap (\varphi_1 + \alpha \leq \arg z \leq \varphi_2 - \alpha), f = b_l\right\}$$

$$+ n\left\{(|z| \leq (1 + 6\alpha)R_j) \cap (\varphi_1 + \alpha \leq \arg z \leq \varphi_2 - \alpha), f = a_0\right\},$$

$$h = \frac{K'}{80(2e+1)(\pi + \alpha')}.$$

Denote by $(\gamma)_1$ the family of all the disks having their centres at a zero of either $f(z) - b_l(l = 1, 2)$ or $f(z) - a_0$ in the region $(|z| \leq (1 + 6\alpha)R_j) \cap (\varphi_1 + \alpha \leq \arg z \leq \varphi_2 - \alpha)$ and the same radius $5\alpha R_j h/(n_1 + 1)$.

Let z_1 be the intersection point of $|z| = R_j$ and the bisector of G_{ν_0}, and let Γ_j be the disk $|z - z_1| \leq \alpha R_j$. Denote by $\beta_s(s = 1, 2, \cdots, n_2)$ the zeros of $f(z) - a_0$ in $|z - z_2| \leq 3\alpha R_j$, where z_2 is an arbitrary but fixed point in Γ_j and outside every disk in $(\gamma)_1$. According to the Boutroux-Cartan theorem, we have

$$\prod_{s=1}^{n_2} |z - \beta_s| > (5\alpha R_j h)^{n_2}$$

for the points outside at most n_2 disks with the total sum of radii not exceeding $10eh\alpha R_j$. Denote by $(\gamma)_2$ the family of these disks.

Since the total sum of the diameters of the disks in $(\gamma)_1$ or $(\gamma)_2$ is less than or equal to

$$10h\alpha R_j + 20eh\alpha R_j = \frac{\alpha K'}{8(\pi + \alpha)} R_j,$$

there is a point z_3 in Γ_j but outside every disk in $(\gamma)_1 \cup (\gamma)_2$ such that

$$\log |f(z_3) - a_0| < -\eta U(R_j).$$

Applying the Poisson-Jensen's formula, we have

$$
\begin{aligned}
\log \frac{1}{|f(z_3) - a_0|} &\leq \frac{3\alpha R_j + 2\alpha R_j}{3\alpha R_j - 2\alpha R_j} m\left(3\alpha R_j, z_2, \frac{1}{f - a_0}\right) \\
&\quad + \sum \log \left|\frac{(3\alpha R_j)^2 - \bar{\beta}_s z_3}{3\alpha R_j(z_3 - \beta_s)}\right| \\
&\leq 5m\left(3\alpha R_j, z_2, \frac{1}{f - a_0}\right) + n\left(3\alpha R_j, z_2, \frac{1}{f - a_0}\right) \log \frac{2}{h} \\
&< \left(5 + 4\log \frac{2}{h}\right) T\left(4\alpha R_j, z_2, \frac{1}{f - a_0}\right),
\end{aligned}
$$

where $m(3\alpha R_j, z_2, 1/(f - a_0))$ and $T(4\alpha R_j, z_2, 1/(f - a_0))$ denote $m\left(3\alpha R_j, \frac{1}{f(z + z_2) - a_0}\right)$ and $T\left(4\alpha R_j, \frac{1}{f(z + z_2) - a}\right)$ respectively.

Furthermore Lemma 6.2 gives us

$$T\left(4\alpha R_j, z_2, \frac{1}{f - a_0}\right)$$

$$\leq T\left(4\alpha R_j, z_2, \frac{\dfrac{1}{f - a_0} - \dfrac{1}{b - a_0} - \dfrac{1}{b_2 - a_0}}{\dfrac{1}{f - a_0} - \dfrac{1}{b_2 - a_0}} \cdot \frac{1}{b_1 - a_0}\right)$$

$$+ \log\left|\frac{1}{f(z_2) - a_0} - \frac{1}{b_2 - a_0}\right| + D$$

$$< C\left(\sum_{l=1}^{2} n(5\alpha R_j, z_2, f = b_l) + 1\right)$$

$$\times \left\{ \log \frac{\left(\displaystyle\sum_{l=1}^{2} n(5\alpha R_j, z_2, f = b_l) + 1\right)}{h/(n_1 + 1)} + \log 10 \right\}$$

$$+ 2\log^{+} \frac{1}{|f(z_2) - a_0|} + D,$$

where D is a constant, which may vary with each occurrence.

In view of the fact that

$$\sum_{l=1}^{2} n(5\alpha R_j, z_2, f = b_l)$$

$$\leq \sum_{l=1}^{2} n\left\{ (|z| \leq (1 + 6\alpha)R_j) \cap (\varphi_1 + \alpha \leq \arg z \leq \varphi_2 - \alpha), f = \beta_l \right\}$$

$$< (1 + 6\alpha)^\tau R_j^\tau, \quad j > j_0,$$

and

$$n_1 < R_j^{\lambda+1}, \quad j > j_0,$$

we have

$$C\left(\sum_{l=1}^{2} n(5\alpha R_j, z_2, f = b_l) + 1\right) \left\{ \log \frac{\left(\displaystyle\sum_{l=1}^{2} n(5\alpha R_j, z_2, f = b_l) + 1\right)}{\dfrac{h}{n_1 + 1}} \right.$$

$$\left. + \log 10 \right\} + D < \frac{\eta}{4\left(5 + 4\log \dfrac{2}{h}\right)^{2N_1 + N_2}} U(R_j).$$

$$(6.4.6)$$

Thus

$$\log \frac{1}{|f(z_2) - a_0|} > \frac{\eta}{4\left(5 + 4\log \frac{2}{h}\right)} U(R_j). \qquad (6.4.7)$$

Now let us start from the disk Γ_j and constuct a chain Γ_Σ of at most $2N$ disks by moving its centre successively on $|z| = R_j$ with the argument increasing (or decreasing) each time by the amount of α to cover the region

$$\left(R_j - \frac{\sqrt{3}}{2}\alpha R_j \leq |z| \leq R_j + \frac{\sqrt{3}}{2}\alpha R_j\right) \cap (\varphi_1 + 8\alpha \leq \arg z \leq \varphi_2 - 8\alpha).$$

In a similar way, we can prove that

$$\log \frac{1}{|f(z) - a_0|} > \frac{\eta}{\left\{4\left(5 + 4\log \frac{2}{h}\right)\right\}^{2N_1}} U(R_j),$$

i.e.,

$$|f(z) - a_0| < e^{-\frac{\eta}{\{4(5+4\log(2/h))\}^{2N_1}} U(R_j)}, \qquad (6.4.8)$$

whenever the point $z \in \Gamma_j$, but outside the disks in $(\gamma)_1$. Since the total sum of the diameters of the disks in $(\gamma)_1$ does not exceed $10h\alpha R_j < \alpha R_j/8$, it follows from the maximum modulus theorem that (6.4.8) holds everywhere in the region

$$\left(R_j - \frac{\alpha R_j}{2} \leq |z| \leq R_j + \frac{\alpha R_j}{2}\right) \cap \left(\varphi_1 + \frac{33}{4}\alpha \leq \arg z \leq \varphi_2 - \frac{33}{4}\alpha\right).$$

Let $z_0 = R_j e^{i\theta_0}$ be an arbitrary point on

$$(|z| = R_j) \cap \left(\varphi_1 + \frac{33}{4}\alpha \leq \arg z \leq \varphi_2 - \frac{33}{4}\alpha\right).$$

We construct two concentric disks

$$\Gamma_j' : |z - z_0| \leq \frac{\alpha R_j}{2}, \quad \Gamma_j'' : |z - z_0| \leq \frac{\alpha R_j}{2Q}.$$

Starting from Γ_j' and moving its centre along the segment

$$L_1 : \{re^{i\theta_0} : R_j \leq r \leq (Q + 2\alpha)R_j\}$$

in succession with its distance increasing by $\alpha R_j/2$ every time, we obtain $N_2 = [2(Q-1)/\alpha] + 4Q$ such disks to cover L_1. Similarly,

$$L_2 : \{re^{i\theta_0} : \left(\frac{1}{Q} - 2\alpha\right)R_j \leq r \leq R_j\}$$

can be covered by N_2 disks obtained by moving Γ''_j along L_2 with the distance of its centre increasing by $\alpha R_j/2Q$ every time.

Therefore we have

$$\log|f(z) - a_0| < -\frac{\eta}{\left\{4(5 + 4\log\frac{2}{h})\right\}^{2N_1+N_2}}U(R_j) \qquad (6.4.9)$$

in the region

$$D'_j : \left\{\left(\frac{1}{Q} - \alpha\right)R_j \leq |z| \leq (Q + \alpha)R_j\right\} \cap (\varphi_1 + 9\alpha \leq \arg z \leq \varphi_2 - 9\alpha).$$

Let ζ be an arbitrary point in

$$D_j : \left\{\frac{R_j}{Q} \leq |z| \leq QR_j\right\} \cap (\varphi_1 + 10\alpha \leq \arg z \leq \varphi_2 - 10\alpha).$$

It is clear that the disk $|z - \zeta| \leq (\alpha/2Q)R_j$ is contained in D'_j. Thus, applying the Cauchy inequality yields

$$|f^{(k)}(\zeta)| \leq k!\frac{\max_{z \in D'_j}|f(z) - a_0|}{\left(\frac{\alpha R_j}{2Q}\right)^k}, \qquad k = 1, 2, \cdots. \qquad (6.4.10)$$

Consequently, (6.4.3) follows from (6.4.9) and (6.4.10).   □

Lemma 6.5   *Let G be the region $(t_1 < |z| < t_2) \cap (\varphi_1 < \arg z < \varphi_2)$ $(0 < t_1 < t_2 < \infty, 0 \leq \varphi_1 < \varphi_2 < 2\pi + \varphi_1)$, $\Gamma_\nu(\nu = 1,2)$ the common part of the boundary of G and $|z| = t_\nu$, and $\omega(z, \Gamma_\nu, G)(\nu = 1,2)$ the harmonic measure of Γ_ν with respect to G. Then the following inequalities hold:*

$$\omega(z, \Gamma_1, G) \leq \frac{4\left(\frac{t_1}{r}\right)^{\frac{\pi}{\varphi_2-\varphi_1}}}{\pi\left[1 - \left(\frac{t_1}{r}\right)^{\frac{2\pi}{\varphi_2-\varphi_1}}\right]} \qquad (6.4.11)$$

and

$$\omega(z, \Gamma_2, G) \leq \frac{4\left(\frac{r}{t_2}\right)^{\frac{\pi}{\varphi_2-\varphi_1}}}{\pi\left[1 - \left(\frac{r}{t_2}\right)^{\frac{2\pi}{\varphi_2-\varphi_1}}\right]}. \qquad (6.4.12)$$

Proof.   Since an estimate of $w(z, \Gamma_1, G)$ can be reduced to an estimate of $w(z, \Gamma_2, G)$ by a transformation of the variable z, it is suffices to make an estimate of $w(z, \Gamma_2, G)$.

It is obvious that

$$w(z, \Gamma_2, G) \leq w(z, \Gamma_2, G'),$$

where G' denotes the region $(|z| < t_2) \cap (\varphi_1 < \arg z < \varphi_2)$. Since the transformation

$$\zeta = \xi + i\eta = \left\{ \frac{z}{t_2} e^{-i\frac{\varphi_1+\varphi_2}{2}} \right\}^{\frac{\pi}{\varphi_2-\varphi_1}}$$

maps G' and Γ_2 onto

$$D : (|\zeta| < 1) \cap \left(-\frac{\pi}{2} < \arg \zeta < \frac{\pi}{2} \right)$$

and

$$S : (|\zeta| = 1) \cap \left(-\frac{\pi}{2} < \arg \zeta < \frac{\pi}{2} \right),$$

respectively, we have

$$w(z, \Gamma_2, G') = w(\zeta, S, D).$$

If θ denotes the supplement of the angle subtended at the point ζ by to the diameter $(\xi = 0) \cap (-i \leq \eta \leq i)$, then we have

$$w(\zeta, S, D) = \frac{2\theta}{\pi}.$$

Consequently

$$w(z, \Gamma_2, G) \leq \frac{2}{\pi} \left\{ \arctg \frac{\xi}{1+\eta} + \arctg \frac{\xi}{1-\eta} \right\} \leq \frac{2}{\pi} \left(\frac{\xi}{1+\eta} + \frac{\xi}{1-\eta} \right)$$

$$= \frac{4\xi}{\pi(1-\eta^2)} \leq \frac{4|\zeta|}{\pi(1-|\zeta|^2)} = \frac{4\left(\frac{r}{t_2}\right)^{\frac{\pi}{\varphi_2-\varphi_1}}}{\pi\left[1 - \left(\frac{r}{t_2}\right)^{\frac{2\pi}{\varphi_2-\varphi_1}}\right]}. \quad \square$$

Lemma 6.6   Let $f(z)$ be an entire function of order $\lambda(0 < \lambda < \infty)$, $\lambda(r)$ a precise order of $f(z)$ with $U(r) = r^{\lambda(r)}$, and suppose $f(z)$ has no Borel direction in the angles, $G_1 : \varphi_1 < \arg z < \varphi_2$ and $G_2 : \varphi_3 < \arg z < \varphi_4$ $(0 \leq \varphi_1 < \varphi_2 < \varphi_3 < \varphi_4 < \varphi_1 + 2\pi)$. If there exist two distinct finite *complex numbers $a_\nu (\nu = 1, 2)$, positive numbers η and K', and a sequence of positive numbers R_j tending to infinity such that*

$$\text{mes } E\{\varphi : \varphi_1 < \varphi < \varphi_2, \log |f(R_j e^{i\varphi}) - a_1| < -\eta U(R_j)\} > K' \quad (6.4.13)$$

and

$$\text{mes } E\{\varphi : \varphi_3 < \varphi < \varphi_4, \log |f(R_j e^{i\varphi}) - a_2| < -\eta U(R_j)\} > K', \quad (6.4.14)$$

then

$$\varphi_3 - \varphi_2 \geq \frac{\pi}{\lambda}$$

and

$$(\varphi_1 + 2\pi) - \varphi_4 \geq \frac{\pi}{\lambda}.$$

Proof.   Using the method of reductional absurdum, let us assume that say $\varphi_3 - \varphi_2 < \pi/\lambda$, will lead to a contradiction.

Take a sufficiently small positive number α such that

$$\varphi_3 - \varphi_2 + 20\alpha < \frac{\pi}{\lambda}.$$

Applying Lemma 6.4 to $f(z)$ in $G_\nu(\nu = 1, 2)$ for all sufficiently large j, we have

$$\log |f(R_j e^{i(\varphi_2 - 10\alpha)}) - a_1| < -\frac{\eta}{\left\{4\left(5 + 4\log \frac{2}{h}\right)\right\}^{2N_1 + N_2}} U(R_j), \quad (6.4.15)$$

and

$$\log |f(R_j e^{i(\varphi_3 + 10\alpha)}) - a_2| < -\frac{\eta}{\left\{4\left(5 + 4\log \frac{2}{h}\right)\right\}^{2N_1 + N_2}} U(R_j), \quad (6.4.16)$$

where N_1, N_2 and h are defined by (6.4.4) and (6.4.5). Moreover, the inequality

$$\log |f'(z)| < -\frac{\eta}{\left\{4\left(5 + 4\log \frac{2}{h}\right)\right\}^{2N_1 + N_2}} U(R_j) + \log \frac{2Q}{\alpha R_j} \quad (6.4.17)$$

holds on the segments

$$\left(\frac{R_j}{Q} \leq |z| \leq QR_j\right) \cap (\arg z = \varphi_2 - 10\alpha)$$

and

$$\left(\frac{R_j}{Q} \le |z| \le QR_j\right) \cap (\arg z = \varphi_3 + 10\alpha)$$

whenever Q satisfies $1 < Q < 1/4\alpha$.

Let N be a positive number and $e^N > 36$. Denote by D the region

$$\left\{\left(\frac{R_j}{e^N} < |z| < e^N R_j\right) \cap (\varphi_2 - 10\alpha < \arg z < \varphi_3 + 10\alpha)\right\}$$

and by Γ its boundary, and write

$$\left.\begin{array}{l}
\Gamma_1 = \Gamma \cap \left(\frac{R_j}{36} \le |z| \le 36R_j\right), \quad \Gamma_2 = \Gamma \cap (|z| = e^N R_j), \\[3mm]
\Gamma_3 = \Gamma \cap \left(|z| = \frac{R_j}{e^N}\right), \quad \Gamma_4 = \Gamma - \left(\bigcup_{\nu=1}^{3} \Gamma_\nu\right).
\end{array}\right\} \tag{6.4.18}$$

Then, for $R_j e^{i\varphi} \in D$, we have

$$\log |f'(R_j e^{i\varphi})| \le \sum_{\nu=1}^{4} \omega(R_j e^{i\varphi}, \Gamma_\nu, D) \cdot \max_{z \in \Gamma_\nu}(\log |f'(z)|). \tag{6.4.19}$$

Now let us give a proper estimate for every factor of the right-hand side of (6.4.19).

Let D' be the region

$$\left(\frac{R_j}{36} < |z| < 36R_j\right) \cap (\varphi_2 - 10\alpha < \arg z < \varphi_3 + 10\alpha),$$

Γ' its boundary, $\Gamma'_2 = \Gamma' \cap (|z| = 36R_j)$ and

$$\Gamma'_3 = \Gamma' \cap \left(|z| = \frac{R_j}{36}\right).$$

It is clear that

$$\sum_{\nu=2}^{4} \omega(R_j e^{i\varphi}, \Gamma_\nu, D) \le \sum_{\mu=2}^{3} \omega(R_j e^{i\varphi}, \Gamma'_\mu, D').$$

By Lemma 6.5 and $(\varphi_3 + 10\alpha) - (\varphi_2 - 10\alpha) \le 2\pi$, we have

$$\sum_{\mu=2}^{3} \omega(R_j e^{i\varphi}, \Gamma'_\mu, D') \le 2\frac{4\left(\frac{1}{36}\right)^{\frac{1}{2}}}{\pi\left(1 - \frac{1}{36}\right)} < \frac{1}{2};$$

hence

$$\omega(R_j e^{i\varphi}, \Gamma_1, D) > \frac{1}{2}.$$

Again, by Lemma 6.5, we have

$$\omega(R_j e^{i\varphi}, \Gamma_2, D) \leq \frac{4\left(\dfrac{1}{e^N}\right)^{\frac{\pi}{\varphi_3 - \varphi_2 + 20\alpha}}}{\pi\left\{1 - \left(\dfrac{1}{e^N}\right)^{\frac{2\pi}{\varphi_3 - \varphi_2 + 20\alpha}}\right\}} < 2e^{\frac{-\pi N}{\varphi_3 - \varphi_2 + 20\alpha}}$$

and

$$\omega(R_j e^{i\varphi}, \Gamma_3, D) \leq \frac{4\left(\dfrac{1}{e^N}\right)^{\frac{\pi}{\varphi_3 - \varphi_2 + 20\alpha}}}{\pi\left\{1 - \left(\dfrac{1}{e^N}\right)^{\frac{2\pi}{\varphi_3 - \varphi_2 + 20\alpha}}\right\}} < 2e^{\frac{-N}{2}}.$$

Choosing $Q = 36$ and $Q = e^N$ in (6.4.17) in succession, we have

$$\max_{z \in \Gamma_1} \log |f'(z)| < -K^* U(R_j) \quad \left(K^* = \frac{\eta}{\left\{4\left(5 + 4\log\dfrac{2}{h}\right)\right\}^{2N_1 + N_2}}\right)$$

and

$$\max_{z \in \Gamma_4} \log |f'(z)| < 0$$

for sufficiently large j.

Since $f(z)$ is an entire function of finite order, we have

$$\max_{|z|=r} \log |f'(z)| \leq \frac{2r + r}{2r - r} m(2r, f') = 3m(2r, f')$$

$$\leq 3\left\{m(2r, f) + m\left(2r, \frac{f'}{f}\right)\right\} < 4m(2r, f).$$

Thus

$$\max_{z \in \Gamma_2} \log |f'(z)| < 4m(2e^N R_j, f) < 8(2e^N)^\lambda U(R_j)$$

and

$$\max_{z \in \Gamma_3} \log |f'(z)| < 4m(2e^{-N} R_j, f) < 5U(R_j),$$

provided j is sufficiently large.

Then, substituting all these estimates into (6.4.19), we obtain

$$\log |f'(R_j e^{i\varphi})| < \left\{-\frac{K^*}{2} + 2^{\lambda+4} e^{-N\left(\frac{\pi}{\varphi_3 - \varphi_2 + 20\alpha} - \lambda\right)} + 10e^{-\frac{N}{2}}\right\} U(R_j).$$

Because $\pi/(\varphi_3 - \varphi_2 + 20\alpha) - \lambda > 0$, we can choose N beforehand to satisfy

$$-\frac{K^*}{2} + 2^{\lambda+4}e^{-N(\frac{\pi}{\varphi_3 - \varphi_2 + 20\alpha} - \lambda)} + 10e^{-\frac{N}{2}} < -\frac{K^*}{4}.$$

Therefore

$$\log|f'(R_j e^{i\varphi})| < -\frac{K^*}{4}U(R_j).$$

Integrating f' along the arc

$$(|z| = R_j) \cap (\varphi_2 - 10\alpha \le \arg z \le \varphi_3 + 10\alpha),$$

we deduce that

$$|f(R_j e^{i(\varphi_3+10\alpha)}) - f(R_j e^{i(\varphi_2-10\alpha)})| < 2\pi R_j e^{-\frac{K^*}{4}U(R_j)} \to 0, \quad j \to \infty.$$
$$(6.4.20)$$

This contradicts (6.4.15) and (6.4.16).   □

6.4.2   Deficient values and Borel directions of an entire function

In this subsection our aim is to prove the following theorem.

Theorem 6.7   *Let $f(z)$ be an entire function of order λ, where $0 < \lambda < \infty$. If p is the number of its finite deficient values and q is the number of its Borel directions, then $p \le q/2$.*

Proof.   According to Theorem 3.8, every entire function of finite positive order must have at least one Borel direction. Thus $q \ge 1$.

When $q = +\infty$, the conclusion of Theorem 6.7 is obvious.

When $q = 1$, Theorem 6.5 concludes that for a meromorphic function of finite positive order, the number of its deficient values does not exceed the number of its Borel directions. Now however $f(z)$ is entire, and infinity is a deficient value. Thus $f(z)$ has no finite deficient value.

When $2 \le q < \infty$, let us prove that the negation of our theorem will lead to a contradiction. In this case, we can find $[q/2] + 1$ distinct finite deficient values a_ν ($\nu = 1, 2, \cdots, [q/2] + 1$) of $f(z)$ with

$$\delta(a_\nu, f) = \delta_\nu > 0.$$

Set $\delta = \min_{1 \le \nu \le [q/2]+1} \delta_\nu$ and denote by $B_m : \arg z = \varphi_m (m = 1, 2, \cdots, q,$
$0 \le \varphi_1 < \varphi_2 < \cdots < \varphi_q < \varphi_{q+1} = 2\pi + \varphi_1)$, the q Borel directions of $f(z)$.

By Lemma 6.1, there is a sequence of positive numbers R_j, $\lim\limits_{j\to\infty} R_j = \infty$, such that all the sets $E_{j\nu}$ of values $\varphi(0 \le \varphi < 2\pi)$ for which

$$\log \frac{1}{|f(R_j e^{i\varphi}) - a_\nu|} > \frac{\delta}{2^{\lambda+4}} U(R_j),$$

$$\left(\nu = 1, 2, \cdots, \left[\frac{q}{2}\right] + 1; \ j \text{ sufficiently large}\right),$$

(6.4.21)

have measure larger than a positive number $K = K(\delta, [q/2] + 1, \lambda)$, where $U(r) = r^{\lambda(r)}$ and $\lambda(r)$ is a precise order of $f(z)$.

Among the q angles $\varphi_m < \arg z < \varphi_{m+1}$ $(m = 1, 2, \cdots, q)$, there exists at least one angle, say $G_1 : \varphi_{m_1} < \arg z < \varphi_{m_1+1}$, and a sequence R_{j1} of R_j tending to infinity, with the property that the set of values φ satisfying $\varphi_{m_1} < \varphi < \varphi_{m_1+1}$ and

$$\log \frac{1}{|f(R_{j1} e^{i\varphi}) - a_1|} > \frac{\delta}{2^{\lambda+4}} U(R_{j1}),$$

has measure greater than K/q.

Taking α, $0 < \alpha < k/20q$, and applying Lemma 6.4 to $f(z)$, a_1, G_1, the sequence (R_{j_1}) and positive numbers $K' = K/q$ and $Q = 2$, we have

$$\log \frac{1}{|f(R_{j1} e^{i\varphi}) - a_1|} > \frac{\delta}{2^{\lambda+4} \left\{ 4 \left(5 + 4 \log \frac{2}{h} \right) \right\}^{3N_1}} U(R_{j1}) \qquad (6.4.22)$$

on the arc $\{R_{j1} e^{i\varphi} : \varphi_{m_1} + 10\alpha < \varphi < \varphi_{m_1+1} - 10\alpha\}$, where

$$N_1 = \left[\frac{\pi}{\alpha}\right] + 1, \quad h = \frac{K}{80(2e+1)(\pi+\alpha)q}.$$

In the angle G_1, the set of values φ satisfying at least one of the following inequalities has measure less than or equal to $20\alpha < K/q$:

$$\log \frac{1}{|f(R_{j1} e^{i\varphi}) - a_\nu|} > \frac{\delta}{2^{\lambda+4}} U(R_{j1}), \quad j > j_0, \nu = 2, 3, \cdots, \left[\frac{q}{2}\right] + 1,$$

Thus, among the $q - 1$ angles $\varphi_m < \arg z < \varphi_{m+1} (m = 1, 2, \cdots, m_1 - 1, m_1+1, \cdots, q)$, we can find $G_2 : \varphi_{m_2} < \arg z < \varphi_{m_2+1}$ and a subsequence R_{j2} of R_{j1} tending to infinity, such that the set of values φ satisfying $\varphi_{m_2} < \varphi < \varphi_{m_2+1}$ and

$$\log \frac{1}{|f(R_{j2} e^{i\varphi}) - a_2|} > \frac{\delta}{2^{\lambda+4}} U(R_{j2}),$$

has measure larger than K/q.

Similarly, applying Lemma 6.4 to $f(z)$, a_2, G_2, (R_{j2}), and $K' = K/q$, we have

$$\log \frac{1}{|f(R_{j2}e^{i\varphi}) - a_2|} > \frac{\delta}{2^{\lambda+4}\left\{4(5 + 4\log \frac{2}{h})\right\}^{3N_1}} U(R_{j2}) \qquad (6.4.23)$$

on $\{R_{j2}e^{i\varphi} : \varphi_{m_2} + 10\alpha < \varphi < \varphi_{m_2+1} - 10\alpha\}$.

Now we claim that G_1 and G_2 cannot be adjacent angles. In fact, let us assume otherwise, say $\varphi_{m_2} = \varphi_{m_1+1}$. Then we can take two angles $G_1' : \varphi_{m_1} < \arg z < \varphi_{m_1+1} - \varepsilon$ and $G_2' : \varphi_{m_2} + \varepsilon < \arg z < \varphi_{m_2+1}$, where $\varepsilon = \min(K/(2q), \pi/(4\lambda))$. Applying Lemma 6.6 to $f(z)$, two distinct finite complex numbers a_1 and a_2, the angles G_1' and G_2', $\eta = \delta/2^{\lambda+4}$ and $K' = K/2q$, we see that the magnitude of the angle between any side of G_1' and any side of G_2' is not less than π/λ, i.e.,

$$2\varepsilon = (\varphi_{m_2} + \varepsilon) - (\varphi_{m_1+1} - \varepsilon) \geq \frac{\pi}{\lambda}.$$

This contradicts the choice of ε.

In $G_1 \cup G_2$, the set of values φ satisfying one of the inequalities

$$\log \frac{1}{|f(R_{j2}e^{i\varphi}) - a_\nu|} > \frac{\delta}{2^{\lambda+4}} U(R_{j2}), \quad j > j_0, \nu = 3, 4, \cdots, \left[\frac{q}{2}\right] + 1$$

has measure less than or equal to $40\alpha < 2K/q$. Among $q - 2$ angles $\varphi_m < \arg z < \varphi_{m+1}(m \neq m_1, m_2)$, there exists at least one, say G_3 : $\varphi_{m_3} < \arg z < \varphi_{m_3+1}$, having a similar property with a subsequence (R_{j3}) of (R_{j2}) and the complex number a_3. For the same reason, G_3 can not be a neighbor of G_1 or G_2.

Following this procedure on, we can obtain $[q/2]+1$ angles $G_\nu : \varphi_{m_\nu} < \arg z < \varphi_{m_\nu+1}(\nu = 1, 2, \cdots, [q/2]+1)$, corresponding to $[q/2]+1$ deficient values $a_\nu(\nu = 1, 2, \cdots, [q/2] + 1)$, and a subsequence $(R_{j,[q/2]+1})$ tending to infinity such that the inequality

$$\log \frac{1}{|f(R_{j,[\frac{q}{2}]+1}e^{i\varphi}) - a_\nu|} > \frac{\delta}{2^{\lambda+4}\left\{4(5 + 4\log \frac{2}{h})\right\}^{3N_1}} U\left(R_{j,[\frac{q}{2}]+1}\right),$$

$$\nu = 1, 2, \cdots, \left[\frac{q}{2}\right] + 1$$

$$(6.4.24)$$

holds on the arc $\{R_{j,[q/2]+1}e^{i\varphi}: \varphi_{m_\nu} + 10\alpha < \varphi < \varphi_{m_\nu+1} - 10\alpha\}$. None of these $[q/2] + 1$ angles G_ν is a neighbor of the others. On the other hand, since the finite plane is divided into q angles $\varphi_m < \arg z < \varphi_{m+1}(m = 1, 2, \cdots, q)$ by the q Borel directions B_m, there are at most $[q/2]$ of them having the above-mentioned property. This contradiction shows that Theorem 6.7 is true.   □

Remark.   *The estimate $p \leq q/2$ of Theorem 6.7 is sharp in the sense that given a positive integer n, there is an entire function of finite positive order such that it has exactly n finite deficient values and $2n$ Borel directions.*

In fact, when $n = 1$, the function e^z has order one, a unique finite deficient value zero and two Borel directions, namely, the positive and negative imaginary axes. When $n \geq 2$, we shall see that the function

$$f(z) = \int_0^z e^{-t^n} dt$$

provides such an example. On the one hand, since $f(z)$ tends to

$$\lim_{R \to \infty} \int_0^{Re^{i\frac{2\pi l}{n}}} e^{-t^n} dt = e^{i\frac{2\pi l}{n}} \int_0^\infty e^{-t^n} dt = a_l$$

uniformly in the angle

$$\left| \arg z - \frac{2\pi l}{n} \right| \leq \frac{\pi}{2n} - \varepsilon, \quad l = 1, 2, \cdots, n,$$

where ε is a sufficiently small positive number, and since also

$$f(z) - a_l = -\int_z^\infty e^{-t^n} dt = -\frac{e^{-z^n}}{nz^{n-1}} + \frac{n-1}{n} \int_z^\infty \frac{e^{-t^n}}{t^n} dt$$

$$= -\frac{e^{-z^n}}{nz^{n-1}}(1 + o(1)), \tag{6.4.25}$$

we have

$$m\left(r, \frac{1}{f - a_l}\right) \geq \frac{1}{2\pi}(1 + o(1))r^n \int_{-\frac{\pi}{2n}+\varepsilon}^{\frac{\pi}{2n}-\varepsilon} \cos n\theta d\theta$$

$$= \frac{1}{\pi}(1 + o(1))r^n \frac{\sin\left(\frac{\pi}{2} - n\varepsilon\right)}{n}$$

$$\geq (1 + o(1))\left(\frac{1}{n\pi} - \frac{\varepsilon}{\pi}\right)r^n, \quad l = 1, 2, \cdots, n.$$

Similarly, since $f(z)$ tends to infinity uniformly in the angle

$$\left|\arg z - \frac{(2l-1)\pi}{n}\right| \leq \frac{\pi}{2n} - \varepsilon, \quad l = 1, 2, \cdots, n,$$

and

$$f(z) = \int_0^z e^{-t^n} dt = \frac{e^{-z^n}\{1 + o(1)\}}{nz^{n-1}} + O(1), \qquad (6.4.26)$$

we get

$$m(r, f) = (1 + o(1))\frac{r^n}{\pi}.$$

Therefore, $f(z)$ has order n and n finite deficient values $a_l(l = 1, 2, \cdots, n)$ with $\delta(a_l, f) = 1/n$.

On the other hand, $f'(z) = e^{-z^n}$ has the $2n$ Borel directions given by:

$$\arg z = \frac{(2k-1)}{2n}\pi, \quad k = 1, 2, \cdots, 2n.$$

According to Corollary 2 of Theorem 5.3, these $2n$ rays are also Borel directions of $f(z)$. From (6.4.25) and (6.4.26), it is clear that $f(z)$ has no other Borel direction.

6.4.3 Number of deficient values and order of an entire function.

For an arbitrary entire function of finite positive order, if it has a finite number of Borel directions, then the number of its deficient values has an upper bound depending only on its order.

Theorem 6.8   *Let $f(z)$ be an entire function of order λ, where $0 < \lambda < \infty$. If the number q of its Borel directions is finite, then the number p of finite deficient values of $f(z)$ is less than 2λ.*

Proof.   Denote by $B_m : \arg z = \varphi_m(m = 1, 2, \cdots, q; 0 \leq \varphi_1 < \varphi_2 < \cdots < \varphi_q < \varphi_{q+1} = 2\pi + \varphi_1)$ the q Borel directions of $f(z)$. When $\lambda > 1/2$, the negation of Theorem 6.8 implies the existence of $p'(2\lambda \leq p' < \infty)$ distinct finite deficient values $a_\nu(\nu = 1, 2, \cdots, p')$ of $f(z)$. Then, by the argument in the proof of Theorem 6.7, it can be seen that corresponding to the p' deficient values $a_\nu(\nu = 1, 2, \cdots, p')$, there are p' angles G_ν : $\varphi_{m_\nu} < \arg z < \varphi_{m_\nu+1}$, none of them being a neighbor of the others, and a sequence of positive numbers $R_{jp'}$, tending to infinity, such that the inequality

$$\log \frac{1}{|f(R_{jp'}e^{i\varphi}) - a_\nu|} > \frac{\delta}{2^{\lambda+4}\left\{4\left(5 + 4\log\frac{2}{h}\right)\right\}^{3N_1}} U(R_{jp'}), \quad \nu = 1, 2, \cdots, p'$$

(6.4.27)

holds on the arc $\{R_{jp'}e^{i\varphi} : \varphi_{m_\nu} + 10\alpha < \varphi < \varphi_{m_\nu+1} - 10\alpha\}$, where

$$0 < \alpha < \min_{1 \le m \le q} \left(\frac{K}{20q}, \frac{\varphi_{m+1} - \varphi_m}{40}\right), \qquad N_1 = \left[\frac{\pi}{\alpha}\right] + 1,$$

$$h = \frac{K}{80(2e+1)(\pi+\alpha)q}, \qquad\qquad \delta = \min_{1 \le \nu \le p'}\{\delta(a_\nu, f)\},$$

$$K = K(\delta, p', \lambda), \qquad\qquad U(r) = r^{\lambda(r)}$$

and $\lambda(r)$ is a precise order of $f(z)$ ($K(\delta, p', \lambda)$ is defined in (6.3.2).).

Without loss of generality, we may suppose that

$$0 \le \varphi_{m_1} < \varphi_{m_1+1} < \varphi_{m_2} < \varphi_{m_2+1} < \cdots < \varphi_{m_{p'}}$$

$$< \varphi_{m_{p'}+1} < \varphi_{m_{p'+1}} = 2\pi + \varphi_{m_1}.$$

For otherwise, it is sufficient ot rearrange the order of $a_\nu(\nu = 1, 2, \cdots, p')$. Now applying Lemma 6.6 to $f(z)$, G_ν and $G_{\nu+1}(1 \le \nu \le p', G_{p'+1} = G_1)$, a_ν and $a_{\nu+1}$, with

$$\eta = \frac{\delta}{2^{\lambda+4}\left\{4\left(5 + 4\log\frac{2}{h}\right)\right\}^{3N_1}}$$

and

$$K' = \min_{1 \le m \le q} \left(\frac{\varphi_{m+1} - \varphi_m}{2}\right),$$

we have

$$\varphi_{m_\nu+1} - \varphi_{m_\nu+1} \ge \frac{\pi}{\lambda}, \quad \nu = 1, 2, \cdots, p',$$

$$\sum_{\nu=1}^{p'}(\varphi_{m_\nu+1} - \varphi_{m_\nu+1}) \ge p'\frac{\pi}{\lambda} \ge 2\pi,$$

so that

$$\sum_{\nu=1}^{p'}(\varphi_{m_\nu+1} - \varphi_{m_\nu+1}) + \sum_{\nu=1}^{p'}(\varphi_{m_\nu+1} - \varphi_{m_\nu}) > 2\pi.$$

(6.4.28)

But the left-hand side of (6.4.28) is obviously equal to 2π. This contradiction proves Theorem 6.8 for $\lambda > 1/2$.

When $\lambda \leq 1/2$, let us prove $p = 0$. If $f(z)$ has a finite deficient value a_1, then Lemma 6.1 yields a sequence (R'_j) such that the two sets of values $\varphi(0 \leq \varphi < 2\pi)$ satisfying

$$\log|f(R'_j e^{i\varphi})| > \frac{\delta}{2^{\lambda+4}}U(R'_j) \tag{6.4.29}$$

and

$$\log\frac{1}{|f(R'_j e^{i\varphi}) - a_1|} > \frac{\delta}{2^{\lambda+4}}U(R'_j) \tag{6.4.30}$$

respectively, have measure greater than a positive number K, where $\delta = \min(\delta(\infty, f), \delta(a_1, f)) = \delta(a_1, f)$.

Since $f(z)$ has a finite number of Borel directions $B_m : \arg z = \varphi_m (m = 1, 2, \cdots, q)$, there exist an angle, say $G_1 : \varphi_1 < \arg z < \varphi_2$, and a subsequence (R_j) of (R'_j) such that the set of values $\varphi(\varphi_1 < \varphi < \varphi_2)$ satisfying

$$\log\frac{1}{|f(R_j e^{i\varphi}) - a_1|} > \frac{\delta}{2^{\lambda+4}}U(R_j)$$

has measure greater than K/q. If α is a sufficiently small positive number and Q satisfies $1 < Q < 1/(4\alpha)$, then Lemma 6.4 implies that the inequality

$$\log|f(z) - a_1| < -\frac{\delta}{2^{\lambda+4}\left\{4\left(5 + 4\log\frac{2}{h}\right)\right\}^{2N_1+N_2}}U(R_j) \tag{6.4.31}$$

holds everywhere in the region

$$D_j : \left(\frac{R_j}{Q} \leq |z| \leq QR_j\right) \cap (\varphi_1 + 10\alpha \leq \arg z \leq \varphi_2 - 10\alpha),$$

where j is sufficiently large and N_1, N_2 and h are defined by (6.4.4) and (6.4.5).

Choose a positive number α such that $\varphi_1 + 2\pi - \varphi_2 + 20\alpha < \pi/\lambda$. Let N be a positive number and $e^N > 36$. Denote by D the region

$$\left(\frac{R_j}{e^N} \leq |z| \leq e^N R_j\right) \cap (\varphi_2 - 10\alpha \leq \arg z \leq \varphi_1 + 10\alpha + 2\pi)$$

and by Γ its boundary, and let $\Gamma_\nu(\nu = 1, 2, 3, 4)$ be as in (6.4.18).

If $R_j e^{i\varphi} \in D$, then we have

$$\log|f(R_j e^{i\varphi}) - a_1| \leq \sum_{\nu=1}^{4} \omega(R_j e^{i\varphi}, \Gamma_\nu, D) \cdot \max_{z \in \Gamma_\nu}(\log|f(z) - a_1|), \quad (6.4.32)$$

where

$$\omega(R_j e^{i\varphi}, \Gamma_1, D) > \frac{1}{2},$$

$$\omega(R_j e^{i\varphi}, \Gamma_2, D) < 2e^{-\frac{\pi N}{\varphi_1 - \varphi_2 + 2\pi + 20\alpha}}$$

and

$$\omega(R_j e^{i\varphi}, \Gamma_3, D) < 2e^{-\frac{N}{2}},$$

as shown in the proof of Lemma 6.5.

By choosing $Q = 36$ and $Q = e^N$ in (6.4.31) respectively, we have

$$\max_{z \in \Gamma_1}|f(z) - a_1| < -K^* U(R_j),$$

$$\left(K^* = \frac{\delta}{2^{\lambda+4}\left\{4\left(5 + 4\log\frac{2}{h}\right)\right\}^{2N_1+N_2}}\right)$$

and

$$\max_{z \in \Gamma_4}|f(z) - a_1| < 0,$$

when j is sufficiently large.

Since $f(z)$ is entire and is of order λ, it is clear that

$$\max_{|z|=r}\log|f(z) - a_1| \leq \frac{2r + r}{2r - r}m(2r, f - a_1) = 3m(2r, f) + O(1).$$

Thus

$$\max_{z \in \Gamma_2}\log|f(z) - a_1| \leq 3m(2e^N R_j, f) + O(1) < 4(2e^N)^\lambda U(R_j)$$

and

$$\max_{z \in \Gamma_3}\log|f(z) - a_1| < 3m(2e^{-N} R_j, f) + O(1) < 4U(R_j),$$

provided that j is sufficiently large.

Substituting these estimates into (6.4.32), we deduce that

$$\log|f(R_j e^{i\varphi}) - a_1| < \left\{-\frac{K^*}{2} + 2^{\lambda+3}e^{-N\left(\frac{\pi}{\varphi_1 + 2\pi - \varphi_2 + 20\alpha} - \lambda\right)} + 8e^{-\frac{N}{2}}\right\}U(R_j).$$

In view of the inequality $\pi/(\varphi_1 + 2\pi - \varphi_2 + 20\alpha) > \lambda$, we can choose N sufficiently large in advance so that

$$-\frac{K^*}{2} + 2^{\lambda+3}e^{-N(\frac{\pi}{\varphi_1+2\pi-\varphi_2+20\alpha}-\lambda)} + 8e^{-\frac{N}{2}} < -\frac{K^*}{4}.$$

Therefore

$$\log|f(R_je^{i\varphi}) - a_1| < -\frac{K^*}{4}U(R_j)$$

for $\varphi_2 - 10\alpha < \varphi < \varphi_1 + 10\alpha + 2\pi$.

Combining this fact with (6.4.31), we see that $f(z)$ tends to a_1 uniformly on $|z| = R_j$. However, since (R_j) is a subsequence of (R'_j), we have a contradiction to the fact that the set of values φ satisfying (6.4.29) has a positive measure. $\square$

Chapter 7

The Spread Relation and Its Applications

In the recent developments of the value distribution theory of meromorphic functions, there is an emphasis on the study of deficient values and deficiencies.

Let $f(z)$ be a transcendental meromorphic function. According to §1.4, a complex number a is called a deficient value of $f(z)$, if

$$\delta(a, f) = \lim_{r \to \infty} \frac{m\left(r, \dfrac{1}{f - a}\right)}{T(r, f)} > 0.$$

What does it really mean to say that $\delta(a, f)$ is positive? Roughly speaking, $f(z)$ must be "close to" a on a substantial portion of each circle $|z| = r$, when r is sufficiently large. The present chapter will be devoted to the spread relation which gives a precise answer to the above question, as well as proof of the so-called give some of its applications.

7.1 Sequence of Pólya Peaks and Its Existence

7.1.1 Definition and lemmas. In order to discuss the spread relation, we need to introduce the concept of a sequence of Pólya peaks which plays an important role in the contemporary research of value distribution theory. The following definition was introduced by A.Edrei [3].

Definition 7.1 *Let $f(z)$ be meromorphic and of lower order μ in the finite plane. A sequence of positive numbers (r_j) is said to be a sequence* of Pólya peaks of order μ, if there are three sequences (r'_j), (r''_j) and (ε_j) such that

$$r'_j \to \infty, \quad \frac{r_j}{r'_j} \to \infty, \quad \frac{r''_j}{r_j} \to \infty, \quad \varepsilon_j \to 0 \ (j \to \infty), \qquad (7.1.1)$$

and that the inequalities $r'_j \leq t \leq r''_j$ imply

$$\frac{T(t, f)}{T(r_j, f)} \leq (1 + \varepsilon_j)\left(\frac{t}{r_j}\right)^{\mu}. \qquad (7.1.2)$$

Now we prove two lemmas which are needed in the proof of the existence of sequence of Pólya peaks.

Lemma 7.1 *Let $\varphi(t)$ and $\psi(t)$ be two continuous positive functions defined for $t \geq t_0$. If $\psi(t)$ is non-decreasing,*

$$\varlimsup_{t \to \infty} \varphi(t) = \infty, \qquad (7.1.3)$$

and

$$\lim_{t \to \infty} \frac{\varphi(t)}{\psi(t)} = 0, \qquad (7.1.4)$$

then for any positive number $r_0 \geq t_0$, there exists a value $r > r_0$ such that

$$\varphi(t) \leq \varphi(r), \quad t_0 \leq t < r, \qquad (7.1.5)$$

and

$$\frac{\varphi(t)}{\psi(t)} \leq \frac{\varphi(r)}{\psi(r)}, \quad t \geq r. \qquad (7.1.6)$$

Proof. By (7.1.3), there is a value $u_0 > r_0$ with

$$\varphi(u_0) = \max_{t_0 \leq t \leq u_0} \varphi(t).$$

Since $\varphi(t)/\psi(t)$ is continuous positive function at $t \geq t_0$ and satisfies (7.1.4), there must be a value $u_1 \geq u_0$ such that

$$\frac{\varphi(u_1)}{\psi(u_1)} = \max_{t \geq u_0} \frac{\varphi(t)}{\psi(t)}. \qquad (7.1.7)$$

By the continuity of $\varphi(t)$, there exists a value r such that $u_0 \leq r \leq u_1$ and

$$\varphi(r) = \max_{u_0 \leq t \leq u_1} \varphi(t) = \max_{t_0 \leq t \leq u_1} \varphi(t). \qquad (7.1.8)$$

Hence (7.1.5) holds. By (7.1.7), (7.1.8), $r \leq u_1$ and the fact that $\psi(t)$ is a non-decreasing function, we have for $t \geq u_1$

$$\frac{\varphi(t)}{\psi(t)} \leq \frac{\varphi(u_1)}{\psi(u_1)} \leq \frac{\varphi(r)}{\psi(r)}. \quad \square$$

Lemma 7.2 Let $\varphi(t)$ be a real continuous non-decreasing function defined for $t \geq t_0 > 0$. Suppose that there are two numbers σ and τ such that $0 < \sigma < \tau$,

$$\varlimsup_{r \to \infty} \frac{\varphi(t)}{t^\tau} = \infty, \qquad (7.1.9)$$

and

$$\varliminf_{r \to \infty} \frac{\varphi(t)}{t^\sigma} = 0. \qquad (7.1.10)$$

Then for any positive number $r_0 \geq t_0$, there exists a value $r > r_0$ such that

$$\frac{\varphi(t)}{t^\tau} \leq \frac{\varphi(r)}{r^\tau}, \quad t_0 \leq t \leq r^{\frac{\tau}{\sigma}}. \qquad (7.1.11)$$

Proof. In view of (7.1.9), it is possible to choose $r_1 > r_0$ so that

$$\frac{\varphi(r_1)}{r_1^\tau} = \max_{t_0 \leq t \leq r_1} \left\{ \frac{\varphi(t)}{t^\tau}, 1 \right\}. \qquad (7.1.12)$$

Using (7.1.10), we choose r_2 so that

$$\varphi(r_2) < r_2^\sigma, \quad r_2 > r_1^{\frac{\tau}{\sigma}}. \qquad (7.1.13)$$

Put

$$r_3 = r_2^{\frac{\sigma}{\tau}}, \qquad (7.1.14)$$

and note that

$$t_0 \leq r_0 < r_1 < r_3 < r_2.$$

Since $\varphi(t)/t^\tau$ is continuous for $t \geq t_0$, there exists some r such that $r_1 \leq r \leq r_3$ and

$$\frac{\varphi(r)}{r^\tau} = \max_{r_1 \leq t \leq r_3} \frac{\varphi(t)}{t^\tau}. \qquad (7.1.15)$$

Combining (7.1.12) and (7.1.15), we obtain

$$\frac{\varphi(r)}{r^\tau} \geq \frac{\varphi(t)}{t^\tau}, \quad t_0 \leq t \leq r_3. \tag{7.1.16}$$

In order to prove that (7.1.16) holds also for $r_3 \leq t \leq r_2$, we observe that, since $\varphi(t)$ is non-decreasing, the relations (7.1.13), (7.1.14) and (7.1.12) imply

$$\varphi(t) \leq \varphi(r_2) < r_2^\sigma = r_3^\tau \leq t^\tau \leq t^\tau \frac{\varphi(r_1)}{r_1^\tau}.$$

Hence, using (7.1.15) once more, we have

$$\frac{\varphi(t)}{t^\tau} < \frac{\varphi(r_1)}{r_1^\tau} < \frac{\varphi(r)}{r^\tau}, \quad r_3 \leq t \leq r_2. \tag{7.1.17}$$

Combining (7.1.16) and (7.1.17), we see that (7.1.11) holds for $t_0 \leq t \leq r_2$. Since

$$r_2 = r_3^{\frac{\tau}{\sigma}} \geq r^{\frac{\tau}{\sigma}},$$

we have (7.1.11) for $t_0 \leq t \leq r^{\frac{\tau}{\sigma}}$. □

7.1.2 Existence of sequence of Pólya peaks.

The following theorem on the existence of sequence of Pólya peaks is due to Edrei [3].

Theorem 7.1 *If $f(z)$ is meromorphic and of lower order μ in the finite plane, then there exists a sequence of positive numbers (r_j) such that*

$$\lim_{j \to \infty} \frac{r_j}{j} = \infty,$$

and

$$T(t, f) \leq (1 + \varepsilon_j)\left(\frac{t}{r_j}\right)^\mu T(r_j, f), \quad \frac{r_j}{j} \leq t \leq jr_j, \tag{7.1.18}$$

where $\lim_{j \to \infty} \varepsilon_j = 0$.

Proof.   Choose the sequence of ε_j to be $\varepsilon_j = 1/(\log j)^2$ $(j = 2, 3, \cdots)$. There are two mutually exclusive cases.

(A) The order λ of $f(z)$ equals μ.

Put $r_2 = 1$. In general, after r_{j-1} has been determined, we shall choose r_j. For every fixed j, since

$$\lim_{t \to \infty} \frac{T(t,f)}{t^{\mu - \varepsilon_j}} = \infty, \quad \lim_{t \to \infty} \frac{T(t,f)}{t^{\mu + \varepsilon_j}} = 0,$$

we apply Lemma 7.1 to the functions

$$\varphi(t) = \frac{T(t,f)}{t^{\mu - \varepsilon_j}}, \quad \psi(t) = t^{2\varepsilon_j}.$$

There exists a value $r_j > \max(j^2, r_{j-1})$ such that

$$\frac{T(t,f)}{t^{\mu - \varepsilon_j}} \le \frac{T(r_j,f)}{r_j^{\mu - \varepsilon_j}}; \quad 1 \le t < r_j, \tag{7.1.19}$$

and

$$\frac{T(t,f)}{t^{\mu + \varepsilon_j}} \le \frac{T(r_j,f)}{r_j^{\mu + \varepsilon_j}}, \quad t \ge r_j. \tag{7.1.20}$$

It is clear that

$$\max \left\{ \left(\frac{r_j}{t} \right)^{\varepsilon_j}, \left(\frac{t}{r_j} \right)^{\varepsilon_j} \right\} \le j^{\varepsilon_j} = e^{\frac{1}{\log j}},$$

when $r_j/j \le t \le j r_j$. Thus we have

$$T(t,f) \le e^{\frac{1}{\log j}} \left(\frac{t}{r_j} \right)^{\mu} T(r_j,f), \quad \frac{r_j}{j} \le t \le j r_j, \tag{7.1.21}$$

by virtue of (7.1.19) and (7.1.20). Since $\lim_{j \to \infty} e^{\frac{1}{\log j}} = 1$, the conclusion of Theorem 7.1 is proved in this case.

(B) The order λ of $f(z)$ is greater than μ.

Taking j_0 sufficiently large, we have $\varepsilon_j < \lambda - \mu$ when $j \ge j_0$. We set $r_{j_0} = 1$ and will choose $r_j (j \ge j_0 + 1)$, when r_{j-1} has been determined. Since

$$\overline{\lim_{t \to \infty}} \frac{T(t,f)}{t^{\mu + \varepsilon_j}} = \infty \text{ and } \overline{\lim_{t \to \infty}} \frac{T(t,f)}{t^{\mu + \frac{\varepsilon_j}{2}}} = 0,$$

we apply Lemma 7.2 to $\varphi(t) = T(t,f)$, $\sigma = \mu + \dfrac{\varepsilon_j}{2}$ and $\tau = \mu + \varepsilon_j$. Thus there exists a value $r_j > \max \left(j^{\frac{2\mu + \varepsilon_j}{\varepsilon_j}}, r_{j-1} \right)$ such that

$$\frac{T(t,f)}{t^{\mu + \varepsilon_j}} \le \frac{T(r_j,f)}{r_j^{\mu + \varepsilon_j}}, \quad 1 \le t \le r_j^{\frac{\mu + \varepsilon_j}{\mu + \frac{\varepsilon_j}{2}}},$$

so that (7.1.21) also holds in this case.

7.2 The T^* Function

7.2.1 Definition and continuity of T^* function. In Baernstein's proof [1] of the spread relation, the key point is the introduction of an important function — the T^* function. subsequently, the T^* function has also been applied to the problem of minimum modulus and the estimate of the coefficients of the univalent functions.

Definition 7.2 *Let $f(z)$ be meromorphic and not identical to zero in the finite plane. For a point*

$$z = re^{i\theta}, \quad 0 < r < \infty, \ 0 \le \theta \le \pi,$$

we set

$$m^*(z) = \sup_E \frac{1}{2\pi} \int_E \log |f(re^{i\varphi})| d\varphi, \tag{7.2.1}$$

where the supremum is taken over on all measurable sets E of $(-\pi, \pi)$ with $\operatorname{mes} E = 2\theta$.

Let $N(r, f)$ have its usual meaning, and set

$$T^*(z) = m^*(z) + N(|z|, f). \tag{7.2.2}$$

Then $T^*(z)$ is defined on

$$H = \{z : \operatorname{Im} z \ge 0, \ z \ne 0\}.$$

From the definition, we have

$$T(r, f) = \sup_{0 \le \theta \le \pi} T^*(re^{i\theta})$$

and

$$N(r, f) = T^*(r).$$

The Jensen formula gives us

$$N\left(r, \frac{1}{f}\right) = T^*(re^{i\pi}) + C,$$

where C is a constant depending only on f.

The most important property of $T^*(z)$ is that $T^*(z)$ is a continuous subharmonic function on H. We prove its continuity first and to this end need the following lemma.

Lemma 7.3   *Let $f(z)$ be meromorphic and not identical to zero in the finite plane. If $m^*(z)$ is given by (7.2.1) and $z = re^{i\theta} \in H$, then there exists a measurable set E with $\text{mes}E = 2\theta$, such that*

$$m^*(z) = \frac{1}{2\pi} \int_E \log|f(re^{i\varphi})|d\varphi.$$

Proof.   Lemma 7.3 is obvious for $\theta = 0$ and $\theta = \pi$. If $0 < \theta < \pi$, we observe the distribution function

$$\lambda(t) = \text{mes}\{\varphi : \log|f(re^{i\varphi})| > t\}.$$

It is clear that $\lambda(t)$ is non-increasing in $(-\infty, +\infty)$ and continuous from the right at every point. When t increases from $-\infty$ to $+\infty$, $\lambda(t)$ decreases from 2π to zero. Thus a point t_0 can be found such that $-\infty < t_0 < +\infty$ and

$$\lambda(t_0+) = \lambda(t_0) \leq 2\theta \leq \lambda(t_0-),$$

where

$$\lambda(t_0+) = \lim_{\substack{t \to t_0 \\ t > t_0}} \lambda(t), \quad \lambda(t_0-) = \lim_{\substack{t \to t_0 \\ t < t_0}} \lambda(t).$$

Let

$$A = \{\varphi : \log|f(re^{i\varphi})| > t_0\}, \quad B = \{\varphi : \log|f(re^{i\varphi})| \geq t_0\}.$$

Then we have

$$\text{mes}A = \lambda(t_0), \quad \text{mes}B = \lambda(t_0 - 0), \quad A \subset B.$$

Choose a set E on $[-\pi, \pi]$ with

$$\text{mes}E = 2\theta, \quad A \subset E \subset B.$$

We assert that E meets the requirement of Lemma 7.3.

In fact, if there is another set F with $\text{mes}F = 2\theta$, then we have

$$\int_F \log|f(re^{i\varphi})|d\varphi = \int_F \{\log|f(re^{i\varphi})| - t_0\}d\varphi + 2\theta t_0$$

$$\leq \int_{-\pi}^{\pi} \max\{0, \log|f(re^{i\varphi})| - t_0\}d\varphi + 2\theta t_0$$

$$= \int_E \{\log|f(re^{i\varphi})| - t_0\}d\varphi + 2\theta t_0$$

$$= \int_E \log|(re^{i\varphi})|d\varphi.$$

This proves Lemma 7.3. □

Theorem 7.2 *If $f(z)$ is meromorphic and not identical to zero in the finite plane, then the function $T^*(z)$ defined by (7.2.2) is continuous on $H = \{z : \operatorname{Im} z \geq 0, \ z \neq 0\}$.*

Proof. According to Lemma 7.3, for each point $z = re^{i\theta}$, $0 < r < \infty$, $0 \leq \theta < \pi$, there exists a set E in $[-\pi, \pi]$ with $\operatorname{mes} E = 2\theta$ for which the supremum in (7.2.1) is attained, i.e.,

$$m^*(z) = \frac{1}{2\pi}\int_E \log|f(re^{i\varphi})|d\varphi.$$

Write $f(z) = f_1(z)/f_2(z)$, where $f_1(z)$ and $f_2(z)$ are entire functions and have no common zero. Thus

$$T^*(z) = \frac{1}{2\pi}\int_E \log|f(re^{i\varphi})|d\varphi + N(r, f)$$

$$= \frac{1}{2\pi}\int_E \log\left|\frac{f_1(re^{i\varphi})}{f_2(re^{i\varphi})}\right|d\varphi + N\left(r, \frac{1}{f_2}\right).$$

Since

$$N\left(r, \frac{1}{f_2}\right) = \frac{1}{2\pi}\int_0^{2\pi}\log|f_2(re^{i\varphi})|d\varphi + C,$$

we have

$$T^*(z) = \frac{1}{2\pi}\int_E \log|f_1(re^{i\varphi})|d\varphi + \frac{1}{2\pi}\int_{CE}\log|f_2(re^{i\varphi})|d\varphi + C(f), \quad (7.2.3)$$

where CE denotes the complement of E in $[-\pi, \pi]$.

Now we distinguish between two mutually exclusive cases.

(i) $f_1(z)$ and $f_2(z)$ have no zeros in the finite plane. Taking two finite positive numbers R_1 and R_2 with $R_1 < R_2$, it is obvious that $\log|f_1(z)|$ and $\log|f_2(z)|$ are two continuous functions on $R_1 \leq |z| \leq R_2$. Let $z_1 = r_1 e^{i\theta_1}$

and $z_2 = r_2 e^{i\theta_2}$, where $R_1 \leq r_1 \leq r_2 \leq R_2$. Corresponding to z_1, there exists a set E_1 with mes $E_1 = 2\theta_1$ such that

$$T^*(z_1) = \frac{1}{2\pi} \int_{E_1} \log|f_1(r_1 e^{i\varphi})|d\varphi + \frac{1}{2\pi} \int_{CE_1} \log|f_2(r_1 e^{i\varphi})|d\varphi + C.$$

Choose a suitable small positive number δ. When $|z_2 - z_1| < \delta$, the set E_1 can be combined with or taken off by a small set e to produce a set E_2 with mes $E_2 = 2\theta_2$. We consider only the case when $E_2 = E_1 \cup e$, since the other case is similar. By the definition,

$$T^*(z_2) \geq \frac{1}{2\pi} \int_{E_2} \log|f(r_2 e^{i\varphi})|d\varphi + N(r_2, f)$$

$$= \frac{1}{2\pi} \int_{E_2} \log|f_1(r_2 e^{i\varphi})|d\varphi + \frac{1}{2\pi} \int_{CE_2} \log|f_2(r_2 e^{i\varphi})|d\varphi + C$$

$$= \frac{1}{2\pi} \int_{E_1} \log|f_1(r_2 e^{i\varphi})|d\varphi + \frac{1}{2\pi} \int_{e} \log|f_1(r_2 e^{i\varphi})|d\varphi$$

$$+ \frac{1}{2\pi} \int_{CE_1} \log|f_2(r_2 e^{i\varphi})|d\varphi - \frac{1}{2\pi} \int_{e} \log|f_2(r_2 e^{i\varphi})|d\varphi + C.$$

Thus

$$T^*(z_1) - T^*(z_2) \leq \frac{1}{2\pi} \int_{E_1} (\log|f_1(r_1 e^{i\varphi})| - \log|f_1(r_2 e^{i\varphi})|)d\varphi$$

$$+ \frac{1}{2\pi} \int_{CE_1} (\log|f_2(r_1 e^{i\varphi})| - \log|f_2(r_2 e^{i\varphi})|)d\varphi$$

$$- \frac{1}{2\pi} \int_{e} \log|f_1(r_2 e^{i\varphi})|d\varphi + \frac{1}{2\pi} \int_{e} \log|f_2(r_2 e^{i\varphi})|d\varphi.$$

Since $\log|f_1(z)|$ and $\log|f_2(z)|$ are uniformly continuous on $R_1 \leq |z| \leq R_2$ and

$$|r_2 - r_1| < \delta, \text{ mes } e = |\theta_2 - \theta_1| \leq \frac{|z_2 - z_1|}{R_1} \cdot \frac{\pi}{2} < \frac{\pi}{2R_1}\delta$$

for $|z_2 - z_1| < \delta$, we obtain $T^*(z_1) - T^*(z_2) < \varepsilon$. Similarly, we have $T^*(z_2) - T^*(z_1) < \varepsilon$. This proves the continuity of $T^*(z)$ for this case.

(ii) $f_1(z)$ or $f_2(z)$ has zeros in the finite plane.

For a sufficiently small positive number δ and $R_1 < |z| < R_2$, we define

$$(A_\delta \log|f(z)|) = \frac{1}{\pi\delta^2} \int_0^\delta \int_{-\pi}^\pi \log|f(z + te^{i\omega})|t d\omega dt.$$

Define $A_\delta \log |f_1(z)|$ and $A_\delta \log |f_2(z)|$ similarly. The subharmonicity of $\log |f_1(z)|$ and $\log |f_2(z)|$ implies that

$$A_\delta \log |f_k(z)| \geq \log |f_k(z)|, \quad k = 1, 2.$$

For the simplicity of notations, we write $u_1 = \log |f_1|$, $u_2 = \log |f_2|$, $u = u_1 - u_2$, $T^*(z, u) = T^*(z)$. Thus (7.2.3) becomes

$$T^*(z, u) - C(u) = \frac{1}{2\pi} \int_E u_1(re^{i\varphi}) d\varphi + \frac{1}{2\pi} \int_{CE} u_2(re^{i\varphi}) d\varphi, \quad (7.2.3)'$$

which yields

$$T^*(z, u) - C(u) \leq \frac{1}{2\pi} \int_E A_\delta u_1(re^{i\varphi}) d\varphi + \frac{1}{2\pi} \int_{CE} A_\delta u_2(re^{i\varphi}) d\varphi$$

$$= T^*(z, A_\delta u) - C(A_\delta u).$$

For a point $re^{i\theta}$, $R_1 < r < R_2$, $0 \leq \theta < \pi$, choose a set E' with $\mathrm{mes} E' = 2\theta$. Thus

$$T^*(z, A_\delta u) - C(A_\delta u)$$

$$= \frac{1}{2\pi} \int_{E'} A_\delta u_1(re^{i\varphi}) d\varphi + \frac{1}{2\pi} \int_{CE'} A_\delta u_2(re^{i\varphi}) d\varphi.$$

On the other hand, we see that

$$T^*(z, u) - C(u) = \frac{1}{2\pi} \int_E u_1(re^{i\varphi}) d\varphi + \frac{1}{2\pi} \int_{CE} u_2(re^{i\varphi}) d\varphi$$

$$\geq \frac{1}{2\pi} \int_{E'} u_1(re^{i\varphi}) d\varphi + \frac{1}{2\pi} \int_{CE'} u_2(re^{i\varphi}) d\varphi.$$

Therefore

$$0 \leq T^*(z, A_\delta u) - C(A_\delta u) - T^*(z, u) + C(u)$$

$$\leq \frac{1}{2\pi} \int_{E'} (A_\delta u_1 - u_1)(re^{i\varphi}) d\varphi + \frac{1}{2\pi} \int_{CE'} (A_\delta u_2 - u_2)(re^{i\varphi}) d\varphi$$

$$\leq \frac{1}{2\pi} \int_{-\pi}^{\pi} (A_\delta u_1 - u_1)(re^{i\varphi}) d\varphi + \frac{1}{2\pi} \int_{-\pi}^{\pi} (A_\delta u_2 - u_2)(re^{i\varphi}) d\varphi$$

$$= \frac{1}{2\pi} \{G(r, A_\delta u_1) - G(r, u_1) + G(r, A_\delta u_2) - G(r, u_2)\}, \quad (7.2.4)$$

where

$$G(r, u) = \int_{-\pi}^{\pi} u(re^{i\varphi}) d\varphi.$$

Now we claim that

$$\lim_{\delta \to 0} G(r, A_\delta u_k) = G(r, u_k), \quad k = 1, 2, \tag{7.2.5}$$

uniformly for $r \in (R_1, R_2)$. Together with (7.2.4), this shows that $T^*(z, u) - C(u)$ is the uniform limit as $\delta \to 0$ of $T^*(z, A_\delta u) - C(A_\delta u)$ in the upper half of $R_1 < |z| < R_2$. Since $A_\delta u_1$ and $A_\delta u_2$ are continuous for each δ, the same is true of $T^*(z, A_\delta u) - C(A_\delta u)$ according to case (i) and hence of $T^*(z, u) - C(u)$.

In order to prove (7.2.5), we note that

$$G(r, A_\delta u_k) = \frac{1}{\pi \delta^2} \int_0^\delta t\,dt \int_{-\pi}^{\pi} d\theta \int_{-\pi}^{\pi} u_k(re^{i\theta} + te^{i\omega})d\omega.$$

Thus, for $0 < t \le \delta$,

$$\int_{-\pi}^{\pi} d\theta \int_{-\pi}^{\pi} u_k(re^{i\theta} + te^{i\omega}\,d\omega) = \int_{-\pi}^{\pi} d\theta \int_{-\pi}^{\pi} u_k(re^{i\theta} + te^{i(\omega+\theta)})d\omega$$

$$= \int_{-\pi}^{\pi} G(|r + te^{i\omega}|, u_k)d\omega \le 2\pi \sup\{G(s, u_k) : |s - r| \le \delta\},$$

so

$$G(r, A_\delta u_k) \le \sup\{G(s, u_k) : |s - r| \le \delta\}, \quad k = 1, 2.$$

Since $G(r, u_k) \le G(r, A_\delta u_k)$ and $G(r, u_k)$ is uniformly continuous on $[R_1, R_2]$, the desired statement (7.2.5) follows immediately. This completes the proof of continuity of $T^*(z)$. □

7.2.2 Subharmonicity of the T^* function.   For the proof of subharmonicity, we need a lemma.

Let T be the set $\{e^{i\theta} : 0 \le \theta < 2\pi\}$. For a subset E of T and a positive number ε, we denote by E_ε the set $\{e^{i(\theta+\varepsilon)} : e^{i\theta} \in E\}$ and by $E_{-\varepsilon}$ the set $\{e^{i(\theta-\varepsilon)} : e^{i\theta} \in E\}$.

Lemma 7.4   *If E is a measurable set of T and $0 < \mathrm{mes}\,E < 2\pi$, then there exists a positive number δ such that*

$$\mathrm{mes}\{E_\varepsilon \cap E_{-\varepsilon}\} \le \mathrm{mes}\,E - 2\varepsilon \tag{7.2.6}$$

for $0 < \varepsilon < \delta$.

Proof. If E is an interval, or differs from an interval by a null set, the equality holds in (7.2.6) for all sufficiently small ε. Otherwise, there are points $a_1 < b_1 < a_2 < b_2 < a_1 + 2\pi$ such that a_1 and a_2 are points of density of $CE(= T - E)$ and b_1 and b_2 are points of density of E. We may assume that $0 < a_1$ and $b_2 < 2\pi$. Choose c with $b_1 < c < a_2$ and let χ denote the characteristic function of E. Then for $\varepsilon > 0$,

$$\int_0^c \chi(t+\varepsilon)\chi(t-\varepsilon)dt \leq \int_0^{a_1} \chi(t+\varepsilon)dt + \int_{a_1}^{b_1} \chi(t-\varepsilon)dt + \int_{b_1}^c \chi(t+\varepsilon)dt$$

$$= \int_0^c \chi(t+\varepsilon)dt - \int_{a_1}^{b_1} \chi(t+\varepsilon)dt + \int_{a_1}^{b_1} \chi(t-\varepsilon)dt$$

$$= \int_0^c \chi(t+\varepsilon)dt + \int_{a_1-\varepsilon}^{a_1+\varepsilon} \chi(t)dt - \int_{b_1-\varepsilon}^{b_1+\varepsilon} \chi(t)dt.$$

Choose $\delta > 0$ such that $0 < \varepsilon < \delta$ implies that

$$\int_{a_1-\varepsilon}^{a_1+\varepsilon} \chi(t)dt \leq \frac{\varepsilon}{2}, \quad \int_{b_1-\varepsilon}^{b_1+\varepsilon} \chi(t)dt \geq \frac{3}{2}\varepsilon.$$

Then

$$\int_0^c \chi(t+\varepsilon)\chi(t-\varepsilon)dt \leq \int_0^c \chi(t+\varepsilon)dt - \varepsilon. \tag{7.2.7}$$

Similarly, using

$$\int_c^{2\pi} \chi(t+\varepsilon)\chi(t-\varepsilon)dt$$

$$\leq \int_c^{a_2} \chi(t+\varepsilon)dt + \int_{a_2}^{b_2} \chi(t-\varepsilon)dt + \int_{b_2}^{2\pi} \chi(t+\varepsilon)dt,$$

we obtain

$$\int_c^{2\pi} \chi(t+\varepsilon)\chi(t-\varepsilon)dt \leq \int_c^{2\pi} \chi(t+\varepsilon)dt - \varepsilon, \quad 0 < \varepsilon < \delta. \tag{7.2.8}$$

which combines with (7.2.7), yields, for $0 < \varepsilon < \delta$,

$$\text{mes}(E_\varepsilon \cap E_{-\varepsilon}) = \int_0^{2\pi} \chi(t+\varepsilon)\chi(t-\varepsilon)dt$$

$$\leq \int_0^{2\pi} \chi(t+\varepsilon)dt - 2\varepsilon = \text{mes}E - 2\varepsilon. \quad \square$$

Theorem 7.3   *If $f(z)$ is meromorphic and not identical to zero in the finite plane, then $T^*(z)$, as given by (7.2.2), is subharmonic on $H = \{z : \operatorname{Im} z \geq 0, z \neq 0\}$.*

Proof.   Set $f(z) = f_1(z)/f_2(z)$, where $f_1(z)$ and $f_2(z)$ are entire functions and have no common zeros. Write $u(z) = \log|f(z)|$ and $u_k(z) = \log|f_k(z)|(k = 1, 2)$. For a real number ψ and fixed r and ρ, $0 < \rho < r$, we define

$$r(\psi) = |r + \rho e^{i\psi}|, \quad \alpha(\psi) = \arg(r + \rho e^{i\psi}), \quad |\arg z| < \frac{\pi}{2}.$$

Then

$$r + \rho e^{i\psi} = r(\psi)e^{i\alpha(\psi)}.$$

Note that $r(\psi) = r(-\psi)$ and $\alpha(\psi) = -\alpha(-\psi)$. For any function v we have

$$\int_{-\pi}^{\pi} v(re^{i\varphi} + \rho e^{i\psi})d\psi$$

$$= \int_0^{\pi} \{v(re^{i\varphi} + \rho e^{i(\varphi+\psi)}) + v(re^{i\varphi} + \rho e^{i(\varphi-\psi)})\}d\psi$$

$$= \int_0^{\pi} \{v(r(\psi)e^{i(\varphi+\alpha(\psi))}) + v(r(\psi)e^{i(\varphi-\alpha(\psi))})\}d\psi. \qquad (7.2.9)$$

Fix $re^{i\theta}$ with $r_1 < r < r_2$, $0 < \theta < \pi$. As in (7.2.3), there is a set E with $\operatorname{mes} E = 2\theta$ such that

$$T^*(re^{i\theta}, u) - C(u) = \frac{1}{2\pi} \int_E u_1(re^{i\varphi})d\varphi + \frac{1}{2\pi} \int_F u_2(re^{i\varphi})d\varphi, \qquad (7.2.10)$$

where $F = CE$. Since u_1 and u_2 are subharmonic, we have for sufficiently small ρ and $k = 1, 2$,

$$u_k(re^{i\varphi}) \leq \frac{1}{2\pi} \int_0^{\pi} \{u_k(r(\psi)e^{i(\varphi+\alpha(\psi))}) + u_k(r(\psi)e^{i(\varphi-\alpha(\psi))})\}d\psi.$$

Substituting these inequalities into (7.2.10) and reversing the order of integration yields:

$$T^*(re^{i\theta}, u) - C(u) \leq \left(\frac{1}{2\pi}\right)^2 \int_0^{\pi} \left\{\int_{E_{\alpha(\psi)}} + \int_{E_{-\alpha(\psi)}} u_1(r(\psi)e^{i\varphi})d\varphi\right\}d\psi$$

$$+ \left(\frac{1}{2\pi}\right)^2 \int_0^{\pi} \left\{\int_{F_{\alpha(\psi)}} + \int_{F_{-\alpha(\psi)}} u_2(r(\psi)e^{i\varphi})d\varphi\right\}d\psi.$$

$$(7.2.11)$$

For this E, let δ be as in Lemma 7.4. We assume that ρ is small enough so that $0 \leq \alpha(\psi) < \delta$ and $\theta + \alpha(\psi) \leq \pi$ for $0 \leq \psi \leq \pi$. Fix such a ψ. By Lemma 7.4, we can choose a set C such that

(i) $C \subset (E_{\alpha(\psi)} \cup E_{-\alpha(\psi)}) - (E_{\alpha(\psi)} \cap E_{-\alpha(\psi)})$, and

(ii) if $A = (E_{\alpha(\psi)} \cap E_{-\alpha(\psi)}) \cup C$, then

$$\mathrm{mes}\,A = \mathrm{mes}\,E - 2\alpha(\psi) = 2(\theta - \alpha(\psi)). \tag{7.2.12}$$

Let $B = (E_{\alpha(\psi)} \cup E_{-\alpha(\psi)}) - C$. Then

$$A \cup B = E_{\alpha(\psi)} \cup E_{-\alpha(\psi)}, \quad A \cap B = E_{\alpha(\psi)} \cap E_{-\alpha(\psi)}. \tag{7.2.13}$$

Since

$$\mathrm{mes}\,A + \mathrm{mes}\,B = \mathrm{mes}(A \cap B) + \mathrm{mes}(A \cup B)$$
$$= \mathrm{mes}(E_{\alpha(\psi)} \cap E_{-\alpha(\psi)}) + \mathrm{mes}(E_{\alpha(\psi)} \cup E_{-\alpha(\psi)})$$
$$= \mathrm{mes}\,E_{\alpha(\psi)} + \mathrm{mes}\,E_{-\alpha(\psi)} = 4\theta,$$

it follows that

$$\mathrm{mes}\,B = 2(\theta + \alpha(\psi)). \tag{7.2.14}$$

By (7.2.12), for every integrable function g we have

$$\int_A g(\varphi)d\varphi + \int_B g(\varphi)d\varphi$$
$$= \int_{A \cup B} g(\varphi)d\varphi + \int_{A \cap B} g(\varphi)d\varphi$$
$$= \int_{E_{\alpha(\psi)} \cup E_{-\alpha(\psi)}} g(\varphi)d\varphi + \int_{E_{\alpha(\psi)} \cap E_{-\alpha(\psi)}} g(\varphi)d\varphi \tag{7.2.15}$$
$$= \int_{E_{\alpha(\psi)}} g(\varphi)d\varphi + \int_{E_{-\alpha(\psi)}} g(\varphi)d\varphi.$$

Similarly,

$$\int_{CA} g(\varphi)d\varphi + \int_{CB} g(\varphi)d\varphi = \int_{F_{\alpha(\psi)}} g(\varphi)d\varphi + \int_{F_{-\alpha(\psi)}} g(\varphi)d\varphi. \tag{7.2.16}$$

From (7.2.15) and (7.2.16), we obtain

$$\frac{1}{2\pi}\Big(\int_{E_{\alpha(\psi)}} + \int_{E_{-\alpha(\psi)}}\Big)u_1(r(\psi)e^{i\varphi})d\varphi$$

$$+\frac{1}{2\pi}\Big(\int_{F_{\alpha(\psi)}} + \int_{F_{-\alpha(\psi)}}\Big)u_2(r(\psi)e^{i\varphi})d\varphi$$

$$= \frac{1}{2\pi}\int_A u_1(r(\psi)e^{i\varphi})d\varphi + \frac{1}{2\pi}\int_{CA} u_2(r(\psi)e^{i\varphi})d\varphi$$

$$+\frac{1}{2\pi}\int_B u_1(r(\psi)e^{i\varphi})d\varphi + \frac{1}{2\pi}\int_{CB} u_2(r(\psi)e^{i\varphi})d\varphi. \qquad (7.2.17)$$

According to (7.2.3)′, for $r(\psi)$ and $2(\theta - \alpha(\psi))$, there is a set E' with measure $2(\theta - \alpha(\psi))$ such that

$$T^*(r(\psi)e^{i(\theta-\alpha(\psi))}, u) - C(u)$$

$$= \frac{1}{2\pi}\int_{E'} u_1(r(\psi)e^{i\varphi})d\varphi + \frac{1}{2\pi}\int_{CE'} u_2(r(\psi)e^{i\varphi})d\varphi \qquad (7.2.18)$$

and

$$\frac{1}{2\pi}\int_{E'} u(r(\psi)e^{i\varphi})d\varphi = \max_E \Big\{\frac{1}{2\pi}\int_E u(r(\psi)e^{i\varphi})d\varphi\Big\}, \qquad (7.2.19)$$

where the maximum of the right-hand side of (7.2.19) is taken over all the measurable sets E of $[0, 2\pi)$ with $\mathrm{mes}E = 2(\theta - \alpha(\psi))$.

From (7.2.12), (7.2.19) and (7.2.18), we obtain

$$\frac{1}{2\pi}\int_A u_1(r(\psi)e^{i\varphi})d\varphi + \frac{1}{2\pi}\int_{CA} u_2(r(\psi)e^{i\varphi})d\varphi$$

$$= \frac{1}{2\pi}\int_A u(r(\psi)e^{i\varphi})d\varphi + \frac{1}{2\pi}\int_0^{2\pi} u_2(r(\psi)e^{i\varphi})d\varphi$$

$$\leq \frac{1}{2\pi}\int_{E'} u(r(\psi)e^{i\varphi})d\varphi + \frac{1}{2\pi}\int_0^{2\pi} u_2(r(\psi)e^{i\varphi})d\varphi$$

$$= \frac{1}{2\pi}\int_{E'} u_1(r(\psi)e^{i\varphi})d\varphi + \frac{1}{2\pi}\int_{CE'} u_2(r(\psi)e^{i\varphi})d\varphi$$

$$= T^*(r(\psi)e^{i(\theta-\alpha(\psi))}, u) - C(u). \qquad (7.2.20)$$

Similarly, we have

$$\frac{1}{2\pi}\int_B u_1(r(\psi)e^{i\varphi})d\varphi + \frac{1}{2\pi}\int_{CB} u_2(r(\psi)e^{i\varphi})d\varphi$$

$$\leq T^*(r(\psi)e^{i(\theta+\alpha(\psi))}, u) - C(u). \qquad (7.2.21)$$

Substituting (7.2.20) and (7.2.21) into (7.2.17) yields

$$\frac{1}{2\pi}\Big(\int_{E_{\alpha(\psi)}} + \int_{E_{-\alpha(\psi)}}\Big)u_1(r(\psi)e^{i\varphi})d\varphi$$

$$+\frac{1}{2\pi}\Big(\int_{F_{\alpha(\psi)}} + \int_{F_{-\alpha(\psi)}}\Big)u_2(r(\psi)e^{i\varphi})d\varphi$$

$$\leq T^*(r(\psi)e^{i(\theta-\alpha(\psi))}, u) - C(u)$$

$$+T^*(r(\psi)e^{i(\theta+\alpha(\psi))}, u) - C(u).$$

Comparing this inequality with (7.2.11), we obtain finally

$$T^*(re^{i\theta}, u) \leq \frac{1}{2\pi}\int_{-\pi}^{\pi} T^*(re^{i\theta} + \rho e^{i\psi}, u)d\psi.$$

7.3 The Spread Relation

7.3.1 Definition.   Now we return to the problem of describing precisely the concept of deficiency, mentioned at the beginning of this chapter. To investigate this problem, it is necessary to introduce an important concept due to A. Edrei [3], viz., the spread $\sigma(a)$ of a complex number a.

Definition 7.3   *Let $f(z)$ be meromorphic and of finite lower order μ in the finite plane. Fix a sequence (r_j) of Pólya peaks of order μ of $f(z)$, and let $\Lambda(r)$ be a positive function with*

$$\Lambda(r) = o(T(r, f)), \quad r \to \infty. \tag{7.3.1}$$

For every r, define sets of arguments $E_\Lambda(r, a) \subset (-\pi, \pi)$ by

$$E_\Lambda(r, a) = \begin{cases} (\theta : |f(re^{i\theta}) - a| < e^{-\Lambda(r)}), & \text{when } a \neq \infty, \\ (\theta : |f(re^{i\theta})| > e^{\Lambda(r)}), & \text{when } a = \infty, \end{cases} \tag{7.3.2}$$

and let

$$\sigma_\Lambda(a) = \lim_{j \to \infty} \operatorname{mes} E_\Lambda(r_j, a). \tag{7.3.3}$$

Then

$$\sigma(a) = \inf_{\Lambda} \sigma_{\Lambda}(a), \tag{7.3.4}$$

where the infimum is taken over all functions $\Lambda(r)$ satisfying (7.3.1). $\sigma(a)$ is called the spread of a with respect to $f(z)$.

Below, we shall need the following integral which can be calculated by the residue theorem.

$$\int_0^{\infty} t^{\tau} P(t, 1, \theta)dt = \frac{\sin \tau\theta}{\sin \tau\pi}, \tag{7.3.5}$$

where $0 < \tau < 1$ and $P(t, r, \theta) = (1/\pi) \cdot (r \sin \theta)/(r^2 + 2tr \cos \theta + t^2)$.

7.3.2 Spread relation.   The following theorem was conjectured by A. Edrei [3] in 1965 and proved by A. Baernstein [1] in 1973.

Theorem 7.4   *Let $f(z)$ be a meromorphic function of lower order $\mu(0 < \mu < \infty)$. If a is a deficient value of $f(z)$ with deficiency $\delta(a, f)$, then $\sigma(a)$ as given by Definition 7.3 satisfies*

$$\sigma(a) \geq \min\left\{2\pi, \frac{4}{\mu}\arcsin\sqrt{\frac{\delta(a, f)}{2}}\right\}. \tag{7.3.6}$$

(7.3.6) is called the spread relation.

Proof.   First, we assume that

$$0 < \frac{4}{\mu}\arcsin\sqrt{\frac{\delta(a, f)}{2}} < 2\pi. \tag{7.3.7}$$

We can also assume, without loss of generality, that

$$f(0) = 1, \quad a = \infty. \tag{7.3.8}$$

The general case is reduced to this special case by the following considerations: Suppose $a \neq \infty$. Let

$$g(z) = \frac{1}{f(z) - a},$$

and define $h(z)$ by

$$f(z) = Cz^k h(z),$$

where k is an integer, C is a non-zero constant and $h(0) = 1$. Since $\mu > 0$, it follows that

$$T(r, f) \sim T(r, g) \sim T(r, h), \quad r \to \infty.$$

Thus f, g and h have the same lower order and the same Pólya peaks. With a self-explanatory notation, we have

$$E_\Lambda(r, a; f) = E_\Lambda(r, \infty; g)$$

and

$$E_\Lambda(r, \infty; f) = E_{\Lambda_1}(r, \infty; h),$$

where $\Lambda_1(r) = \Lambda(r) - k \log r - \log |C|$. As $r \to \infty$, we have

$$\Lambda(r) = o(T(r, f)) \iff \Lambda(r) = o(T(r, g))$$
$$\iff \Lambda_1(r) = o(T(r, h)).$$

Hence

$$\sigma(a, f) = \sigma(\infty, g), \quad \sigma(\infty, f) = \sigma(\infty, h).$$

It is also true that $\delta(a, f) = \delta(\infty, g)$, $\delta(\infty, f) = \delta(\infty, h)$. The above facts show that there is no harm in assuming (7.3.8).

Setting

$$\gamma = \frac{2}{\pi\mu} \arcsin\sqrt{\frac{\delta(\infty, f)}{2}},$$

then $0 < \gamma < 1$. Let $\Lambda(r)$ be a positive function satisfying (7.3.1) and put

$$\sigma_j = \text{mes}\, E_\Lambda(r_j, \infty).$$

The spread relation is equivalent to the inequality

$$\varliminf_{j \to \infty} \sigma_j \geq 2\pi\gamma. \tag{7.3.9}$$

Let

$$E_0(r) = \{\theta : |f(re^{i\theta})| \geq 1\}, \quad E^c(r) = \{\theta : 1 \leq |f(re^{i\theta})| \leq e^{\Lambda(r)}\}.$$

We then have

$$T(r_j, f) = \frac{1}{2\pi} \int_{E_0(r_j)} \log|f(r_j e^{i\theta})| d\theta + N(r_j, f)$$

$$= \frac{1}{2\pi} \int_{E_\Lambda(r_j, \infty)} \log|f(r_j e^{i\theta})| d\theta + N(r_j, f)$$

$$+ \frac{1}{2\pi} \int_{E^c(r_j)} \log|f(r_j e^{i\theta})| d\theta$$

$$\leq T^*(r_j e^{\frac{1}{2}i\sigma_j}) + \Lambda(r_j).$$

Dividing by $T(r_j)$ and remembering that $T^*(z) \leq T(|z|, f)$, we find that

$$\lim_{j \to \infty} \frac{T^*(r_j e^{\frac{1}{2}i\sigma_j})}{T(r_j, f)} = 1. \qquad (7.3.10)$$

On the other hand, define

$$v(z) = \begin{cases} 0, & z = 0, \\ T^*(z^\gamma), & z \neq 0, \ \mathrm{Im}\, z \geq 0. \end{cases} \qquad (7.3.11)$$

It follows from Theorem 7.3 that $v(z)$ is subharmonic in $\mathrm{Im}\, z > 0$ and continuous on $\mathrm{Im}\, z \geq 0$, except at the origin where it is not defined. Since $f(0) = 1$, it is clear that putting $T^*(0) = v(0) = 0$ will remove this discontinuity.

Now let us observe the half disk

$$D_R = \{z = re^{i\theta} : 0 < r < R, \ 0 < \theta < \pi\}.$$

With the boundary values of $v(z)$, we construct the Poisson integral $h(z)$ which is harmonic and majorizes $v(z)$ in D_R, i.e.

$$h(z) = \int_{-R}^{R} v(t) A(t, r, \theta, R) dt + \int_{0}^{\pi} v(Re^{i\varphi}) B(\varphi, r, \theta, R) d\varphi,$$

where

$$A(t, r, \theta, R) = \frac{1}{\pi} \cdot \frac{r \sin\theta}{t^2 - 2tr\cos\theta + r^2} - \frac{1}{\pi} \frac{R^2 r \sin\theta}{R^4 - 2rtR^2\cos\theta + r^2t^2},$$

$$B(\varphi, r, \theta, R) = \frac{2Rr\sin\theta}{\pi} \cdot \frac{(R^2 - r^2)\sin\varphi}{|R^2 e^{2i\varphi} - 2rRe^{i\varphi}\cos\theta + r^2|^2}.$$

Using (7.3.11) and the definition of the T^* function, we see that

$$v(t) = T^*(t^\gamma) = N(t^\gamma, f), \qquad\qquad t > 0,$$

$$v(-t) = v(te^{i\pi}) = T^*(t^\gamma e^{i\gamma\pi}) \leq T(t^\gamma, f), \quad t > 0$$

and

$$v(Re^{i\varphi}) = T^*(R^\gamma e^{i\gamma\pi}) \leq T(R^\gamma), \quad 0 \leq \varphi \leq \pi.$$

The Poisson kernels A and B are positive, so we obtain

$$v(re^{i\theta}) \leq \int_0^R N(t^\gamma, f)A(t, r, \theta, R)dt + \int_0^R T(t^\gamma, f)A(-t, r, \theta, R)dt$$

$$+T(R^\gamma, f) \int_0^\pi B(\varphi, r, \theta, R)d\varphi.$$

$$(7.3.12)$$

From

$$|R^2 e^{2i\varphi} - 2rRe^{i\varphi}\cos\theta + r^2|^2 = |(Re^{i\varphi} - re^{i\theta})(Re^{i\varphi} - re^{-i\theta})|^2$$

$$\geq (R - r)^4 > \frac{R^4}{16}, \quad 0 < r < \frac{R}{2},$$

it is easily deduced that

$$B(\varphi, r, \theta, R) < \frac{32r}{\pi R}, \quad 0 < \varphi < \pi, \quad 0 < \theta < \pi, \quad 0 < r < \frac{R}{2}.$$

Since

$$R^4 - 2rtR^2 \cos\theta + r^2 t^2 = |R^2 - rte^{i\theta}|^2 > 0,$$

the second term in the definition of A is positive, and thus, letting

$$P(t, r, \theta) = \frac{1}{\pi} \cdot \frac{r\sin\theta}{t^2 + 2tr\cos\theta + r^2},$$

we have

$$A(t, r, \theta, R) \leq P(t, r, \pi - \theta), \quad A(-t, r, \theta, R) \leq P(t, r, \theta).$$

Using the estimates of A and B in (7.3.12), we obtain the key inequality

$$v(re^{i\theta}) \leq \int_0^R N(t^\gamma, f)P(t, r, \pi - \theta)dt + \int_0^R T(t^\gamma, f)P(t, r, \theta)dt$$

$$+32\left(\frac{r}{R}\right)T(R^\gamma, f), \quad 0 < \theta < \pi, \quad 0 < r < \frac{R}{2}. \qquad (7.3.13)$$

Let (r_j) be a sequence of Pólya peaks of order μ of $f(z)$, and let (r'_j), (r''_j) and (ε_j) be the associated sequences. Put

$$s'_j = (r'_j)^{\frac{1}{\gamma}}, \quad s''_j = (r''_j)^{\frac{1}{\gamma}}, \quad s_j = (r_j)^{\frac{1}{\gamma}}.$$

We have

$$T(t^\gamma, f) < (1 + \varepsilon_j)T(r_j, f)\Big(\frac{t}{s_j}\Big)^{\gamma\mu}, \quad s'_j < t < s''_j. \qquad (7.3.14)$$

Choose j_0 so that $j \geq j_0$ implies $2s'_j < s_j < (s''_j)/2$. Then

$$P(t, s_j, \theta) = \frac{1}{\pi}\text{Im}\Big(\frac{1}{t + s_j e^{-i\theta}}\Big) \leq \frac{1}{\pi} \cdot \frac{1}{|t + s_j e^{-i\theta}|} < \frac{2}{\pi s_j} < \frac{1}{s_j},$$

$$0 < t < s'_j, \quad j \geq j_0.$$

From this and (7.3.14), we find that for $0 < r < \pi$ and $j \geq j_0$

$$\int_0^{s''_j} T(t^\gamma, f)P(t, s_j, \theta)dt = \Big(\int_0^{s'_j} + \int_{s'_j}^{s''_j}\Big)T(t^\gamma, f)P(t, s_j, \theta)dt$$

$$\leq T(r'_j, f)\frac{s'_j}{s_j} + (1 + \varepsilon_j)T(r_j, f)\int_{s'_j}^{s''_j}\Big(\frac{t}{s_j}\Big)^{\gamma\mu}P(t, s_j, \theta)dt$$

$$\leq T(r_j, f)\frac{s'_j}{s_j} + (1 + \varepsilon_j)T(r_j, f)\int_0^{\infty}\Big(\frac{t}{s_j}\Big)^{\gamma\mu}P(t, s_j, \theta)dt.$$

Since

$$\lim_{j\to\infty}\frac{s'_j}{s_j} = \lim_{j\to\infty}\Big(\frac{r'_j}{r_j}\Big)^{\frac{1}{\gamma}} = 0, \quad \gamma\mu = \frac{2}{\pi}\arcsin\sqrt{\frac{\delta(\infty, f)}{2}} \leq \frac{1}{2}$$

and (7.3.5) holds, we have

$$\int_0^{\infty}\Big(\frac{t}{s_j}\Big)^{\gamma\mu}P(t, s_j, \theta)dt = \int_0^{\infty} t^{\gamma\mu}P(t, 1, \theta)dt = \frac{\sin\theta\gamma\mu}{\sin\pi\gamma\mu}.$$

Thus we get finally

$$\int_0^{s''_j} T(t^\gamma, f)P(t, s_j, \theta)dt \leq T(r_j, f)\Big\{\frac{\sin\theta\gamma\mu}{\sin\pi\gamma\mu} + o(1)\Big\}, \quad j \to \infty \quad (7.3.15)$$

uniformly in θ for $0 < \theta < \pi$.

Starting from

$$N(t^\gamma, f) < (1 - \delta(\infty, f) + o(1))T(t^\gamma, f), \quad s'_j < t < s''_j,$$

a similar argument leads to

$$\int_0^{s_j''} N(t^\gamma, f) P(t, s_j, \pi - \theta) dt$$

$$\leq (1 - \delta(\infty, f)) T(r_j, f) \left\{ \frac{\sin(\pi - \theta)\gamma\mu}{\sin \pi\gamma\mu} + o(1) \right\}, \quad j \to \infty, \quad (7.3.16)$$

uniformly in θ for $0 < \theta < \pi$.

When $j \to \infty$, we note that

$$\frac{s_j}{s_j''} T((s_j'')^\gamma, f) = \left(\frac{r_j}{r_j''} \right)^{\frac{1}{\gamma}} T(r_j'', f)$$

$$< (1 + o(1)) T(r_j, f) \left(\frac{r_j}{r_j''} \right)^{\frac{1}{\gamma} - \mu} = o(T(r_j, f)). \quad (7.3.17)$$

In (7.3.13), take $r = s_j$, $R = s_j''$, and use (7.3.15), (7.3.16) and (7.3.17). The result is that

$$v(s_j e^{i\theta}) \leq T(r_j, f) \left\{ \frac{\sin \theta\gamma\mu + (1 - \delta(\infty, f)) \sin(\pi - \theta)\gamma\mu}{\sin \pi\gamma\mu} + o(1) \right\},$$

$$j \to \infty, \quad 0 < \theta < \pi,$$

where the $o(1)$ term is independent of θ.

From the definition of γ, we have

$$1 - \delta = \cos \pi\gamma\mu.$$

Using this and the identity

$$\sin \theta\gamma\mu = \sin\{\pi\gamma\mu - (\pi - \theta)\gamma\mu\}$$

$$= \sin \pi\gamma\mu \cos(\pi - \theta)\gamma\mu - \cos \pi\gamma\mu \sin(\pi - \theta)\gamma\mu,$$

we obtain

$$v(s_j e^{i\theta}) \leq T(r_j, f)\{\cos(\pi - \theta)\gamma\mu + \alpha_j\}, \quad j = 1, 2, \cdots, \quad 0 < \theta < \pi,$$

$$(7.3.18)$$

where (α_j) is a sequence tending to zero.

Let

$$\sigma_j = \text{mes} E_\Lambda(r_j, \infty), \quad J = \{j : \sigma_j \leq 2\pi\gamma\}.$$

If J is a finite set, then (7.3.9) obviously holds. Thus we may assume that J is infinite.

The point

$$(r_j e^{\frac{1}{2}i\sigma_j})^{\frac{1}{\gamma}} = s_j e^{\frac{1}{2}i\frac{\sigma_j}{\gamma}}$$

belongs to the domain of v, i.e., the upper half plane, if and only if $j \in J$, in which case we have

$$T^*(r_j e^{\frac{1}{2}i\sigma_j}) = v(s_j e^{i\frac{\sigma_j}{2\gamma}}), \quad j \in J.$$

Using this in (7.3.10), we obtain

$$\lim_{\substack{j \to \infty \\ j \in J}} \frac{v(s_j e^{\frac{i\sigma_j}{2\gamma}})}{T(r_j, f)} = 1.$$

But (7.3.18) gives

$$v(s_j e^{i\frac{\sigma_j}{2\gamma}}) \le T(r_j, f)\{\cos(\pi - \frac{\sigma_j}{2\gamma})\gamma\mu + \alpha_j\}, \quad j \in J, \quad j \to \infty.$$

Thus

$$\lim_{\substack{j \to \infty \\ j \in J}} \frac{\sigma_j}{2\gamma} = \pi,$$

and the proof is complete in the case of $0 < (4/\mu)\arcsin\sqrt{\delta(a, f)/2} < 2\pi$.

When $\dfrac{4}{\mu}\arcsin\sqrt{\delta(a, f)/2} \ge 2\pi$, take a number d with $0 < d < \delta(a, f)$ such that

$$\frac{4}{\mu}\arcsin\sqrt{\frac{d}{2}} < 2\pi.$$

Put $\gamma = (2/\pi\mu)\arcsin\sqrt{d/2}$. According to the above deductive reasoning, we can prove that

$$\sigma(a) \ge \frac{4}{\mu}\arcsin\sqrt{\frac{d}{2}}.$$

Let d tend to $d_0 = 2\sin^2(\mu\pi/2)$, and we have $\sigma(a) = 2\pi$.   □

Corollary.   Let $f(z)$ be meromorphic and of finite lower order μ in the finite plane. If $f(z)$ has deficient values $a_j(j = 1, 2, \cdots, \nu(f); 1 \le \nu(f) \le \infty)$ and

$$\delta(a_1, f) \ge \delta(a_2, f) \ge \delta(a_3, f) \ge \cdots,$$

with

$$\frac{4}{\mu}\arcsin\sqrt{\frac{\delta(a_1, f)}{2}} < 2\pi, \tag{7.3.19}$$

then

$$\frac{4}{\mu}\sum_{j=1}^{\nu(f)}\arcsin\sqrt{\frac{\delta(a_j, f)}{2}} \le 2\pi. \tag{7.3.20}$$

Proof. When $\nu(f) < \infty$, set

$$d = \min_{\substack{1\le j_1 \ne j_2 \le \nu(f) \\ a_{j_1}, a_{j_2} \ne \infty}} \{|a_{j_1} - a_{j_2}|\},$$

$$A = \max_{\substack{1\le j \le \nu(f) \\ a_j \ne \infty}} \{|a_j|\},$$

$$\Lambda(r) = \max\{1 + \log\frac{2}{d}, \quad \log(A+2)\}.$$

Condition (7.3.19) implies that $\mu > 0$. Thus

$$\Lambda(r) = o\{T(r, f)\}, \quad r \to \infty.$$

We assert that

$$E_\Lambda(r, a_{j_1}) \cap E_\Lambda(r, a_{j_2}) = \emptyset \tag{7.3.21}$$

for $1 \le j_1 \ne j_2 \le \nu(f)$, where $E_\Lambda(r, a)$ is defined by (7.3.2). In fact, if the intersection is non-empty, say

$$\theta \in (E_\Lambda(r, a_{j_1}) \cap E_\Lambda(r, a_{j_2})),$$

then we obtain the inequality

$$d \le |a_{j_1} - a_{j_2}| \le |a_{j_1} - f(re^{i\theta})| + |f(re^{i\theta}) - a_{j_2}|$$
$$< 2e^{-\Lambda(r)} \le 2e^{-(1+\log\frac{2}{d})} < \frac{d}{e},$$

when a_{j_1} and a_{j_2} are finite, and also

$$A + 2 \le e^{\Lambda(r)} < |f(re^{i\theta})| < |a| + 1 \le A + 1,$$

when one of a_{j_1} and a_{j_2} is infinite; both are absurd.

From (7.3.21), (7.3.3) and (7.3.4), we obtain

$$\sum_{j=1}^{\nu(f)} \sigma(a_j) \le 2\pi.$$

Connecting this inequality with Theorem 7.4, (7.3.20) is derived immediately.

When $\nu(f) = \infty$, we have for any positive integer J,

$$\frac{4}{\mu} \sum_{j=1}^{J} \arcsin\sqrt{\frac{\delta(a_j, f)}{2}} \le 2\pi,$$

and hence (7.3.20) holds,   □

(7.3.20) is called the total spread relation.

7.4 Applications of the Spread Relation

7.4.1 Fuchs' theorem.   As an important consequence of the spread relation, we have

Theorem 7.5   *Let $f(z)$ be meromorphic and of finite lower order μ in the finite plane. If $a_j (j = 1, 2, \cdots)$ are the deficient values of $f(z)$ with deficiencies $\delta(a_j, f)$, then the series*

$$\sum_{j=1}^{\infty} \{\delta(a_j, f)\}^{\frac{1}{2}} \tag{7.4.1}$$

is convergent, and its sum does not exceed $\max\{1, (\sqrt{2}/2)\mu\pi\}$.

Proof.   Arrange the deficient values a_j such that $\delta(a_1, f) \ge \delta(a_2, f) \ge \delta(a_3, f) \ge \cdots$. If

$$\frac{4}{\mu} \arcsin\sqrt{\frac{\delta(a_1, f)}{2}} \ge 2\pi,$$

then a_1 is the unique deficient value of $f(z)$ and Theorem 7.5 obviously holds. If

$$\frac{4}{\mu} \arcsin\sqrt{\frac{\delta(a_1, f)}{2}} < 2\pi,$$

then we have

$$\frac{4}{\mu} \sum_{j=1}^{\infty} \arcsin\sqrt{\frac{\delta(a_j, f)}{2}} \leq 2\pi$$

according to the Corollary of Theorem 7.4. Thus

$$\sum_{j=1}^{\infty} \{\delta(a_j, f)\}^{\frac{1}{2}} \leq \sqrt{2} \sum_{j=1}^{\infty} \arcsin\sqrt{\frac{\delta(a_j, f)}{2}} \leq \frac{\sqrt{2}}{2}\mu\pi. \quad \square$$

For the meromorphic functions of finite order, the convergence of the series (7.4.1) was first indicated by O. Teichmuller [1]. However, there was no proof, until W.Fuchs [1] completed it in 1958. W.K.Hayman [2] proved a further result in 1964.

Theorem 7.5'   *Conditions and notations as in Theorem 7.5, the series*

$$\sum_{j=1}^{\infty} \{\delta(a_j, f)\}^{\frac{1}{3}+\varepsilon}$$

converges for any positive number ε. Moreover, if a positive number ε is given, then there exists a meromorphic function $f(z)$ of finite lower order μ in the finite plane such that the series

$$\sum_{j=1}^{\infty} \{\delta(a_j, f)\}^{\frac{1}{3}-\varepsilon}$$

is divergent.

A natural question is whether or not $\sum_{j=1}^{\infty} \{\delta(a_j, f)\}^{\frac{1}{3}}$ converges. Regarding this problem, Petrenko [1] obtained

$$\sum_{j=1}^{\infty} \frac{\{\delta(a_j, f)\}^{\frac{1}{3}}}{\log \frac{e}{\delta(a_j, f)}} < \infty.$$

Bombieri and Ragendda [1] proved: If $\tau(t)$ is a positive function defined in $(0, \infty)$ with $\int^{\infty} \tau(t)/t dt < \infty$, then

$$\sum_{j=1}^{\infty} \{\delta(a_j, f)\tau(\delta(a_j, f))\}^{\frac{1}{3}} < \infty.$$

Finally Weitsman[2] settled this problem completely.

Theorem 7.5″   *Let $f(z)$ be meromorphic and of finite lower order μ in the finite plane. Then*

$$\sum_{j=1}^{\infty} \{\delta(a_j, f)\}^{\frac{1}{3}} < \infty.$$

For entire functions, more can be said. For instance, if $f^{(k)}(z)$ denotes the derivative $(k > 0)$ or the primitive $(k < 0)$ of order k of an entire function $f(z)$ with finite lower order μ, then Yang and Zhang [7] proved that

$$\sum_{k=-\infty}^{\infty} \sum_{a_{kj} \neq 0, \infty} \delta(a_{kj}, f^{(k)})^{\frac{1}{3}} < \infty.$$

Another important problem is the following Arakelyan conjecture [1].

If $f(z)$ is an entire function of finite lower order μ and $a_j (j = 1, 2, \cdots)$ are its deficient values, then

$$\sum_{j=1}^{\infty} \frac{1}{\log \dfrac{1}{\delta(a_j, f)}} < \infty.$$

Recently, Lewis and Wu[1] made significant progress on this problem. They proved

Theorem 7.6   *Let $f(z)$ be an entire function of finite lower order μ and $a_j (j = 1, 2, \cdots)$ its deficient values. Then there exists a positive number γ_0 not depending on f such that*

$$\sum (\delta(a_j, f))^{\frac{1}{3} - \gamma_0} < \infty.$$

However, the Arakelyan conjecture is still open.[1]

7.4.2 Ellipse theorem.   Now we establish a very elementary proposition, by which the famous ellipse theorem will be easily deduced from the total spread relation.

---

1) Drasin recently informed me that the Arakelyan conjecture was disproved by A. Eremenko, who has an example to show that there exist entire functions of finite order for which the nth deficiency $\delta(a_n, f)$ satisfies $\log \dfrac{1}{\delta(a_n, f)} < cn$ for some $c > 0$.

Lemma 7.5   *Let μ be a number with $0 \leq \mu \leq 1$, and A, B be two positive numbers. If $A + B \leq \pi$, then we have*

$$\cos^2 \mu A + \cos^2 \mu B - 2 \cos \mu A \cos \mu B \cos \mu \pi \geq \sin^2 \mu \pi. \qquad (7.4.2)$$

Proof.   We distinguish two cases.
(i) $A + B = \pi$.
In this case, the left-hand side of (7.4.2) equals

$$\cos^2 \mu A + \cos^2 (\mu \pi - \mu A) - 2 \cos \mu A \cos(\mu \pi - \mu A) \cos \mu \pi$$

$$= \cos^2 \mu A + \sin^2 \mu A \sin^2 \mu \pi - \cos^2 \mu A \cos^2 \mu \pi$$

$$= \sin^2 \mu \pi.$$

(ii) $A + B < \pi$.
Without loss of generality, we may assume that $A \geq B$. Then we claim that

$$\cos^2 \mu A + \cos^2 \mu B - 2 \cos \mu A \cos \mu B \cos \mu \pi$$

$$> \cos^2 \mu A + \cos^2 \mu(\pi - A) - 2 \cos \mu A \cos \mu(\pi - A) \cos \mu \pi. \qquad (7.4.3)$$

According to case (i), the right-hand side of (7.4.3) equals $\sin^2 \mu \pi$.
In order to prove (7.4.3), we note that

$$\cos \mu B - \cos \mu(\pi - A) > 0, \qquad (7.4.4)$$

since $0 < \mu B < \mu(\pi - A) < \pi$.
When $\mu(\pi + A) \leq \pi/2$, it is clear that

$$\cos \mu B - \cos \mu(\pi + A) > 0. \qquad (7.4.5)$$

When $\pi/2 < \mu(\pi + A) < 3\pi/2$, (7.4.5) obviously holds. When $\mu(\pi + A) \geq (3/2)\pi$, we have $\mu(\pi + A) \leq 2\pi - \mu B$, so that (7.4.5) is also true.
By (7.4.4) and (7.4.5), we obtain

$$\{\cos \mu B - \cos \mu(\pi - A)\}\{\cos \mu B + \cos \mu(\pi - A)\}$$

$$> \{\cos \mu(\pi + A) + \cos \mu(\pi - A)\}\{\cos \mu B - \cos \mu(\pi - A)\}.$$

Thus (7.4.3) follows immediately. $\square$

Theorem 7.7 Let $f(z)$ be a meromorphic function of finite lower order μ, $0 \le \mu \le 1$. If a and b are two distinct deficient values of $f(z)$ and

$$u = 1 - \delta(a, f), \quad v = 1 - \delta(b, f),$$

then

$$u^2 + v^2 - 2uv \cos \mu\pi \ge \sin^2 \mu\pi. \tag{7.4.6}$$

Furthermore, if $u \le \cos \mu\pi$, then $v = 1$; if $v \le \cos \mu\pi$, then $u = 1$.

Proof. When $\mu = 0$, (7.4.6) obviously holds. In the case of $\mu > 0$, we have

$$\frac{4}{\mu}\arcsin\sqrt{\frac{\delta(a, f)}{2}} + \frac{4}{\mu}\arcsin\sqrt{\frac{\delta(b, f)}{2}} \le 2\pi$$

according to the total spread relation. Setting

$$A = \frac{2}{\mu}\arcsin\sqrt{\frac{\delta(a, f)}{2}}, \quad B = \frac{2}{\mu}\arcsin\sqrt{\frac{\delta(b, f)}{2}},$$

then

$$A > 0, \quad B > 0, \quad A + B \le \pi, \quad \cos \mu A = 1 - \delta(a, f) = u$$

and

$$\cos \mu B = 1 - \delta(b, f) = v.$$

Thus inquality (7.4.6) can be obtained from Lemma 7.5.

When $u \le \cos \mu\pi$, we have $(4/\mu)\arcsin\sqrt{\delta(a, f)/2} \ge 2\pi$. Hence $(4/\mu)$ $\arcsin\sqrt{\delta(b, f)/2} = 0$, i.e. $v = 1$. Similarly, if $v \le \cos \mu\pi$, then $u = 1$. □

Theorem 7.7 states that the point (u, v) cannot be located inside the ellipse $u^2 + v^2 - 2uv \cos \mu\pi = \sin^2 \mu\pi$. Therefore it is called the ellipse theorem, which is due to Edrei and Fuchs [3].

A series of important consequences can be obtained from the ellipse theorem. For instance, we have the following corollaries:

Corollary 1. Let $f(z)$ be meromorphic and of lower order $\mu(\le 1/2)$ in the finite plane. If a is a deficient value of $f(z)$ with $\delta(a, f) \ge 1 - \cos \mu\pi$, then a is the unique deficient value of $f(z)$. In particular, a meromorphic function of lower order zero can have one at most deficient value.

In fact, if $f(z)$ has another deficient value b, then from $u = 1 - \delta(a, f) \leq \cos \mu\pi$ and Theorem 7.7, we have $v = 1$.

Corollary 2. *Let $f(z)$ be a transcendental entire function of finite order. If its lower order μ satisfies $0 \leq \mu \leq 1$, then*

$$\sum_{a \neq \infty} \delta(a, f) \begin{cases} = 0, & \text{when } 0 \leq \mu \leq \dfrac{1}{2}, \\[2mm] \leq 1 - \sin \mu\pi, & \text{when } \dfrac{1}{2} < \mu \leq 1. \end{cases} \tag{7.4.7}$$

Suppose $a_j (j = 1, 2, \cdots, q)$ are q finite distinct complex numbers. The second fundamental theorem of Nevanlinna gives

$$\sum_{j=1}^{q} m\left(r, \frac{1}{f - a_j}\right) < T(r, f) - N\left(r, \frac{1}{f'}\right) + O(\log r).$$

Thus

$$\sum_{j=1}^{q} \lim_{r \to \infty} \frac{m\left(r, \dfrac{1}{f - a_j}\right)}{T(r, f)} \leq \lim_{r \to \infty} \frac{\displaystyle\sum_{j=1}^{q} m\left(r, \dfrac{1}{f - a_j}\right)}{T(r, f)}$$

$$\leq \lim_{r \to \infty} \left\{ 1 - \frac{N\left(r, \dfrac{1}{f'}\right)}{T(r, f)} \right\} + \overline{\lim_{r \to \infty}} \frac{O(\log r)}{T(r, f)}.$$

Since

$$T(r, f') < (1 + o(1))T(r, f), \quad \overline{\lim_{r \to \infty}} \frac{O(\log r)}{T(r, f)} = 0,$$

we have

$$\sum_{j=1}^{q} \delta(a_j, f) \leq \delta(0, f').$$

Hence

$$\sum_{a \neq \infty} \delta(a, f) \leq \delta(0, f'). \tag{7.4.8}$$

Because $f'(z)$ is an entire function of lower order μ, we have $\delta(\infty, f') = 1 \geq 1 - \cos \mu\pi$, when $0 \leq \mu \leq 1/2$. By Corollary 1, $\delta(0, f')$ must equal zero. Thus

$$\sum_{a \neq \infty} \delta(a, f) = 0.$$

When $1/2 < \mu \leq 1$, put $u = 1 - \delta(\infty, f') = 0$ and $v = 1 - \delta(0, f')$. Then (7.4.6) is reduced to $v^2 \geq \sin^2 \mu\pi$. Noting that $0 \leq v \leq 1$, we obtain $v \geq \sin \mu\pi$, i.e., $\delta(0, f') \leq 1 - \sin \mu\pi$. Comparing this with (7.4.8), the conclusion of Corollary 2 follows.

(7.4.7) can be rewritten as

$$\sum_{a \in \check{C}} \delta(a, f) \begin{cases} \leq 1, & \text{when } 0 \leq \mu \leq \dfrac{1}{2}, \\ \leq 2 - \sin \mu\pi, & \text{when } \dfrac{1}{2} < \mu \leq 1. \end{cases} \qquad (7.4.9)$$

A very deep result is that (7.4.9) can be extended to the case of meromorphic functions without any additional conditions. This is a result of Edrei, proved in 1973. We shall introduce it in next section.

7.5 The Deficiency Problem

7.5.1 The deficiency problem. As another application of the spread relation, we shall introduce the deficiency problem for meromorphic functions and settle it in the case of lower order less than one (See Edrei [4].).

Let $f(z)$ be meromorphic and of finite lower order μ in the finite plane. Suppose $a_j(j = 1, 2, \cdots)$ is the sequence of all the deficient values of $f(z)$, arranged so that

$$\delta(a_1, f) \geq \delta(a_2, f) \geq \delta(a_3, f) \geq \cdots > 0. \qquad (7.5.1)$$

The number of deficient values of $f(z)$ is denoted by $\nu(f)(0 \leq \nu(f) \leq \infty)$, and the total deficiency $\Delta(f)$ is defined to be

$$\Delta(f) = \sum_{j=1}^{\nu(f)} \delta(a_j, f).$$

Definition 7.4 (The Deficiency problem)   *Let $\mathcal{F}_\mu$ be the class of all meromorphic functions of lower order exactly equal to $\mu < \infty$.*

(1) Give a precise determination of

$$\Omega(\mu) = \sup_{f \in \mathcal{F}_\mu} \Delta(f).$$

(2) *Define a function $f \in \mathcal{F}_\mu$ to be extremal if*

$$\Delta(f) = \Omega(\mu). \qquad (7.5.2)$$

Do extremal functions exist? If so, what properties characterize extremal functions?

Since $a_j (j = 1, 2, \cdots, \nu(f))$ are the deficient values of $f(z)$, we have by the spread relation,

$$\sum_{j=1}^{\nu(f)} \frac{4}{\mu} \arcsin \sqrt{\frac{\delta(a_j, f)}{2}} \le 2\pi. \qquad (7.5.3)$$

Here we assume that $(4/\mu)\arcsin\sqrt{\delta(a_1, f)/2} < 2\pi$. Otherwise, $\delta(a_1, f) > 2\sin^2(\mu\pi/2) = 1 - \cos\mu\pi$. Thus a_1 is the unique deficient value of $f(z)$ by Corollary 1 of Theorem 7.7.

Now we want to find a precise upper bound of

$$\Delta(f) = \sum_{j=1}^{\nu(f)} \delta(a_j, f)$$

for the case

$$\frac{4}{\pi} \sum_{j=1}^{\nu(f)} \arcsin \sqrt{\frac{\delta(a_j, f)}{2}} \le 2\mu.$$

In order to simplify this problem, let

$$s_j = \frac{4}{\pi} \arcsin \sqrt{\frac{\delta(a_j, f)}{2}}, \quad j = 1, 2, \cdots, \nu(f)$$

i.e., $\delta(a_j, f) = 2\sin^2(s_j\pi/4)$. Therefore, the above problem can be rewritten in the following form:

If $0 \le s_j \le 1 (j = 1, 2, \cdots, \nu(f))$ and $\displaystyle\sum_{j=1}^{\nu(f)} s_j \le 2\mu$, we want to find a precise upper bound for

$$\sum_{j=1}^{\nu(f)} 2\sin^2 \frac{s_j\pi}{4}.$$

In the next subsection, we shall pose a general problem in convex programming.

7.5.2 A problem in convex programming. Let $x = \varphi(s)$ be a real continuous function on $[0, 1]$ satisfying the following conditions:

(1) $\varphi(0) = 0$, $\varphi(1) = 1$,

(2) $\varphi'(s)$ and $\varphi''(s)$ exist in $(0, 1)$ and they are strictly positive and continuous in this interval.

Denote by $\psi(x)$ the inverse function

$$s = \psi(x) = \varphi^{-1}(x), \quad 0 \leq x \leq 1.$$

Problem P.   Let

$$s_1, s_2, \cdots, s_k, \quad 2 \leq k < \infty$$

be subject to the constraints

$$0 \leq s_j \leq 1, \quad j = 1, 2, \cdots, k, \quad \sum_{j=1}^{k} s_j \leq H, \quad H > 0.$$

1) Find

$$\Lambda = \sup\left(\sum_{j=1}^{k} \varphi(s_j) \right),$$

where the supremum is taken over all points satisfying the constraints.

2) What can be said about an optimum point $s = (s_1, s_2, \cdots, s_k)$, which is a point such that the coordinates s_j satisfy not only the constraints, but also

$$\sum_{j=1}^{k} \varphi(s_j) = \Lambda.$$

The solution of Problem P depends on the following lemma.

Lemma 7.6   *If $s = (s_1, s_2, \cdots, s_k)$ is optimum, it has at most one coordinate which is not 0 or 1.*

Proof.   If this were not so, two coordinates of s, say s_1 and s_2, would be such that

$$0 < s_1 < 1, \quad 0 < s_2 < 1, \quad s_1 + s_2 \leq H - \sum_{j=3}^{k} s_j,$$

$$\varphi(s_1) + \varphi(s_2) = \Lambda - \sum_{j=3}^{k} \varphi(s_j). \tag{7.5.4}$$

The point
$$s(t) = (s_1 + t, s_2 - t, s_3, \cdots, s_k)$$
satisfies the constraints, provided $|t|$ is sufficiently small. Hence

$$\varphi(s_1 + t) + \varphi(s_2 - t) + \sum_{j=3}^{k} \varphi(s_j) \leq \Lambda. \tag{7.5.5}$$

Comparing (7.5.4) and (7.5.5), the function
$$G(t) = \varphi(s_1 + t) + \varphi(s_2 - t)$$
has a relative maximum at $t = 0$ and consequently
$$G''(0) \leq 0.$$

On the other hand,
$$G''(0) = \varphi''(s_1) + \varphi''(s_2) > 0.$$

We get a contradiction and the proof of Lemma 7.6 is complete.

Lemma 7.7   *There always exists an optimum point.*
(i) If
$$k \leq H,$$
then
$$\Lambda = k. \quad s_1 = s_2 = \cdots = s_k = 1.$$

(ii) If
$$H < k,$$
then,
$$\Lambda = [H] + \varphi(H - [H]),$$
where $[H]$ denotes the greatest integer contained in H. In this case,
(a) exactly $[H]$ coordinates of s are equal to 1;
(b) one coordinate is equal to $H - [H]$;
(c) all other coordinates, if they exist, are equal to zero.

Lemma 7.7 follows immediately from Lemma 7.6.

Lemma 7.8   *Let the quantities x_j $(j = 1, 2, \cdots, k; 2 \leq k \leq \infty)$ be subject to the constraints*

$$0 \leq x_j \leq 1, \quad 1 \leq j \leq k, \quad \sum_{j=1}^{k} \psi(x_j) \leq H < \infty. \tag{7.5.6}$$

(i) *If $k \leq H$, then*

$$\sum_{j=1}^{k} x_j \leq k.$$

(ii) *If $H < k < \infty$, then*

$$\sum_{j=1}^{k} x_j \leq [H] + \varphi(H - [H]),$$

and equality holds if and only if
  (a) *exactly $[H]$ of the x are equal to 1;*
  (b) *one x is $\varphi(H - [H])$;*
  (c) *all other x, if they exist, are equal to zero.*

(iii) *If $k = \infty$ and $x_j > 0$ for all $j > 1$, then*

$$\sum_{j=1}^{\infty} x_j < [H] + \varphi(H - [H]).$$

Proof.   Assertions (i) and (ii) constitute an obvious restatement of the preceding solution and require no proof.

To prove assertion (iii) select $l \geq H + 1$, large enough to imply that

$$R_{l+1} = \sum_{j=l+1}^{\infty} \psi(x_j) < 1. \qquad (7.5.7)$$

This is possible because of the convergence of the series in (7.5.6).

For a positive integer $N \geq l + 1$, since

$$0 \leq x_j \leq 1, \quad j = l+1, \ l+2, \cdots, N$$

and (7.5.7) holds, we obtain by (ii) of Lemma 7.8,

$$\sum_{j=l+1}^{N} x_j \leq \varphi(R_{l+1}),$$

and hence

$$\sum_{j=l+1}^{\infty} x_j \leq \varphi(R_{l+1}).$$

Noting that $s = \psi(x)$ is an increasing function on $[0, 1]$, we have

$$\psi\left(\sum_{j=l+1}^{\infty} x_j\right) \le R_{l+1} = \sum_{j=l+1}^{\infty} \psi(x_j) < 1. \tag{7.5.8}$$

Let

$$\xi_{l+1} = \sum_{j=l+1}^{\infty} x_j < 1. \tag{7.5.9}$$

Combining (7.5.6), (7.5.8) and (7.5.9), we find that

$$\sum_{j=1}^{l} \psi(x_j) + \psi(\xi_{l+1}) \le H.$$

In this inequality all the x and ξ_{l+1} are strictly positive and there are $l + 1 \ge H + 2$ such quantities. Hence by assertion (ii) of Lemma 7.8, equality cannot be attained (when equality is attained, there are $[H]+1 \le l$ positive quantities). i.e.,

$$\sum_{j=1}^{l} x_j + \xi_{l+1} < [H] + \varphi(H - [H]).$$

The proof of (iii) is derived. □

7.5.3 Settlement of the deficiency problem in the case of $\mu < 1$.
Now we shall prove the following theorem:

Theorem 7.8 *Let $f(z)$ be meromorphic and of lower order $\mu(1/2 < \mu \le 1)$ in the finite plane. Then*

$$\Delta(f) = \sum_{j=1}^{\nu(f)} \delta(a_j, f) \le 2 - \sin \mu\pi. \tag{7.5.10}$$

with equality possible if and only if

$$\nu(f) = 2, \quad \delta_1 = 1, \quad and \quad \delta_2 = 1 - \sin \mu\pi.$$

Proof. The function

$$x = \varphi(s) = 2\sin^2 \frac{s\pi}{4}$$

is defined on $[0, 1]$, with $\varphi(0) = 0$ and $\varphi(1) = 1$. Moreover, $\varphi'(s)$ and $\varphi''(s)$ are continuous and strictly positive in $(0, 1)$. Its inverse can be denoted by

$$s = \psi(x) = \frac{4}{\pi}\arcsin\sqrt{\frac{x}{2}}.$$

Using the total spread relation, the deficiencies $\delta(a_j, f)(j = 1, 2, \cdots, \nu(f))$ of $f(z)$ must satisfy

$$\frac{4}{\pi}\sum_{j=1}^{\nu(f)}\arcsin\sqrt{\frac{\delta(a_j, f)}{2}} \leq 2\mu.$$

Since

$$0 < \delta(a_j, f) < 1, \quad j = 1, 2, \cdots, \nu(f)$$

and

$$\sum_{j=1}^{\nu(f)}\psi(\delta(a_j, f)) \leq 2\mu,$$

we have by Lemma 7.8

$$\sum_{j=1}^{\nu(f)}\delta(a_j, f) \leq [2\mu] + \frac{2\sin^2(2\mu - [2\mu])\pi}{4}$$

$$= 1 + 2\sin^2\frac{(2\mu - 1)\pi}{4} = 2 - \sin\mu\pi,$$

where equality is possible only if

$$\nu(f) = 2, \quad \delta(a_1, f) = 1, \quad \delta(a_2, f) = 1 - \sin\mu\pi. \quad \square$$

Equality in (7.5.10) can actually be attained. For $\mu = 1$, the function e^z provides such an example. For $1/2 < \mu < 1$, it is sufficient to note the function

$$f(z) = \prod_{k=1}^{\infty}\left(1 + \frac{z}{k^{\frac{1}{\mu}}}\right). \tag{7.5.11}$$

Obviously $\delta(\infty, f) = 1$. We shall prove in the next lemma that both the order and lower order of $f(z)$ are equal to μ, and $\delta(0, f) = 1 - \sin\mu\pi$.

Lemma 7.9 *Let $f(z)$ be given by (7.5.11) with $1/2 < \mu < 1$. Then for a sufficiently small positive number δ, we have*

$$\log|f(re^{i\theta})| = \pi\frac{\cos\mu\theta}{\sin\mu\pi}r^\mu + o(r^\mu) \tag{7.5.12}$$

uniformly in θ for $|\theta| < \pi - \delta$.

Proof. From (7.5.11), we have

$$\log f(z) = \sum_{k=1}^{\infty} \log\left(1 + \frac{z}{k^{\frac{1}{\mu}}}\right) = \int_{\frac{1}{2}}^{\infty} \log\left(1 + \frac{z}{t}\right) dn(t, f = 0)$$

$$= n(t, f = 0)\log\left(1 + \frac{z}{t}\right)\Big|_{\frac{1}{2}}^{\infty} + z\int_{\frac{1}{2}}^{\infty} \frac{n(t, f = 0)dt}{t(t + z)}.$$

Since $\mu < 1$, we have

$$\lim_{t\to\infty} n(t, f = 0)\log\left(1 + \frac{z}{t}\right) = \lim_{t\to\infty} t^{\mu}\log\left(1 + \frac{z}{t}\right) = 0.$$

Thus

$$\log f(z) = z\int_{\frac{1}{2}}^{\infty} \frac{n(t, f = 0)dt}{t(t + z)}. \tag{7.5.13}$$

For any positive number ε, there is a sufficiently large positive number t_0 such that

$$|n(t, f = 0) - t^{\mu}| < \varepsilon t^{\mu}, \quad t > t_0.$$

Hence

$$\left|\log f(z) - z\int_0^{\infty} \frac{t^{\mu}}{t(t + z)}dt\right|$$

$$\leq |z|\int_{\frac{1}{2}}^{t_0} \frac{n(t, f = 0)}{t|t + z|}dt + |z|\int_0^{t_0} \frac{t^{\mu}dt}{t|t + z|} + \varepsilon|z|\int_{t_0}^{\infty} \frac{t^{\mu}dt}{t|t + z|}. \tag{7.5.14}$$

We easily see that

$$|z|\int_{\frac{1}{2}}^{t_0} \frac{n(t, f = 0)}{t|t + z|}dt \leq \frac{|z|}{|z| - t_0}\int_{\frac{1}{2}}^{t_0} \frac{n(t, f = 0)}{t}dt = O(1),$$

$$|z|\int_0^{t_0} \frac{t^{\mu}dt}{t|t + z|} = O(1)$$

and

$$z\int_0^{\infty} \frac{t^{\mu}dt}{t(t + z)} = \frac{\pi z^{\mu}}{\sin \mu\pi} \tag{7.5.15}$$

from the residue theorem.

If $z = re^{i\theta}$ and $|\theta| \leq \pi/2$, then from

$$|t + z| \geq \sqrt{t^2 + r^2} > \frac{t + r}{2},$$

we have

$$|z| \int_0^\infty \frac{t^\mu dt}{t|t+z|} \leq 2r \int_0^\infty \frac{t^\mu dt}{t(t+r)} = \frac{2\pi r^\mu}{\sin \mu \pi}.$$

If $\pi/2 < |\theta| < \pi - \delta$, then we have

$$|t+z| \geq \frac{t+r}{2} \sin \frac{\pi - |\theta|}{2} > \frac{t+r}{2} \sin \frac{\delta}{2},$$

so that

$$|z| \int_0^\infty \frac{t^\mu dt}{t|t+z|} < \frac{2r}{\sin \dfrac{\delta}{2}} \int_0^\infty \frac{t^\mu dt}{t(t+r)} = \frac{2\pi r^\mu}{\sin \dfrac{\delta}{2} \sin \mu \pi}.$$

Lemma 7.9 follows from (7.5.14) and these estimates.

For the function $f(z)$ given by (7.5.11), we now calculate its characteristic function $T(r, f)$.

Choose a sufficiently small positive number δ such that $\delta < \pi - \dfrac{\pi}{2\mu}$.
Obviously, we have

$$\left| m(r, f) - \frac{1}{2\pi} \int_{-\pi+\delta}^{\pi-\delta} \log^+ |f(re^{i\theta})| d\theta \right| \leq (\log^+ M(r, f)) \frac{\delta}{\pi}.$$

But (7.5.11), (7.5.13) and (7.5.15) give

$$\log M(r, f) = \log f(r) = r \int_{\frac{1}{2}}^\infty \frac{n(t, f = 0)}{t(t+r)} dt = (1 + o(1)) \frac{\pi r^\mu}{\sin \mu \pi}.$$

Thus

$$\left| m(r, f) - \frac{1}{2\pi} \int_{-\pi+\delta}^{\pi-\delta} \log^+ |f(re^{i\theta})| d\theta \right| \leq \delta \cdot O(r^\mu).$$

From Lemma 7.9, we have

$$\frac{1}{2\pi} \int_{-\pi+\delta}^{\pi-\delta} \log^+ |f(re^{i\theta})| d\theta$$

$$= \frac{\pi r^\mu}{\sin \mu \pi} \cdot \frac{1}{2\pi} \int_{-\frac{\pi}{2\mu}}^{\frac{\pi}{2\mu}} \cos \mu\theta d\theta + o(r^\mu)$$

$$= \frac{r^\mu}{\mu \sin \mu \pi} + o(r^\mu).$$

Hence

$$T(r, f) = \frac{r^\mu}{\mu \sin \mu \pi} + o(r^\mu), \tag{7.5.16}$$

and the order of $f(z)$ is equal to μ. Since

$$n(r, f = 0) = (1 + o(1))r^\mu, \quad r \to \infty,$$

we have

$$N(r, f = 0) = \left(\frac{1}{\mu} + o(1)\right)r^\mu, \quad r \to \infty.$$

Therefore,

$$\delta(0, f) = 1 - \sin \mu\pi. \quad \square$$

Theorem 7.9 *If $f(z)$ is meromorphic and of lower order $\mu(0 < \mu \le 1/2)$ in the finite plane, then*

$$\Delta(f) = \sum_{j=1}^{\nu(f)} \delta(a_j, f) \le 1. \tag{7.5.17}$$

Furthermore, $\Delta(f) \ge 1 - \cos \mu\pi$ only if $\nu(f) = 1$.

Proof. We distinguish two cases.

(i) $f(z)$ has a deficient value a_1 with $\delta(a_1, f) \ge 1 - \cos \mu\pi$. By Corollary 1 of Theorem 7.7, a_1 is the unique deficient value of $f(z)$, and there is nothing to prove.

(ii) $1 - \cos \mu\pi > \delta(a_1, f) \ge \delta(a_2, f) \ge \cdots$.

We define quantities $d_j (j = 1, 2, \cdots)$ by

$$\delta(a_j, f) = (1 - \cos \mu\pi)d_j = \left(2 \sin^2 \frac{\mu\pi}{2}\right)d_j.$$

Thus

$$0 \le d_j < 1, \quad j = 1, 2, \cdots, \nu(f).$$

Consider the function

$$x = \varphi(s) = \left\{\frac{\sin \dfrac{\mu s\pi}{2}}{\sin \dfrac{\mu\pi}{2}}\right\}^2, \quad 0 \le s \le 1,$$

which is twice continuously differentiable on $[0, 1]$, $\varphi(0) = 0$ and $\varphi(1) = 1$. $\varphi'(s)$ and $\varphi''(s)$ are strictly positive in $(0, 1)$. Its inverse is

$$s = \varphi^{-1}(x) = \psi(x) = \frac{2}{\pi\mu} \arcsin\left(x^{\frac{1}{2}} \sin \frac{\mu\pi}{2}\right).$$

The total spread relation takes the form

$$\sum_{j=1}^{\nu(f)} \psi(d_j) \le 1.$$

Hence Lemma 7.8 gives us

$$\sum_{j=1}^{\nu(f)} d_j \le 1,$$

so

$$\sum_{j=1}^{\nu(f)} \delta(a_j, f) \le 1 - \cos \mu\pi.$$

Since $d_1 < 1$, equality cannot be attained.   □

The deficiency problem is still open in the case of $\mu > 1$. Drasin and Weitsman [1] posed the following conjecture.

Let $\lambda \ge 1$,

$$\Lambda_1(\lambda) = 2 - \frac{2\sin\frac{\pi}{2}(2\lambda - [2\lambda])}{[2\lambda] + 2\sin\frac{\pi}{2}(2\lambda - [2\lambda])},$$

$$\Lambda_2(\lambda) = 2 - \frac{2\cos\frac{\pi}{2}(2\lambda - [2\lambda])}{[2\lambda] + 1},$$

and

$$\Lambda(\lambda) = \max\{\Lambda_1(\lambda), \ \Lambda_2(\lambda)\}.$$

Then, for any meromorphic function of order λ, we have

$$\sum \delta(a, f) \le \Lambda(\lambda).$$

On the other hand, Drasin and Weitsman [1] proved: Given $\lambda > 1$, there exists a meromorphic function of order λ such that $\Sigma\delta(a, f) = \Lambda(\lambda)$. Therefore, the upper bound $\Lambda(\lambda)$ is precise if the above conjecture is verified.

To end this chapter, we formulate several very important results on deficient values.

The inverse problem of the Nevanlinna deficiency relation was settled in the case of entire functions by Fuchs and Hayman [1], and completely by Drasin [3].

Theorem 7.10 (Drasin)   *Let (a_j) be an arbitrary sequence of distinct complex numbers and (δ_j) and (θ_j) two sequence of non-negative numbers, $1 \le j \le N \le +\infty$, subject to*

$$0 < \delta_j + \theta_j \le 1$$

and

$$\sum_{j=1}^{N} \delta_j + \sum_{j=1}^{N} \theta_j \le 2.$$

Then there exists a meromoprhic function $f(z)$ in the finite plane such that

$$\delta(a_j, f) = \delta_j, \quad \theta(a_j, f) = \theta_j, \quad (j = 1, 2, \cdots, N)$$

and

$$\delta(a, f) = 0, \quad \theta(a, f) = 0, \quad (a \ne a_j),$$

where $\theta(a, f)$ is the index of multiplicity (See p. 32).

It is clear that the meromorphic function in Theorem 7.10 is usually of infinite order. As for the case of finite order, Eremenko [3] gave a complete solution in 1986.

Theorem 7.11 (Eremenko)   *Let (a_j), $1 \le j \le N \le \infty$ be an arbitrary sequence of distinct complex numbers and (δ_j) a sequence of positive numbers satisfying*

$$0 < \delta_j < 1, \quad j = 1, 2, \cdots, N,$$

$$\Delta = \sum_{j=1}^{N} \delta_j < 2,$$

$$\sum_{j=1}^{N} \delta_j^{\frac{1}{3}} < +\infty. \qquad (7.5.18)$$

Then there exists a meromorphic function $f(z)$ of finite order in the plane such that $\delta(a_j, f) = \delta_j (j = 1, 2, \cdots, N)$ and $\delta(a, f) = 0$ $(a \ne a_j)$.

Condition (7.5.18) is necessary, in view of Theorem 7.5″.

Around 1930, F. Nevanlinna posed the following conjecture.

Let $f(z)$ be meromorphic and of finite lower order μ in the finite plane. If the total deficiency of $f(z)$ equals two, then we have

(i) The order λ of $f(z)$ equals μ and has the form $\dfrac{n}{2}$, where n is a positive integer.

(ii) Every deficiency of $f(z)$ must be a multiple of $\dfrac{1}{\mu}$.

(iii) Every deficient value of $f(z)$ must also be its asymptotic value.

The conjecture was proved in the case of entire functions, by Pfluger [1], Edrei and Fuchs [2]. The general case, on the other hand, is very difficult. In 1987, Drasin [5] published a very important paper in which he presented a complete verification of the conjecture of F. Nevanlinna.

Theorem 7.12 (Drasin)   *The conjecture of F. Nevanlinna is correct.*

The author omits the proofs of Theorems 7.10–7.12 with great regret. The interested reader is encouraged to study the original papers.

Bibliography

Ahlfors, L. V.
[1] Untersuchungen zur Theorie der konformen Abbildung und der ganzen Funktionen, *Acta Soc. Sci. Fenn.*, **1** (1930), 1–40.
[2] *Complex Analysis*, 3rd ed. McGraw-Hill, New York, 1979.

Anderson, J. M. and Baernstein, A.
[1] The size of the set on which a meromorphic function is large, *Proc. London Math. Soc.*, **36** (1978), 518–539.

Anderson, J. M., Baker, I. N. and Clunie, J.
[1] The distribution of values of certain entire and meromorphic functions, *Math. Z.*, **178** (1981), 509–525.

Anderson, J.M., Barth, K. F. and Brannan, D. A.
[1] Research problems in complex analysis, *Bull. London Math. Soc.*, **9** (1977), 129–162.

Anderson, J. M. and Clunie, J.
[1] Entire functions of finite order and lines of Julia, *Math. Z.*, **112** (1969), 59–73.

Arakelyan, N. U.
[1] Entire functions of finite order with a set of infinite deficient values (in Russian), *Dokl, USSR*, **170** (1966), 999–1002.

Baernstein, A.
[1] Proof of Edrei's spread conjecture, *Proc. London Math. Soc.*, **26** (1973), 418–434.
[2] Integral means, univalent functions and circular symmetrization, *Acta Math.*, **133** (1974), 139–169.
[3] A generalization of the $\cos \pi \rho$ theorem, *Trans. Amer. Math. Soc.*, **193** (1974), 181–197.

Baker, I. N.
[1] The distribution of fix-points of entire functions, *Proc. London Math. Soc.*, **16** (1966), 493–506.
[2] The domains of normality of an entire function, *Ann. Acad. Sci. Fenn.*, Series A, I. Math., **1** (1975), 277–283.

Balaguer, F. S.
[1] Directions of Borel-Valiron of maximum kind common to an entire function and its successive derivatives and integrals, *Mem. Acad. Ci. Madrid*, **5** (1956), 1–51.

Barsegyan, G.A.
[1] On the relation between the behaviour of asymptotic values and a-points of meromorphic functions (in Russian), *Akad. Nauk Armyan SSR Dokl.*, **18** (1983), 124–133.

Barth, K.F., Brannan, D.A. and Hayman W. K.
[1] Research problems in complex analysis, *Bull. London Math. Soc.*, **16** (1984), 490–517.

Begehr, H.
[1] Über Defektbegriffe in der Theorie der meromorphen Funktionen, *Math. Zeit.*, **116** (1970), 349–354.

Biernacki, M.
[1] Sur les directions de Borel des fonctions méromorphes, *Acta Math.*, **56** (1930), 197–204.

Bombieri, E. and Ragendda, P.

[1] Sulle deficienze delle funzioni meromorfe di ordine inferiere finite, *Rend. Sem. Fac. Sci. Univ. Cagliari*, **37** (1967), 23–38.

Borel, E.

[1] Sur les zéros des fonctions entières, *Acta Math.*, **20** (1897), 357–396.

Brannan, D.A. and Hayman, W. K.

[1] Research problems in complex analysis, *Bull. London Math. Soc.*, **21** (1989), 1–35.

Campbell, D. M., Clunie, J. G. and Hayman, W.K.

[1] *Research problems in complex analysis, in Aspects of Contemporary Complex Analysis* (edited by D. A. Brannan and J. G. Clunie), Academic Press, London, 1980, 527–572.

Cartan, H.

[1] Sur les systèmes de fonctions holomorphes à variété lacunaires et leurs applications, *Ann. École Norm. Sup.*, **45** (1928), 255–346.

[2] Sur les zéros des combinaisons linéaires de p fonctions holomorphes données, *Mathematica (cluj)*, **7** (1933), 80–103.

Cartwright, M. L.

[1] *Integral functions*, Cambridge Univ. Press, Cambridge, 1956.

Chen, Huaihui

[1] Singular directions corresponding to Hayman's inequality (in Chinese), *Adv. in Math.* (Beijing), 16 (1987), 73–80.

[2] The singular direction of a meromorphic function of order zero corresponding to Hayman's inequality (in Chinese), *Acta Math. Sinica*, **30** (1987), 234–237.

Chern, S. S.

[1] Complex analytic mappings of Riemann surfaces I, *Amer. J. Math.*, **82** (1960), 323–337.

Chuang, Chitai

[1] Un théorème relatif aux directions de Borel des fonctions méromorphes d'ordre fini, *C. R. Acad. Sci., Paris*, **204** (1937), 951–952.

[2] Sur la comparaison de la croissance d'une fonction méromorphe et de celle de sa dérivée, *Bull. Sci. Math.*, **75** (1951), 171–190.

[3] Une généralisation d'une inégalité de Nevanlinna, *Sci. Sinica*, **13** (1964), 887–895.

Chuang, Chitai and Yang, Lo

[1] Distribution of the values of meromorphic functions, *Contemporary Math., Amer. Math. Soc.*, **48** (1985), 21–63.

Clunie, J.

[1] On a result of Hayman, *J. London Math. Soc.*, **42** (1967), 389–392.

Clunie, J. and Hayman, W. K.

[1] *Proceedings of the symposium on complex analysis at Canterbury* (1973), London Math. Soc. Lecture Notes, no. 12, Cambridge, 1974.

Collingwood, E.F.

[1] Sur quelques théorèmes de M. R. Nevanlinna, *C. R. Acad. Sci.*, **179** (1924), 955–957.

Dai, Chongji and Jin, Lu

[1] Number of deficient values of a class of meromorphic functions, *Kodai Math. J.*, **10** (1987), 74–82.

Dinghas, A.

[1] *Wertverteilung meromorpher Funktionen in ein - und mehrfach zusammenhängenden Gebieten*, Lecture Notes in Math., no. 783, Springer–Verlag, 1980.

Drasin, D.

[1] Normal families and the Nevanlinna theory, *Acta Math.*, **122** (1969), 231–263.

[2] An introduction to potential theory and meromorphic functions, Complex Analysis and its Applications, *IAEA, Vienna*, 1 (1976), 1–93.

[3] The inverse problem of the Nevanlinna theory, *Acta Math.*, **138** (1977), 83–151.

[4] Quasi-conformal modifications of functions having deficiency sum two, *Ann. of Math.*, **114** (1981), 493–518.

[5] Proof of a conjecture of F. Nevanlinna concerning functions which have deficiency sum two, *Acta Math.*, **158** (1987), 1–94.

Drasin, D. and Hayman, W. K.

[1] Value distribution of functions meromorphic in an angle, *Proc. London Math. Soc.*, (3) **48** (1984), 319–340.

Drasin, D. and Weitsman, A.

[1] Meromorphic functions with large sums of deficiencies, *Advances in Math.*, **15** (1974), 93–126.

[2] On the Julia directions and Borel directions of entire functions, *Proc. London Math. Soc.*, **32** (1976), 199–212.

Drasin, D., Weitsman, A., Yang, Lo and Zhang, Guanghou

[1] Deficient values of entire functions and their derivatives, *Proc. Amer. Math. Soc.*, **82** (1981), 607–612.

Edrei, A.

[1] Meromorphic functions with three radially distributed values, *Trans. Amer. Math. Soc.*, **78** (1955), 276–293.

[2] The deficiencies of meromorphic functions of finite lower order, *Duke Math. J.*, **31** (1964), 1–21.

[3] Sums of deficiencies of meromorphic functions, I, II, *J. d' Analyse Math.*, **14** (1965), 79–107; **19** (1967), 53–74.

[4] Solution of the deficiency problem for functions of small lower order, *Proc. London Math. Soc.*, **26** (1973), 435–445.

Edrei, A. and Fuchs, W. H. J.

[1] On the growth of meromorphic functions with several deficient values, *Trans Amer. Math. Soc.*, **93** (1959), 292–328.

[2] Valeurs déficientes et valeurs asymptotiques des fonctions méromorphes, *Comment. Math. Helv.*, **33** (1959), 258–295.

[3] The deficiencies of meromorphic functions of order less than one, *Duke Math. J.*, **27** (1960), 233–249.

[4] Bounds for the number of deficient values of certain classes of meromorphic functions, *Proc. London Math. Soc.*, **12** (1962), 315–344.

[5] On meromorphic functions with regions free of poles and zeros, *Acta Math.*, **108** (1962), 113–145.

[6] Asymptotic behavior of meromorphic functions with extremal spread I, II, *Ann. Acad. Sci. Fenn.*, Series A. I. Math., **2** (1976), 67–111; **3** (1977), 141–168.

Edrei, A., Fuchs, W. H. J. and Hellerstein, S.

[1] Radial distribution and deficiencies of the values of a meromorphic function, *Pacific J. Math.*, **2** (1961), 135–151.

Eremenko, A. E.

[1] On the natural asymptotic curves of meromorphic functions, *Complex Variables*, **4** (1985), 305–309.

[2] On deviations of meromorphic functions of finite lower order. *Amer. Math.Soc. Transl.*, (2) **131** (1986), 45–54.

[3] Inverse Problem of value-distribution theory for meromorphic functions of finite order, *Sibersky Math. J.*, **27** (1986), 87–102.

Eremenko, A. E. and Sodin, M. L.

[1] Proof of a basic theorem of Littlewood on the value distribution of entire functions (in Russian), *Izv. Akad. Nauk SSSR Ser. Mat.*, **51** (1987), 421–428.

Essén, M. R.

[1] *The cos$\pi\lambda$ Theorem*, Lecture Notes in Math., no. 467, Springer-Verlag, 1975.

Fenton, P.

[1] Entire functions having asymptotic functions, *Bull. Austral. Math. Soc.*, **27** (1983), 321–328.

Frank, G.

[1] Eine Vermutung von Hayman über Nullstellen meromorpher Funktionen, *Math. Zeit.*, **149** (1976), 29–36.

[2] Über die Nullstellen meromorpher Funktionen und deren Ableitungen, *Math. Ann.*, **255** (1977), 145–154.

[3] Über die Nullstellen von linearen Differentialpolynomen mit meromorphen Koeffizienten, preprint.

Frank, G. and Hennekemper, W.

[1] Einige Ergebnisse über die Wertverteilung meromorpher Funktionen und ihrer Ableitungen, *Resultate der Math.*, **4** (1981), 39–54.

Frank, G. and Mues, E.

[1] Differentialpolynome, 1979, Oberwolfach (unpublished).

Frank, G and Weissenborn, G.

[1] Rational deficient functions of meromorphic functions, *Bull. London Math. Soc.*, **18** (1986), 29–33.

[2] On the zeros of linear differential polynomials of meromorphic functions, *Complex Variables*, **12** (1989), 77–81.

Fuchs, W. H. J.

[1] A theorem on the Nevanlinna deficiencies of meromorphic functons of finite order, *Ann. of Math.*, **68** (1958), 203–209.

[2] Developments in the classical Nevanlinna theory of meromorphic functions, *Bull. Amer. Math. Soc.*, **73** (1967), 275–291.

[3] Topics in Nevanlinna theory, Proc. of the NRL conference on classical function theory, Naval Research Lab., Washington, 1970, 1–32.

[4] A Phragman-Lindelof theorem conjectured by D. J. Newman, *Trans. Amer. Math. Soc.*, **267** (1981), 285–293.

[5] The development of the theory of deficient values since Nevanlinna, *Ann. Acad. Sci. Fenn.*, Ser. A. *I. Math.*, **7** (1982), 38–44.

Fuchs, W. H. J. and Hayman, W. K.

[1] An entire function with assigned deficiencies, studies in mathematical analysis and related topics, *Essays in honor of George Polya*, Stanford Univ. Press, 1962, 117–125.

Gauthier, P. M.

[1] Cercles de remplissage and asymptotic behaviour, *Can. J. Math.*, **21** (1969), 447–455.

Goldberg, A. A.

[1] On the deficiencies of meromorphic functions (in Russian), *Dokl. Akad. Nauk. SSSR*, **98** (1954), 893–895.

[2] On the sets of deficient values for meromorphic functions of finite order (in Russian), *Ukrainian Math. J.*, **11** (1959), 438–443.

[3] Sets on which the modulus of an entire function is bounded below (in Russian), *Siberian Math. J.*, **20** (1979), 512–518.

Goldberg, A. A. and Eremenko, A. E.

[1] On asymptotic curves of entire functions of finite order (in Russian), *Mat. Sb. (N. S.)*, **109** (151) (1982), 555–581.

Goldberg, A. A. and Ostrovskii, I. V.

[1] *The distribution of values of meromorphic functions* (in Russian), Nauka Moscow, 1970.

Goldberg, A. A., Eremenko, A. E. and Sodin, M. L.

[1] Exceptional values in the sense of R. Nevanlinna and in the sense of S. P. Petrenko (in Russian), *Teor. Funktsii Funktsional Anal. i Prilozhen*, **47** (1987), 41–51 and **48** (1987), 58–70.

Goluzin, G. M.

[1] *Geometric Theory of Functions of a Complex Variable* (in Russian), 2nd ed, Nauka Moscow, 1966.

Gong, Xianghong and Gu, Yongxing
[1] On Hayman directions, *Sci. Sinica*, Series A, **31** (1988), 1053–1064.
Griffiths, P. A.
[1] *Entire Holomorophic Mappings in one and Several Complex Variables*, Ann. of Math. Studies, no. 85, 1976, Princeton Univ. Press.
Gross, F.
[1] *Proceedings of the NRL Conference on Classical Function Theory*, Naval Research Lab., 1970, Washington.
[2] *Factorization of Meromorphic Functions*, Naval Research Lab., 1972, Washington.
Gu, Yongxing
[1] Sur les familles normales de fonctions méromorphes, *Sci. Sinica*, **21** (1978), 431–445.
[2] A criterion for normality of meromorphic functions (in Chinese), **Sci. Sinica**, special issue (I) (1979), 267–274.
Hayman, W. K.
[1] Picard values of meromorphic functions and their derivatives, *Ann. of. Math.*, **70** (1959), 9–42.
[2] *Meromorphic Functions*, Clarendon Press, Oxford, 1964.
[3] On the characteristic of functions meromorphic in the plane and of their integrals, *Proc. London Math. Soc.*, **14** (1965), 93–128.
[4] *Research Problems in Function Theory*, Athlone Press (Univ. of London), 1967.
[5] Some achievements of Nevanlinna theory, *Ann. Akad. Sci. Fenn.*, Ser. A, I. *Math.*, **7** (1982), 65–71.
[6] Value distribution and exceptional sets, *Sém. Math. Sup., Montréal*, 1982, 79–148.
Hayman, W. K. and Kennedy, P. B.
[1] *Subharmonic Functions* (I), Academic Press, London, 1976.
Hayman, W. K. and Miles, J.
[1] On the growth of a meromorphic function and its derivatives, *Complex Variables*, **12** (1989), 245–260.
Hayman, W. K. and Rossi, J. F.
[1] Characteristic, maximum modulus and value distribution, *Trans. Amer. Math. Soc.*, **284** (1984), 651–664.
Hayman, W. K. and Yang, Lo
[1] Growth and values of functions regular in an angle, *Proc. London Math. Soc.*, **44** (1982), 193–214.
Hellerstein, S.
[1] On a class of meromorphic functions with deficient zeros and poles, *Pacific J. Math.*, **13** (1963), 115–124.
Hellerstein, S. and Shea, D. F.
[1] Bounds for the deficiencies of meromorphic functions of finite order, *Proc. Symposia in pure Math.*, **11** (1968), 214–239.
[2] Minimal deficiencies for entire functions with radially distributed zeros, *Proc. London Math. Soc.*, **37** (1978), 35–55.
Hellerstein, S., Shen, Li Chien, Williamson, J.
[1] Reality of the zeros of an entire function and its derivatives, *Trans. Amer. Math. Soc.*, **275** (1983), 319–331.
Hellerstein, S. and Williamson, J.
[1] Entire functions with negative zeros and a problem of R. Nevanlinna, *J. d'Analyse Math.*, **22** (1969), 233–267.
[2] Derivatives of entire functions and a question of Pólya I, II, *Trans. Amer. Math. Soc.*, **227** (1977), 227–249; **234** (1977), 497–503.
Hennekemper, W.
[1] Über die Wertverteilung von $(f^{(k+1)})^{(k)}$, *Math. Z.*, **177** (1981), 375–380.

Hiong, King-lai

[1] Sur les fonctions entières et les fonctions méromorphes d'ordre infini, *J. Math. pures et appl.*, **14** (1935), 233–308.

[2] Sur les fonctions holomorphes dont les dérivées admettent une valeur exceptionnelle, *Ann. École Norm. Sup.*, **72** (1955), 165–197.

[3] *Sur les Fonctions Méromorphes et les Fonctions Algébroïdes*, Mém. Sci. Math., Fasc. **139**, Paris, 1957.

Julia, G.

[1] Sur quelques propriétés nouvelles des fonctions entières ou méromorphes, *Ann. École Norm. Sup.*, **36** (1919), 93–125; 37 (1920), 165–218.

Kirwan, W. E. and Zalcman, L. (editors)

[1] *Advances in Complex Function Theory*, Proc. of Seminars held at Maryland Univ. (1973/74), Lecture Notes in Math., no. 505, Springer, 1976.

Laine, I., Lehto, O. and Sorvali, T. (Editors)

[1] *Complex Analysis*, Joensuu, 1978, Proc., Lecture Notes in Math., no. 747, Springer, 1978.

Laine, I. and Rickman, S. (Editors)

[1] *Value distribution theory*, Lecture Notes in Math., no. 981, Springer, 1983.

Langley, J. K.

[1] The distribution of zeros of certain differential polynomials, *J. London Math. Soc.*, (2) **29** (1984), 485–498.

[2] On differential polynomials and results of Hayman and Doeringer, *Math. Z.*, **189** (1984), 1–11.

[3] On normal families and a result of Drasin, *Proc. of the Royal Soc. of Edinburgh*, **98A** (1984), 385–393.

Lewis, J. L. and Wu, J. M.

[1] On conjectures of Arakelyan and Littlewood, *J. d'Analyse Math.*, **50** (1988), 259–283.

Li, Xianjin

[1] Proof of Hayman's conjecture on normal families, *Sci. Sinica*, **28** (1985), 596–603.

Lin, Qun and Dai, Chongji

[1] On a conjecture of Shah concerning small functions, *Kexue Tongbao (Bull. of Sci.)*, **31** (1986), 220–224.

Marty, F.

[1] Recherches sur la répartion des valeurs d'une fonction méromorphe, *Ann. Fac. Sci. Univ. Toulouse*, **23** (1931), 183–261.

Miles, J.

[1] Some examples of the dependence of the Nevanlinna deficiency upon the choice of origin, *Proc. London Math. Soc.*, (3) 47 (1983), 145–176.

Miles, J. and Williamson, J.

[1] A characterisation of the exponential function, *J. London Math. Soc.*, (2) **33** (1986), 110–116.

Milloux, H.

[1] Le théorème de Picard, suites de fonctions holomorphes; fonctions méromorphes et fonctions entières, *J. de Math.*, **3** (1924), 345–401.

[2] *Extension d'un Théorème de M. R. Nevanlinna et Applications*, Act. Sci. et Ind., **888** (1940).

[3] Sur les directions de Borel des fonctions entières, de leurs derivées et de leurs integrales, *J. d'Analyse Math.*, 1 (1951), 244–330.

Miranda, C.

[1] Sur un nouveau critère de normalité pour les familles des fonctions holomorphes, *Bull. Sci. Math. France*, **63** (1935), 185–196.

Montel, P.

[1] *Leçons sur les Familles Normales de Fonctions Analytiques et leurs Applications*, Coll. Borel, 1927.

Mues, E.

[1] Über ein Defekt und Verzweigungsrelation für die Ableitung Meromorpher Funktionen, *Manuscripta Math.*, **5** (1971), 275–297.

[2] Zur Wertverteilung von Differentialpolynomen, *Archiv der Math.*, **32** (1979), 55–67.

[3] Über ein Problem von Hayman, *Math. Zeit.*, **164** (1979), 239–259.

Mues, E. and Steinmetz, N.

[1] The theorem of Tumura-Clunie for meromorphic functions, *J. London Math. Soc.*, **23** (1981), 113–122.

Nevanlinna, R.

[1] Zur Theorie der meromorphen Funktionen, *Acta Math.*, **46** (1925), 1–99.

[2] *Le Théorème de Picard-Borel et la Théorie des Fonctions Méromorphes*, Coll. Borel, 1929.

[3] *Analytic Functions*, Springer-Verlag, New York, 1970.

Osgood, C. F.

[1] Sometimes effective Thue-Siegel-Roth-Schmidt-Nevanlinna bounds, or better, *J. Number Theory*, **21** (1985), 347–389.

Oshkin, I. B.

[1] On a condition for the normality of families of holomorphic functions (in Russian), *Uspekhi, Mat. Nauk*, **37** (1982), 221–222.

Ostrowski, A.

[1] Über Folgen analytischer Funktionen und einige Verschärfungen des Picardschen Satzes, *Math. Zeit.*, **24** (1926), 215–258.

Ostrowski, E. B.

[1] Connection of the growth of meromorphic functions with their value-distribution on argument (in Russian), *Izv. Akad. Nauk SSSR Ser. Mat.*, **25** (1961), 277–328.

Ozawa, M.

[1] *Modern Theory of Functions I, Theory of Value-Distribution* (in Japanese), Tokyo, 1976.

Pang, Xuecheng

[1] Bloch's principle and normal criterion, *Sci. Sinica*, Series A, **32** (1989), 782–791.

Petrenko, F. B.

[1] Some estimate for deficiencies of meromorphic functions (in Russian), *Sibir. Math. J.*, **7** (1966), 1319–1336.

Pfluger, A.

[1] Zur Defektrelation ganzer Funktionen endlicher Ordnung, *Comment Math. Helv.*, **19** (1946), 91–104.

Polya, G.

[1] Untersuchungen über Lücken und Singularitäten von Potenzreihen, *Math. Zeit.*, **29** (1929), 549–640.

Rauch, A.

[1] Extension de théorème relatifs aux directions de Borel des fonctions méromorphes, *J. Math. pures et Appl.*, **12** (1933), 109–171.

[2] Cas où une direction de Borel d'une fonction entière $f(z)$ d'ordre fini est aussi direction de Borel pour $f'(z)$, *C. R. Acad. Sci.*, **199** (1934), 1014–1016.

Rossi, J. and Weitsman, A.

[1] A unified approch to certain questions in value distribution theory, *J. London Math. Soc.*, (2) 28 (1983), 310–326.

Schottky, F.

[1] Über den Picardschen Satz und die Borelschen Ungleichungen, *S. B. Preuss Akad. Wiss*, (1904), 1244–1263.

Schwick, W.

[1] Normality criteria for families of meromorphic functions, *J. d'Analyse Math.*, **52** (1989), 241–289.

Shea, D.F.
[1] On the Valiron deficiencies of meromorphic functions of finite order, *Trans. Amer. Math. Soc.*, **124** (1966), 201–227.

Shea, D. F. and Sons, L. R.
[1] Value distribution theory for meromorphic functions of slow growth in the disk, Houston *J. Math.*, **12** (1986), 249–266.

Shiffman, B.
[1] A general second main theorem for meromorphic functions on $\mathbb{C}^n$, *Amer. J. Math.*, **106** (1984), 509–531.

Sons, L. R.
[1] Value distribution for unbounded functions in the unit disc, *Complex Variables.*, **7** (1987), 337–341.

Steinmetz, N.
[1] Über eine Verallgemeinerung des zweiten Nevanlinnaschen Hauptsatzes, *J. Reine Angew. Math.*, **368** (1986), 134–141.

Stoll, W.
[1] *Introduction to value distribution theory of meromorphic maps*, Lecture Notes in Math., no. 950, Springer, 1982, 210–359.
[2] *Value Distribution Theory for Meromorphic Maps.* Vieweg, Braunschweig, 1985.

Teichmüller, O.
[1] Vermutungen und Sätze über die Wertverteilung gebrochener Funktionen endlicher Ordnung, *Deutsche Math.*, **4** (1939), 163–190.

Titchmarsh, E. C.
[1] *The Theory of Functions*, 2nd edition, Oxford, 1939.

Toppila, S.
[1] On the characteristic of meromorphic functions and their derivatives, *J. London Math. Soc.*, (2) 25 (1982), 261–272.
[2] On the Nevanlinna characteristic of functions of small lower order, *Rev. Roumaine Math. Pures Appl.*, **32** (1987), 929–939.

Tsuji, M.
[1] *Potential Theory in Modern Function Theory*, Chelsea Publ. Corp., New York, 1975.

Valiron, G.
[1] *Lectures on the General Theory of Integral Functions*, Edouard Privat, Toulouse, 1923.
[2] Recherches sur le théorème de Borel dans la théorie des fonctions méromorphes, *Acta Math.*, **52** (1928), 67–92.
[3] *Directions de Borel des Fonctions Méromorphes*, Mémor. Sci. Math., fasc. 89, Paris, 1938.
[4] Sur les valeurs déficientes des fonctions méromorphes d'ordre nul, *C. R. Acad. Sci. Paris*, **230** (1950), 40–42.

Weitsman, A.
[1] Meromorphic functions with maximal deficiency sum and a conjecture of F. Nevanlinna, *Acta Math.*, **123** (1969), 115–139.
[2] A theorem on Nevanlinna deficiencies, *Acta Math.*, **128** (1972), 41–52.

Whittaker, J. M.
[1] The order of the derivative of a meromorphic function, *J. London Math. Soc.*, **11** (1936), 82–87.

Winkler, J.
[1] Zur Existenz ganzer Funktionen bei vorgegebener Menge der Nullstellen und Einsstellen, *Math. Zeit.*, **168** (1979), 77–85.

Wittich, H.
[1] *Neuere Untersuchungen über eindeutige analytische Funktionen*, Springer-Verlag, 1955.

Wu, H.
[1] *The Equidistribution Theory of Holomorphic Curves*, Ann. of Math. Studies **64**, Princeton Univ. Press, 1970.

Wu, J. -M. G.

[1] Length of paths for subharmonic functions, *J. London Math. Soc.* (2) 32 (1985), 497–505.

Yang, Chungchun

[1] On deficiencies of differential polynomials, *Math. Zeit.*, 116 (1970), 197–204 and 125 (1972), 107–112.

[2] On meromorphic functions with three almost linear values, *Math. Zeit.*, 123 (1971), 131–138.

[3] *Analysis of one Complex Variable* (editor), Word Sci., Singapore, 1987.

Yang, Lo

[1] Multiple values of meromorphic functions and of their combinations (in Chinese), *Acta Math. Sinica*, 14 (1964), 428–437.

[2] Common Borel directions of a meromorphic function and its derivatives, *Sci. Sinica*, Special Issue (II) (1979), 91–104.

[3] Angular distribution and multiple values between entire functions and their derivatives, *Sci. Sinica*, 23 (1980), 16–39.

[4] Deficient functions of meromorphic functions, *Sci. Sinica*, 24 (1981), 1179–1189.

[5] Meromorphic functions and their derivatives, *J. London Math. Soc.*, 25 (1982), 288–296.

[6] Value distribution of meromorphic functions and their derivatives, *Sci. Sinica*, Series A, 25 (1982), 572–582.

[7] *Value Distribution Theory and its new Research* (in Chinese), Science Press, Beijing, 1982.

[8] A fundamental inequality and its application, *Chin. Ann. of Math.*, Series B, 4 (1983), 339–346.

[9] Normal families and differential polynomials, *Sci. Sinica*, Series A, 26 (1983), 673–686.

[10] Normal families and fix-points of meromorphic functions, *Indiana Univ. Math. J.*, 35 (1986), 179–191.

[11] Deficient values and angular distribution of entire functions, *Trans. Amer. Math. Soc.*, 308 (1988), 583–601.

[12] Growth and angular distribution of entire functions, *Complex Variables*, 13 (1989), 155–160.

[13] Precise fundamental inequalities and sum of deficiencies, *Sci. Sinica*, Series A, 34 (1991), 157–165.

[14] Precise estimate of total deficiency of meromorphic derivatives, *J. d'Analyse Math.*, 55 (1990), 287–296.

[15] Deficient values and deficient functions, International Symposium in Memory of Hua Loo Keng, Vol. II Analysis, Springer-Verlag and Sci. Press, 1991, 313–321.

[16] Recent results and problems in the theory of value distribution, preprint.

Yang, Lo and Shiao, Shiouzhi

[1] Sur les points de Borel des fonctions méromorphes et de leur derivées, *Sci. Sinica*, 14 (1965), 1556–1573.

Yang, Lo and Zhang, Guanghou

[1] Recherches sur la normalité des familles de fonctions analytiques à des valeurs multiples, I. Un nouveau critère et quelques applications, *Sci. Sinica*, 14 (1965), 1258–1271; II Géneralisations, ibid. 15 (1966), 433–453.

[2] Sur la distribution des directions de Borel des fonctions méromorphes, *Sci. Sinica*, 16 (1973), 465–482.

[3] Recherches sur le nombre des valeurs deficientes et le nombre des directions de Borel des fonctions méromorphes, *Sci. Sinica*, 18 (1975), 23–37.

[4] On the number of deficient values of entire functions, *Acta Math. Sinica*, 18 (1975), 35–53.

[5] Sur la construction des fonctions méromorphes ayant des directions singulières données, *Sci. Sinica*, 19 (1976), 445–459.

[6] Deficient values of extremal entire functions, *Sci. Sinica* 20 (1977), 421–435.

[7] Deficient values and asymptotic values of entire functions, *Sci. Sinica*, Special Issue II (1979), 190–203.

[8] Distribution of zeros and poles of meromorphic functions and their filling disks, *Sci. Sinica*, Series A, **25** (1982), 371–383.

Yang, Lo and Zhang, Qingde
[1] New singular direction of meromorphic functions, *Sci. Sinica*, Series A, **27** (1984), 352–366.

Yi, HongXun
[1] On a theorem of Tumura and Clunie for a differential polynomial, *Bull. London Math. Soc.*, **20** (1988), 593–596.

Zalcman, L.
[1] A heuristic principle in complex function theory, *Amer. Math. Monthly*, **82** (1975), 813–817.

Index